27
3
RI

AF744280

7000545586

UWE BRISTOL
WITHDRAWN
LIBRARY SERVICES

OCEANOGRAPHY AND OCEAN ENGINEERING

SEA LEVEL RISE, COASTAL ENGINEERING, SHORELINES AND TIDES

OCEANOGRAPHY AND OCEAN ENGINEERING

Additional books in this series can be found on Nova's website
under the Series tab.

Additional E-books in this series can be found on Nova's website
under the E-books tab.

OCEANOGRAPHY AND OCEAN ENGINEERING

SEA LEVEL RISE, COASTAL ENGINEERING, SHORELINES AND TIDES

LINDA L. WRIGHT
EDITOR

UWE Bristol
FR-51763
10 MAY 2012
FET
Library Services

Nova Science Publishers, Inc.
New York

Copyright © 2011 by Nova Science Publishers, Inc.

All rights reserved. No part of this book may be reproduced, stored in a retrieval system or transmitted in any form or by any means: electronic, electrostatic, magnetic, tape, mechanical photocopying, recording or otherwise without the written permission of the Publisher.

For permission to use material from this book please contact us:
Telephone 631-231-7269; Fax 631-231-8175
Web Site: http://www.novapublishers.com

NOTICE TO THE READER

The Publisher has taken reasonable care in the preparation of this book, but makes no expressed or implied warranty of any kind and assumes no responsibility for any errors or omissions. No liability is assumed for incidental or consequential damages in connection with or arising out of information contained in this book. The Publisher shall not be liable for any special, consequential, or exemplary damages resulting, in whole or in part, from the readers' use of, or reliance upon, this material. Any parts of this book based on government reports are so indicated and copyright is claimed for those parts to the extent applicable to compilations of such works.

Independent verification should be sought for any data, advice or recommendations contained in this book. In addition, no responsibility is assumed by the publisher for any injury and/or damage to persons or property arising from any methods, products, instructions, ideas or otherwise contained in this publication.

This publication is designed to provide accurate and authoritative information with regard to the subject matter covered herein. It is sold with the clear understanding that the Publisher is not engaged in rendering legal or any other professional services. If legal or any other expert assistance is required, the services of a competent person should be sought. FROM A DECLARATION OF PARTICIPANTS JOINTLY ADOPTED BY A COMMITTEE OF THE AMERICAN BAR ASSOCIATION AND A COMMITTEE OF PUBLISHERS.

Additional color graphics may be available in the e-book version of this book.

LIBRARY OF CONGRESS CATALOGING-IN-PUBLICATION DATA

Sea level rise, coastal engineering, shorelines and tides / [edited by]
Linda L. Wright.
p. cm.
Includes index.
ISBN 978-1-61728-655-1 (hardcover)
1. Sea level. 2. Coastal engineering. I. Wright, Linda L., 1974-
GC89.S428 2010
627'.58--dc22
2010025396

Published by Nova Science Publishers, Inc. † New York

Contents

PREFACE

Current sea level rise is due significantly to global warming, which will increase sea level over the coming century and longer periods. Increasing temperatures result in sea level rise by the thermal expansion of water and through the addition of water to the oceans from the melting of continental ice sheets. Values for predicted sea level rise over the course of this century typically range from 90 to 880 mm. This new book presents research data on sea level rise, coastal engineering, shorelines and tides, including the tides and topography of the Strait of Gibraltar; tidal energy use for the future and sea level variation along the Estonian coast of the Baltic Sea. Also discussed herein is the potential impacts of sea-level rise on the Atlantic and Gulf of Mexico coasts and coastal management of shorelines in Colombia, South America.

The Strait of Gibraltar is the unique and narrow connection between the Mediterranean basin and the open Atlantic Ocean. One of the main features of this place appears as the strong topographic constriction that happens both in the horizontal (minimum width of 14 km) and vertical (minimum depth of 250 meters) dimension in conjunction with the presence of a two-layered counter-current circulation and strong inequalities of the tidal range at both sides. These differences in tidal amplitude induce intense barotropic and baroclinic currents which interact with the sharp sea-bottom topography in the main sill of the Strait creating internal waves in the Atlantic-Mediterranean Interface (AMI). These intense and periodic undulatory processes created in the western side of the Strait modify the main along-strait circulation in a number of different ways. On one hand, they can mix the Atlantic and Mediterranean waters creating pulsating upwelling events which greatly alter the biogeochemical properties of the incoming surface layer. Also, during the formation of the train of internal waves over the Sill there is a creation of alternating bands with different vertical and horizontal velocities (associated to the underlying troughs and crests of the waves) that could accumulate biological material in narrow bands across the main channel of the Strait. Finally, they could also generate horizontal divergences being responsible for the suction of coastal waters which are then transported to the inner Mediterranean basin along with the main flux.

All of these processes acting together significantly influence the biogeochemical budget of the Mediterranean basin, so a great research effort has been made in recent years to elucidate the fundamental mechanisms controlling these hydrological processes and how they control and shape the biogeochemical patterns in the area. In Chapter 1 a comprehensive review of the most recent findings is presented along with some directions for future research for the enhancement of the knowledge of the tidal influence on the inter-basin biogeochemical budget.

Energy provides an essential impetus to support the development of countries in the whole world, with the increasing demand for energy, the traditional fossil energy sources will eventually be depleted in the foresee future, such as coal, oil, natural gas, and a large number consumption of fossil energy leads the enormous environmental issues, such as acid rain, greenhouse effect and pollution of river and soil, in order to the world's sustainable development, in order to future generations of breeding, renewable energy research has been regarded by governments and the general public, such as wind, solar, ocean energy, biomass, etc., renewable and sustainable energy has been applied to the daily life of ordinary people. But for now, the application of ocean energy is still relatively backward compare with the application of wind energy and solar power, and the fundamental reason is that the application of ocean energies need huge investment and the technologies either are not mature enough or are too expensive, and the gains of project are limited, and the investment has great risk, so the normal companies are unwilling to invest in this area, and the support of government is necessary. With the improved of ocean energy developed technology and the increase of traditional energy prices and ordinary people aware the important of ocean energy application, the ocean energy in the future can enter a new era. Tide is due to the earth and the celestial movement and the interaction, which lead the seawater fluctuations. In hundreds years ago, the tide is used to grind corn by the ancient human in order to reduce the labor intensity of people and livestock, and the modern use of tide is tidal energy generation. According to the statistics data, the total theoretical potential of tidal energy in the whole world is more than 3 billion kW, and the potential development of tidal energy is about 200TWh. At the same time, the total installed capacity of electric power around the world in 2004 is about 4 billion kilowatts, so the development of tidal energy has great potential and bright prospects in future. Chapter 2 analyses the formation of tide, distribution, power generation principle and application examples in the world, and forecast the prospects of tidal energy power generation and the various problems for tidal energy development.

The statistical distributions of wave pressure, orbital velocity and wave period in the vicinity of the seabed are systematically determined from the comprehensive field wave data collected at 0.5m above the seabed in finite coastal water depth. The zero-crossing method is used to analyse the field data to obtain wave pressure and orbital velocity amplitudes of individual waves. It is found that the distributions of instantaneous wave pressure and orbital velocity are identical and perfectly follow the Gaussian distribution as commonly assumed. The distributions of wave pressure and orbital velocity amplitudes are also found to be almost identical and more closely follow the Weibull and modified Rayleigh distributions than the Rayleigh distribution. The distribution of wave periods is shown to follow the normal distribution. The joint distribution of wave pressure and period is also discussed in Chapter 3. Several useful formulas are derived to compute characteristic wave pressure, orbital velocity and wave period, and the expected values of maximum wave pressure and orbital velocity. These derived formulas may be also applied to calculate characteristic wave heights based on linear wave theory.

The aim of Chapter 4 is to present an overview and statistical analysis of the sea level data obtained from the main Estonian coastal tide gauges (Pärnu, Tallinn, Ristna, Narva-Jõesuu) over the period 1842–2009. Long-term variations of both mean and extreme sea level values are studied in the eastern section of the semi-enclosed and nearly tideless Baltic Sea.

After correcting the sea level series to spatially varying postglacial land uplift rates (0–2.5 mm/yr), the series display increasing (1–3 mm/yr) trends, which are roughly equal to or insignificantly higher than the recent global sea level rise estimates. The increase is larger in winter, which is in accordance with similar seasonal structures of the NAO index trends. However, the remarkably steep rise in annual maximum sea levels (3–13 mm/yr) could be explained by the local response to the changing regional wind climate. Due to its windward location, the sea level variations in the semi-enclosed study area are sensitive to the ongoing intensification of cyclones and prevailing west winds. Maximum value analysis revealed that in the south-westerly exposed Pärnu Bay, two storm surge events (253 in 1967 and 275 cm in 2005) can not be predicted by means of return statistics. The parameters of maximum expected storm surges could be estimated on the basis of hydrodynamic modelling.

In Chapter 5 the authors made projections of relative sea-level rise, horizontal inundation, and the associated impacts on people and infrastructure in the Atlantic and the Gulf of Mexico coast of the United States. The authors first estimated the relative sea-level rise by 2100 in the study area by using (1) rates of subsidence in different localities derived from historical tide gauge observations and (2) rates of global sea-level rise from the semi-empirical model developed by Rahmstorf (2007). The range of sea-level rise projections mainly reflects equal contributions of spatial variability (due to subsidence) and the range of projected greenhouse gas scenarios, the range of projected temperature change (from global climate models) for a given scenario, and uncertainty in the semi-empirical model itself. Using the USGS 30-m National Elevation Dataset, projected sea-level rise, and local subsidence estimates, they mapped the horizontal inundation in the region, and examine the different land use categories in the inundated zones. Finally, they overlaid the inundation maps with census block data to determine the number of people and housing units located in the area vulnerable to future sea-level rise. Analysis results were summarized by state to indicate spatial variability and facilitate information usage by stake holders. In total, 31-53 thousand km^2 of land will be inundated depending on different sea-level rise scenarios; 3.8 - 6.6 million people and 1.8 - 3.2 million housing units are at risk of inundation by future sea-level rise. The risk is not distributed evenly. In general, northern states are less vulnerable because of relatively small future sea-level rise and the steep coastal terrain, whereas southern states are more vulnerable because of the higher sea-level rise, low elevation and flat terrain profile. However, some northern states such as New York and New Jersey have highly developed coast. As a result, even moderate amount of inundation area may lead to large number of people and housing units at risk of future sea-level rise.

Chapter 6 discusses shorelines, which are essential resources for almost all South American countries. However, the lack of adequate planning and management are accelerating their destruction. In Colombia, significant issues for coastal management and sustainable development include overpopulation, sewage discharges/microbial contamination, destruction of mangrove forests, industrial/port development, shrimp farming, recreation projects, sedimentation and erosion; among others. Moreover, it is quite clear that several shoreline cities, in particular at the north coast of Colombia where the most populated are found, must act quickly to start developing viable and long lasting strategies to adapt to the impacts of climate change. Similarly, it is quite urgent to promote actions to balance the conflicting interests of development with the preservation of the shoreline natural resources. This includes creating strong education programs, shoreline use/development ordinances that are environmentally-friendly and derived from technical advisory committees, campaigns to massify the use of eolic and solar energy, and conservation plans to protect and enhance the environment while encouraging public access and adequate use of the shorelines for current and future generations.

Intertidal sandy beach is full of life, as described in Chapter 7. Despite its deserted look and constantly changing conditions, intertidal sandy beach hosts diverse forms of life including meiofaunal or bacterial population, as well as single cell eukaryotes called protists. Either photosynthetic or non-photosynthetic, protists play important roles as the primary producer or the primary predator. The life as a beach protist is different from planktonic one, directly affected by the constant tidal activity that correspondingly changes physical conditions including surface and ground water level, salinity, temperature, light intensity, oxygen availability. As a result, beach protists community is different from the planktonic one even if it's from the same geographic locality. Beach protists community significantly differ within a single beach due to different conditions in microenvironments and zonation across the shore face. Until recently, protists in intertidal sandy beach received little attention, which makes this accessible environment one of the last frontier in exploration of microbial biodiversity. Interestingly, this unique also environment seems to be an experimental field of evolutionary process. The biodiversity of beach protists includes evolutionary intriguing protists, of which the most interesting is *Hatena arenicola. Hatena arenocola* is at the intermediate stage of acquiring photosynthetic ability via endosymbiosis of another microalga, demonstrating otherwise in-observable evolutionary process in real time.

As presented in Chapter 8, an expression developed for beach run-up heights for long gravity waves based on work energy theorem concept of classical physics is validated against real historical tsunami data including the 26th December 2004 Indian Ocean tsunami. It can also be applied for pilot estimation of storm surge inundation approximately as tsunamis and storm surges differ mainly in their generation mechanisms but behave in a similar way in shallow waters (long-period gravity waves). The time required by a wave to travel from shallow depths (d_1 m) to 0 m depth and shore slopes show significant nonlinear functional relationships (power series, $R^2 > 0.99$). The average deceleration ($\bar{a}$) estimated by Monte-Carlo method (Composite mid-point rule) rather than constant deceleration (a), when a tsunami wave travels from 1 m to 0 m depth seems to be more effective in estimating near shore tsunami heights for steep shore slopes (>0.3). Storm surge beach run-up height depends on the height near coastline, on the shore slopes and shallow water bottom topography over which the wave travels (as that determines the greater momentum of terminal speed) and slopes on land. Inundation is estimated for 61 cyclonic events in the Bay of Bengal during the period, 1972 - 2006.

ADCP velocity data collected in the two main sills of the Strait of Gibraltar (Camarinal and Espartel sills) have been used for analysing the vertical structure of main tidal constituents (M_2, S_2, O_1 and K_1) currents in this area. In Chapter 9, two different periods (winter and summer) were considered in correspondence to seasonal variations in density profiles. Amplitudes and phases of the various tidal constituents have been compared for both periods and locations. Barotropic and baroclinic parts of the tidal currents have been extracted using the dynamical mode decomposition technique and the relative importance of each mode has been established in terms on the energy associated. In Espartel sill, the barotropic mode is more energetic in wintertime for all constituents except for K_1. Baroclinic modes have smaller contribution to total energy. Over Camarinal sill, barotropic mode accounts for more than 90% of total energy in all the tidal constituents, the highest value (97%) observed for M_2.

Electromagnetic fluctuations in the ocean have external sources like ionospheric–magnetospheric current systems, and purely internal oceanic sources associated with interaction between water velocity fields and the geomagnetic field. The oscillations of the telluric field originated by tides were first predicted by Faraday in 1832. The physical phenomena that explain those oscillations are connected to the tidal water movement in the earth's geomagnetic field. This phenomenon has been used to estimate average water mass transport in straits, channels, throats or even in the open ocean, using submarine cables. However, the telluric field induced by tides that spreads out far inland could be used to characterize tidal phenomenon. This result opened the possibility to estimate mass transport alongshore associated with tidal flow using onshore measurement of the telluric field.

A review of the fundaments, processing and interpretation of telluric oscillations with tide origin is presented and discussed in Chapter 10. The paper presents results obtained from the analysis of data collected in two different systems located in Portugal: 1) a submarine cable crossing the channel at the entrance of a lagoon (Aveiro, Portugal) and, 2) two antennas installed close to the coast line (Portuguese west coast). Spectral analysis of the data revealed that measured voltages are dominated by semidiurnal M_2, S_2/K_2 periods. Values of 720 $m^3\ s^{-1}\ mV^{-1}$ and of 3.0×10^4 and 4.25×10^3 were estimated for the coefficient relating voltage and water transport at Aveiro channel and at the two different sites in the coast line, respectively. The results show that it is possible to indirectly measure the water transport by tidal flow measuring onshore differences of electrical potential. The method shows great potential for low cost accurate long-term monitoring of integral water transport.

Nowadays several global tidal models, most of them freely available for the community, provide tidal sea level heights practically everywhere around the World Ocean, constituting powerful tools for scientists, oceanographers and users of different areas of geosciences. In addition, most of these global models are frequently used to provide tidal heights at the open boundaries of regional high resolution coastal models. Chapter 11 is an evaluation of the performance of three of the most recognized global tidal models: FES2004 (Lyard et al., 2006), GOT 4.7, an update of GOT00 (Ray, 1999) and TPXO 7.2, an update of TPXO 6 (Egbert et al., 1994; Egbert and Erofeeva, 2002). The tidal harmonic constants from these models are compared to the ones obtained from ten sea level data series gathered at tidal stations located along the Argentine coast (from the Río de la Plata estuary to Tierra del Fuego) and from twenty data series of sea level heights obtained on the crossing of the satellite tracks from the Ocean Topography Experiment (Topex/Poseidon: T/P) on the Argentine continental shelf (in the period 1992–2002). Four semidiurnal constituents (M_2, S_2, N_2 and K_2) and four diurnal ones (K_1, O_1, P_1 and Q_1) are studied and the corresponding harmonic vectorial differences of observed sea level and altimeter data series are presented and discussed. The main differences between harmonic amplitudes and phases obtained from observations and from global models are detected in coastal waters of the Río de la Plata (a shallow area characterized by a micro tidal regime). Such differences, in general, significantly decrease towards the south, where the tidal range is greater than 10 m and the tide is practically semidiurnal.

Coral reef topographies and reef growth patterns are influenced from relative sea-level histories. Several types of reef growth patterns responding to the relative sea-level histories, e.g. balanced aggrading/onlapping, seaward prograding, back stepping, etc., have been identified in previous studies. Recently, Nakamura and Nakamori (*Coral Reefs* 2007, 26, 741–755) developed a geochemical model for coral reef formation based on diffusion-limited and light-enhanced calcification, and the model reconstructed well the reef topography and Holocene reef-growth history. In Chapter 12, we modified the model, and simulated it on four scenarios of relative sea-level histories. The simulation result on the first scenario, which is similar with global sea level history between 8,000 years B.P. and present, is very similar with balanced aggrading/onlapping type of Holocene fringing reefs. The simulations on the scenarios of stable sea level and gradual sea level falling are well-reconstructed seaward prograding type reefs. The result of the simulation on the scenario of faster sea-level rising is similar with back-stepping type reefs. Therefore, the reef topographies and growth pattern responding to relative sea level histories simulated by the model were in general well-reconstructed concerning Holocene reefs observed in nature.

As explained in Chapter 13, altimetry measurements over the Ionian region have been used to analyse the negative sea level trend over the Ionian basin in the last decades. The apparent decreasing trend should be better understood as an abrupt sea level drop in 1998 probably linked to changes in the surface circulation in the Ionian basin induced by the Eastern Mediterranean Transient, which changed from anticyclonic to cyclonic about March 1998. From then onwards, a rising rate of 7.9 ± 0.9 mm/year is observed over the basin.

Recently generalised extreme value distribution is widely used for extreme event modeling and ocean engineering applications. Hence, in Chapter 14 the characteristic function (CF) of generalized extreme value distribution is derived and the moment generating function (MGF) is deduced from it. It generates all the moments of the distribution and satisfies the tests to verify a function to be a characteristic function. Expressions for mean, variance, skewness and kurtosis are obtained from MGF. The GEV is suggested as the sampling distribution of the newly defined significant wave height distribution by the method of characteristic function. The CF of Frechet distribution, a special case of GEV is also derived. The MGF deduced from it generates all the moments of the distribution. The characteristic function has many useful and important properties which gives it a central role in statistical theory.

A topic of current interest in the analysis of sea-level states is to investigate the occurrence of future rare events which is essential for the prediction of flooding risks, coastal management and in the design of coastal defences and offshore structures. Nowadays, it is widely believed that the frequency of such rare events is increasing as a result of climatic and other changes, although they are hard to predict and their effects are, yet, poorly understood. Recent developments in multivariate statistical techniques for discrimination, clustering and dimension reduction for time series, have the potential to aid on the construction of new tools and models for forecasting the occurrence and impact of such future rare events. In studies of regional sea-level variability, tidal measurements are often analyzed individually for characterizing sea-level variability at each location. Marginal analysis, however, is in itself insufficient to come with an accurate description of regional sea-level variability. An alternative approach is to consider simultaneously the whole data set of sea-level records from a given region, and characterize regional variability in terms of locations exhibiting similar behavior through clustering techniques. Cluster analysis is a useful approach for characterizing regional variability of locations exhibiting similar behavior in terms of, for example, short-term or long-term predictions of extreme values. In Chapter 15, time series clustering is applied to the analysis of long tide gauge records from the Baltic Sea. In order to describe the regional variability of Baltic sea-level, tide gauge measurements are clustered on the basis of their corresponding predictive distributions for 25-, 50- and 100-years return values. This is relevant for the design of marine systems and coastal structures, which requires a good knowledge of the most severe sea-level conditions that they need to withstand during their lifetime, and also for describing and understanding the variability of extreme sea heights in a climate change context.

Chapter 16 presents recent developments concerning application of neural-networkbased nonlinear modeling and data analysis techniques in coastal engineering, with particular attention to analysis of the hydrodynamics of the coastal zone. The chapter provides a concise description of the theoretical basis of artificial neural network (ANN) methods (Section 3.), as well as examples of their application and a discussion of their strengths and limitations. The methods presented include: gap-filling and interpolation (Section 4.1.); pattern recognition with nonlinear principal component analysis (Section 4.2.); nonlinear regression (Section 5.1.); and hybrid ANN and physically-based models (Section 5.2.). Cross-shore and alongshore current data from a field experiment in North Carolina, USA (Section 2.) are used to illustrate possibilities and limitations of some of these data-analysis and modeling techniques. The chapter concludes with a summary of recent developments and a brief description of some other predictive-learning methods with potential applicability in coastal engineering.

In: Sea Level Rise, Coastal Engineering, Shorelines and Tides ISBN: 978-1-61728-655-1
Editor: Linda L. Wright, pp. 1-46 © 2011 Nova Science Publishers, Inc.

Chapter 1

THE STRAIT OF GIBRALTAR: TIDES, TOPOGRAPHY AND ASSOCIATED BIOLOGICAL EFFECTS

D. Macías[1,2], F. Echevarría[1], M. Bruno[3] and C.M.García[1]
[1]Departamento de Biología. Área de Ecología. Facultad de Ciencias del Mar y Ambientales. Universidad de Cádiz. Puerto Real. Cádiz. SPAIN
[2]Integrative Oceanography Division. Scripps Institution of Oceanography University of California at San Diego. La Jolla. California. USA
[3]Departamento de Física Aplicada. Facultad de Ciencias del Mar y Ambientales. Universidad de Cádiz. Puerto Real. Cádiz. SPAIN

Abstract

The Strait of Gibraltar is the unique and narrow connection between the Mediterranean basin and the open Atlantic Ocean. One of the main features of this place appears as the strong topographic constriction that happens both in the horizontal (minimum width of 14 km) and vertical (minimum depth of 250 meters) dimension in conjunction with the presence of a two-layered counter-current circulation and strong inequalities of the tidal range at both sides. These differences in tidal amplitude induce intense barotropic and baroclinic currents which interact with the sharp sea-bottom topography in the main sill of the Strait creating internal waves in the Atlantic-Mediterranean Interface (AMI). These intense and periodic undulatory processes created in the western side of the Strait modify the main along-strait circulation in a number of different ways. On one hand, they can mix the Atlantic and Mediterranean waters creating pulsating upwelling events which greatly alter the biogeochemical properties of the incoming surface layer. Also, during the formation of the train of internal waves over the Sill there is a creation of alternating bands with different vertical and horizontal velocities (associated to the underlying troughs and crests of the waves) that could accumulate biological material in narrow bands across the main channel of the Strait. Finally, they could also generate horizontal divergences being responsible for the suction of coastal waters which are then transported to the inner Mediterranean basin along with the main flux.

All of these processes acting together significantly influence the biogeochemical budget of the Mediterranean basin, so a great research effort has been made in recent years to elucidate the fundamental mechanisms controlling these hydrological processes and how they control and shape the biogeochemical patterns in the area. In this chapter a comprehensive review of the most recent findings is presented along with some directions for future research for the enhancement of our knowledge of the tidal influence on the inter-basin biogeochemical budget.

1. Introduction

The Strait of Gibraltar is quite a unique place from various points of view including the social, economic and oceanographic dimensions. From an anthropogenic perspective, it was considered to be the end of the known world for a long time and even today, it is perceived as a natural border between the developed European continent and developing Africa. It is also a particular place for migrating species as it constitutes the natural pathway for multiple animals including birds, fishes and marine mammals moving across continents or between marine basins.

From a purely geographic perspective, the Strait of Gibraltar is the Mediterranean's only communication with the world ocean, thus its significance. It extends about 60 kilometers in a southwest-northeast, 15 km wide at its narrowest section (Tarifa narrows, red line in Figure 1) and only 280m deep at its main sill (Figure 1).

Coastal topography is quite complicated in the Strait as it presents both capes and bights with sharp latitudinal and longitudinal gradients, which interact with the main currents creating a very complex and variable circulation pattern. Bottom topography highlights the presence of two sills. The main sill, known as Camarinal Sill, is very shallow and defines the smallest section that plays like a bottleneck for water exchange between the basins. The second, Spartel Sill, is deeper and placed to the west on the channel that runs in a southerly direction along Majuan ridge by which the main part of the Mediterranean water drainage takes place. Between the two sills is the Tangier basin (with a maximum depth of over 600 m), which is a small reservoir of significant importance for tidal dynamics. To the east of the Camarinal Sill, the bottom sharply falls to 900 m in the eastern side of the Strait (Figure 1).

The mean water circulation through the Strait of Gibraltar is mainly created and maintained by the negative hydrological budget of the Mediterranean Sea where evaporative losses exceed the freshwater income by both riverine discharges and direct rainfall. These hydrological conditions induce a well-known inverse estuarine circulation along the main channel of the Strait (Lacombe and Richez, 1982; Armi and Farmer, 1988; Hopkins, 1999). This circulation is basically composed by a surface eastward flow of nutrient-poor, open-ocean Atlantic waters and a deep outflow (i.e. westward) of nutrient-rich Mediterranean waters which leads to a natural tendency to oligotrophy in the entire Mediterranean basin. The biogeochemical budget of the Mediterranean Sea depends on the water exchange through the Strait of Gibraltar, as well as atmospheric and river inputs (Béthoux et al., 1998). Hence, the study of the balance of water and elements trough the Strait as well as its dynamics has implications not only at the regional level, but, also for large basin scale budget calculations (Packard et al., 1988; Minas et al., 1991).

Over annual timescales the water exchange through the Strait of Gibraltar can be regarded as a nearly constant inflow of Atlantic waters towards the Mediterranean in the upper layer (Atlantic Jet, AJ thereinafter) and a nearly constant outflow of deep Mediterranean waters toward the below Atlantic (i.e. the antiestuarine circulation presented above).

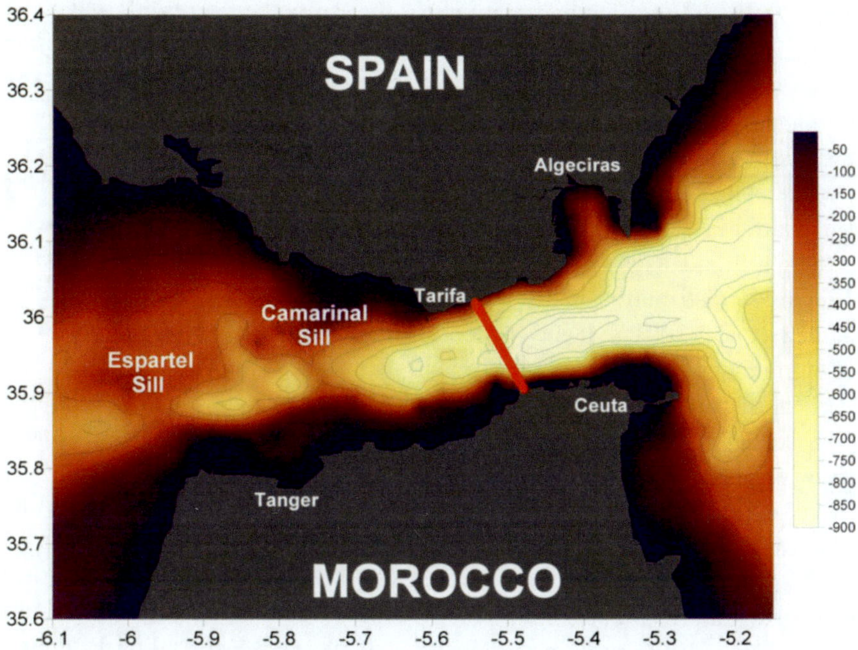

Figure 1. Strait of Gibraltar. Isohalines from 900 meters each 100 meters. Main cities and bottom signatures are shown. Red line marks the position of the Tarifa Narrows.

The magnitude of these virtually regular flows has been estimated by direct measurement (e.g. Bryden and Kinder, 1988; Pettigrew, 1989; Bryden et al., 1994; García-Lafuente et al., 2000; Tsimplis and Bryden, 2000) and numerical models (Wu and Haines, 1996; Sein et al., 1998; Hopkins, 1999; Sanino et al., 2002) and are considered to be basically dependent on the climatic conditions over the Mediterranean area and of the Strait's main geographical characteristics (Bryden and Kinder, 1991; García Lafuente and Criado, 2001).

However, this description of the along-strait circulation is a simplification of the real one given that at least three different water masses can be observed in the region: Surface Atlantic Water (SAW), North Atlantic Central Water (NACW) (which together constitute the AJ) and Mediterranean Outflowing Water (MOW) (Gascard and Richez, 1985); the amount of each one is strongly dependent on multiple factors including tidal height (Gascard and Richez, 1985) and the along-strait position (Bray et al., 1995).

Another consequence of tidally-intensified currents comes from their interaction with the bottom topography. As stated above, bottom depth changes dramatically near the Camarinal Sill (Figure 1). The interaction between flows and this sharp topography leads to the Atlantic Mediterranean Interface (AMI) changing its position abruptly too. This is particularly true during certain phases of the tidal cycle, when the abrupt changes of the AMI are related either to the formation of internal hydraulic jumps (Boyce, 1975; Armi and Farmer, 1985), a phenomenon that prevails during moderate to strong (spring) tides, or arrested internal waves (Bruno et al., 2002), which are more the usual during weak (neap) tides. Such undulatory processes enhance interfacial mixing (Wesson and Gregg, 1994) and can inject deep, nutrient-rich water into the upper layer of Atlantic water. The upwelled inorganic nutrients are advected towards the Mediterranean Sea in the upper layer, enhancing, as a result, the primary production in the Alboran Sea to the east of the Strait. The turbulence-favouring abrupt nature of hydraulic jumps relative to the smoother arrested waves suggests a fortnightly cycle for mixing and thus, for the exchange of dissolved substances. The simultaneous occurrence of enhanced mixing and strong tidal currents can give rise to positive correlations between nutrient concentration and tidal flows, in which case, the mean flux into the Mediterranean needs to be revised estimated above by assuming steady flows.

Also, water composition of the AJ has been described as being influenced by the tidal forcing. For example, the quantity of the less-abundant NACW is clearly linked with tidal amplitude (Gascard and Richez, 1985, Macías et al., 2006) or wind regime (Gómez et al., 2004). Water mass distribution in the AJ and biological patterns are highly related in this region (Gómez et al., 2001; Macías et al., 2006, 2008) characterised by a discontinuous [horizontal entrainment] transport of chlorophyll patches into the Alborán Sea.

The vertical distribution of chlorophyll is also expected to be linked to the dynamics of the water masses, which show clear vertical segregation, with SAW at the surface, MOW in the deeper layers and NACW at mid-depths. Thus, three important contact zones can be defined, SAW-NACW, SAW-MOW, and NACW-MOW, where according to previous observations (Macías et al., 2006) using *in vivo* chl *a* fluorescence, deep chlorophyll maximum (DCM) are likely to occur. Those DCMs have also been characterised by their biogeochemical signature (Macías et al., 2008) with a great diversity associated also with the tidal intensity. These differences in composition of the DCM according to their origin and position could be of particular relevance for the pelagic ecosystem of the Alboran Sea as the discontinuous entrance of biogeochemical material should affect the composition and behaviour of the planktonic community.

In the present chapter we will (1) describe the main characteristics of the flow and tidal dynamic within the Strait of Gibraltar and in the nearby areas of both the Atlantic Ocean and the Mediterranean Sea. Afterwards we will present some of the most prominent biogeochemical signatures in the region associated with the tidal forcing including (2) along-strait dynamic associated with interfacial mixing events, (3) diversity and dynamic of the planktonic community composition in the main channel and, finally, (4) coastal-channel interactions in the Camarinal Sill region and associated effects on the biogeochemical budget between basins.

2. Flows, Tides and Associated Hydrology in the Strait of Gibraltar

As already pointed out in the introduction to the chapter, the water flux through the Strait of Gibraltar is quite complex and it is possible to identify at least four main components (Lacombe and Richez, 1982; Candela, 1991): a tidal, mainly barotropic flow, with magnitudes up to 2.5ms^{-1} (Candela et al., 1990); a barotropic subinertial component (with periodicity ranging from days to several months) driven by atmospheric pressure fluctuations within the Mediterranean and with magnitudes close to 0.4ms^{-1} (Candela et al., 1989); a long-term baroclinic component driven by the internal pressure gradient due to the density difference between the Mediterranean and the Atlantic Waters, with magnitudes of about 0.5 ms^{-1} (Bryden et al., 1994); and subtidal currents that are associated large amplitude internal waves induced by the interaction of tidal flows with vertical stratification and bottom topography which are mainly generated around Camarinal Sill (Armi and Farmer, 1988; Bruno et al., 2002; Vázquez et al., 2006).

Therefore, from a physical point of view the Strait is a very energetic system with long-term, subinertial, tidal and subtidal currents all being of significant amplitude. In this next section of the chapter each of the afore mentioned temporal scales of the flux will be presented and analysed.

2.1. Long-Term Dynamics

Long-term currents are generally characterised by a two-layer flow with the Atlantic water flowing in the upper layer towards the Mediterranean and the Mediterranean water doing that in the lower layer towards the Atlantic. The reason for this two-layer exchange is to compensate for the negative budget of water due to strong evaporation in the Mediterranean Sea. The upper layer flow (Atlantic inflow) is driven by a sea level difference between western and eastern sides of the strait of about 15 cm (García-Lafuente, 2008) while the lower layer flow (Mediterranean outflow) is driven by the density difference between the two basins.

Surface Atlantic Water (SAW) and, to a lesser extent, North Atlantic Central Water (NACW), whose salinity varies between 36 and 36.5 compose the Atlantic inflow that originates in the Gulf of Cádiz. Most Mediterranean outflow consists of Levantine Intermediate Water (LIW), which is formed in winter in the eastern Mediterranean (Rhodes Basin) and returns as an intermediate counter-current through the Strait of Sicily (Pettigrew, 1989), reaching the Strait of Gibraltar once it skirts the European coast in an anti-clockwise motion. Western Mediterranean Deep Water (WMDW) is the second kind of outflowing water and forms in winter in the Gulf of Lion. It is colder, less saline and slightly denser than LIW and is located beneath it. The volume of WMDW formed depends on the harshness of winter. Consequently it shows a high inter-annual variability in the Strait (García Lafuente et al., 2007). Due to the strong vertical mixing affecting to LIW and WMDW they are indistinguishable when they cross the Strait towards the Gulf of Cadiz.

In the Strait, salinity is the variable that separates the upper and lower layer flows. It is usual to take the surface of 37.5 as a reference. The time averaged position of this surface, commonly known as interface, slopes up along the axis of the Strait from 200-250 m offshore of Cape Spartel to less than 100 m at the eastern boundary. There is also a cross-strait slope induced by the Earth's rotation that raises the mean position of the interface from South to North.

Long-term flows through the Strait show a clear seasonal variability that is clearly linked to the formation of WMDW (García-Lafuente, 2008), as it affects the outflow characteristics, being slightly increased and colder in late spring and diminishing at the end of the year. Furthermore, seasonal warming of surface waters, which causes a marked density contrast in the summer, enhances the Atlantic inflow in late summer.

2.2. Subinertial Dynamics

On subinertial time-scale (days to a few months), the variation in atmospheric pressure over the Mediterranean basin is the main cause of fluctuations in the current intensity through the Strait. The physical mechanism is related to the isostatic response of sea level variations to the atmospheric pressure fluctuations over the western Mediterranean (Crepon, 1965; Candela et al., 1989; García-Lafuente et al., 2002). Also, subinertial flows are almost 180º out of phase with respect to the atmospheric pressure fluctuations for oscillation periods ranging from 3 to 80 days. In addition to these meteorologically forced flows, there is a significant part of the subinertial fluctuations, showing a baroclinic behaviour with opposite current directions below and above the interface, that is responsible for the temporal variations of the vertical shear of currents, which shows its maximum/minimum during neap/spring tides. Bryden et al. (1994) pointed out variation in tidal mixing intensity between spring and neap tides cycles might be a likely mechanism of explaining this behaviour.

2.3. Tidal Dynamics

It is customary in tidal analysis to distinguish between barotropic and baroclinic components of the tide. The barotropic part is characterised by the vertical displacement of the sea level (surface tide) and the currents forced by the horizontal gradient. On the other hand, the baroclinic tide is characterised by the vertical displacement of the isopicnal surfaces (internal tide) and the currents originated by the induced horizontal gradients of density.

Concerning the barotropic or surface tide, the Strait of Gibraltar links the large tidal range of the Atlantic Ocean (exceeding 3 m during spring tides) with that of the Mediterranean, where tides are, in general, practically non-existent. In the Strait the dominant oscillations are of semidiurnal period, with amplitude of 1 m at the Western side decreasing down to 0.3 m at the Eastern side (Candela et al., 1990). Once inside the Mediterranean it becomes null in Alicante at the Eastern end of the Alboran Sea basin.

Barotropic tidal currents are directed towards the Atlantic between low and high tide, carrying the water masses demanded in order to adjust the level of the oceanic high tide. During the ebb the tidal currents flow towards the Mediterranean, draining water to adjust the low tide in the oceanic side of the Strait.

Since the amplitude of the tidal currents is considerably greater than the mean currents, a periodical reversion of the current direction would be expected, at all depths of the water column. However, it does not occur at all of the places along the Strait due to the interaction of intense tidal flow with the abrupt bottom topography and the strong vertical stratification of the water column, which originates an important internal tide. As a result of this interaction, the surface current in the easternmost part of the Strait is never reversed by tidal currents and the same occurs with the Mediterranean outflow in the westernmost area.

2.4. Subtidal Dynamics. Large Amplitude Internal Waves

The interaction of the barotropic tidal flow with the main sill (Camarinal Sill) topography and the stratified water column, primarily causes internal tides that by non-linear processes and non-hydrostatic dispersion evolve into large-amplitude (more than 100m) internal waves exhibiting much shorter period and wavelength than those related to the basic tidal variability (Richez, 1994; Bruno et al., 2002).

As discussed by Bruno et al. (2002), once internal waves are formed they are trapped on the lee side of the sill because of the establishment of critical or supercritical conditions over the sill and that is an important factor explaining the growth in amplitude that internal waves experience here. In practise, critical/supercritical conditions mean that celerity of the internal undulations is equal/less than a depth-averaged current intensity flowing in the opposite direction. Some of the internal undulations generated around Camarinal Sill during the flooding phase of tidal currents, are also propagated towards the Atlantic. However, they are almost imperceptible because (i) they are rapidly advected towards the Atlantic (having no time to grow) and (ii) because of the deeper position of the interface west of Camarinal Sill.

The release of the internal waves towards the Mediterranean begins with the establishment of subcritical conditions over the sill (i.e internal wave celerity is greater than depth-averaged current). It happens almost at the beginning of the inflow phase of the barotropic tidal currents (towards the Mediterranean Sea). In Figure 2 the different stages of the internal wave generation are illustrated.

These internal waves have been found to be one of the major contributors to the mixing between the Atlantic and Mediterranean layers within the Strait (Wesson and Gregg, 1994; Macias et al., 2006), being able to have significant remote effects on the hydrography of the Alboran Sea (Vázquez et al., 2006). This fact confers special importance to the study of internal wave phenomena in the Strait of Gibraltar.

Vázquez et al. (2008) have inferred on an empirical basis, using a long record of ADCP profiles over Camarinal Sill, that large amplitude internal waves are generated when barotropic tidal currents during flood tide (flow towards the Atlantic) reach an intensity of 1 m/s. Also, the release of the internal waves takes place when intensity is reduced down to 0.5 m/s. This result enables the use of tidal current predictions over the Camarinal Sill to forecast occurrence of large amplitude internal waves around the Camarinal Sill. However, the same authors have found that the empirical model fails in some occasions due to the effect of subinertial flows. They concluded that modification of the hydraulic conditions over the sill produced by subinertial flow variations (forced by atmospheric pressure fluctuations in the western Mediterranean) can produce an activation/inhibition of internal wave events during neap/ spring tides. This could create the possibility of activating these internal wave events during neap tides, when predicted hydraulic conditions over the sill do not favour their occurrences, and to inhibit their generation during spring tide conditions, when large internal wave events are expected.

Once internal waves are released, they travel towards the Alboran Sea. The time spent to arrive to the Alboran Sea is significantly affected by the diurnal inequality of tidal currents and also by subinertial flow variations that must exert some effect at a longer time scale. Sánchez-Garrido et al. (2008) have reported due to the diurnal inequality, the travel time of internal waves between Camarinal Sill and the eastern side of the Strait ranges from 8 to 11 hours.

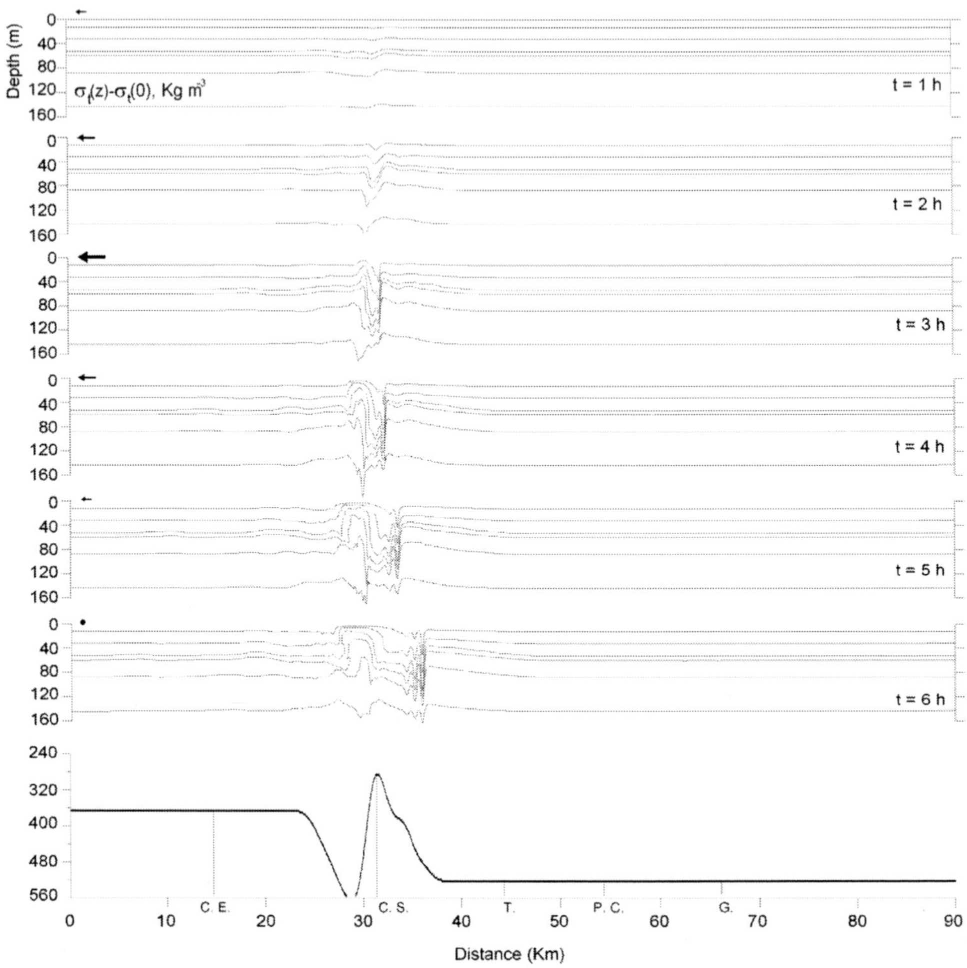

Figure 2. Vertical sections of density due to internal tide as simulated with a non-hydrostatic numerical model in an idealised Strait of Gibraltar. Arrows on the left upper corner of each section indicate the direction and intensity of barotropic tidal currents at Camarinal Sill (CS). At time t=2 h internal tide begins to be generated at CS left flank of the perturbation propagates against the barotropic flow. At time t=3 h perturbation is trapped at Camarinal Sill due to the critical conditions there and a hydraulic jump (internal bore) is generated. At t= 4 h, hydraulic jump disintegrates into several undulations and finally at t=5 h internal perturbations are released towards the Alboran Sea.

3. Along-Strait Dynamics. Mixing and Advection

As stated in the previous sections of the chapter, the most important and energetic processes in the Strait happen in the West-East direction along its main channel including the two-layer circulation and the intense interfacial mixing (between Mediterranean and Atlantic layers) in the internal-waves generation area (i.e. the Camarinal Sill). Thereby, it seems reasonable to make a first approach to the general dynamics of the area by focusing in the along-strait processes of water flow and mixing events.

Although the reported water flux values across the Strait differ from each other, a reasonable value of general agreement is around 0.8 Sv (1 Sv = $1\text{x}10^6$ m^3 /s), the inflow being around 5% greater than the outflow in order to compensate for evaporative losses in the Mediterranean Sea. These mean fluxes (in and out of the Mediterranean basin) correspond to 300 times the mean flux approximately of the Nile River, which implies that, every second, roughly 600 Nile Rivers are being interchanged through the Strait of Gibraltar.

This long-term average circulation pattern exhibits large fluctuations at different time scales as presented in the previous section of the chapter. Seasonal and subinertial (meteorologically-induced) fluctuations of, typically, 0.1 Sv and 0.5 Sv respectively have been reported (Candela, 1990; García-Lafuente et al., 2002), but the main source of variability is tidal. There are significant differences in tidal amplitude between the Western and Eastern parts of the strait, and this induces barotropic and baroclinic tidal currents along the main channel (Lacombe and Richez, 1982) whose amplitude can be up to 4 Sv during spring tides, more than four times greater in magnitude than the time-averaged flow (García-Lafuente and Vargas, 2003).

An interesting and curious fact pointed out by Bryden et al. (1994) is that tides contribute to the mean exchange through the positive correlations between the position of the interface separating Mediterranean and Atlantic waters (AMI) and the strength of the tidal currents. Bryden et al. (1994) showed that, on average, almost half of the flow exchange measured in the main sill of Camarinal, occurs by virtue of this correlation. The analysis of Vargas et al. (2006) confirmed this mechanism (which they named tidal-rectification of flows) they went further and showed that tidal rectification dominates the exchange during spring tides and is negligible during neap tides. The question remains open as to whether or not the exchange of other substances also follows a pulsating pattern related to the fortnightly cycle of tides, an issue that becomes more complex if mixing is taken into account.

These reported values of interchanged flows have their reflection on the biogeochemical budget between both marine basins. For example the nitrate concentration of the exchanged waters is estimated as 1.2 mmol N/m^3 in the inflowing Atlantic waters and 9.6 mmol N/m^3 in the outflowing Mediterranean waters (Gómez et al., 2000b; Minas et al., 1991; Dafner et al., 2003). With these concentrations and using the flow estimates of Bascheck et al. (2001), the nutrient fluxes towards the Mediterranean Sea and towards the Atlantic Ocean would be 972 and 7296 mol N/s, respectively. Thus, the Mediterranean Sea would export a net amount of 6324 mol N/s (or 2914 ton N/year) to the global ocean through the Strait of Gibraltar. However, this is only an average calculation, which should be greatly influenced by the intense and quick hydrological (mainly tidally-related) processes happening within the Strait.

Due to the high frequency of such processes it is difficult to study them by using the classic approach of vessel-based field-data sampling because of the time needed to take and process the samples. Thereby, alternative methods should be used to adequately describe such quick processes. In this sense two alternative but complementary approaches have been taken during the last half-decade of oceanographic research in the area. On the one hand, a specific-designed boat-based sampled was conducted by performing observation on a fixed position within the main channel of the Strait along several tidal cycles. This sampling strategy allows to register the characteristics of the Atlantic Jet (AJ) coming through the Strait along the tidal cycle and relate them with the timing and amplitude of the tide.

Another alternative is to use numerical hydrological-biogeochemical coupled models. There have been several attempts to simulate the water circulation through the Strait of Gibraltar using hydrodynamic models (e.g. Wang, 1989 and 1993, Brandt et al., 1996, Sein et al., 1998). These models have different levels of complexity ranging from the simplest one-dimensional two-layer model to 3D coarse resolution models. Recently, high spatial-resolution models allowing for realistic bottom topography-flow interaction have been developed: both two-dimensional, two-layer types (Izquierdo et al., 2001; Castro et al., 2004) and three-dimensional ones (Sannino et al., 2002, 2004). All of them have some degree of success in simulating the short-scale undulatory phenomena over the Camarinal Sill but they are either unable to deal with mixing (two-layer models) or mixing was not specifically addressed in the study (Sannino et al., 2004).

Less attention has been paid to the development of physical-biological coupled models, however, which explore the effect of the strong advection and mixing processes on biogeochemical exchanges and the behavior of the pelagic ecosystem in the Strait and adjacent marine regions. A conceptual model of the plankton distribution in the Strait was proposed by Gómez et al. (2000a) and Echevarría et al. (2002). These papers related the quasi-permanent enrichment of phytoplankton biomass in the North-Eastern area of the Strait to mixing processes over the Camarinal Sill and the subsequent eastwards advection of the water masses. However, these authors did not carry out numerical calculations to test this conceptual model.

The circulation and its variability are known to modify the distribution patterns of biological variables in the Strait (Gómez et al., 2001; Macías et al., 2006) and in its neighbour areas (Mercado et al., 2005). Therefore, the correct simulation of the pelagic community in this area must be achieved by means of a physical-biological coupled model in which the physical part can resolve short time-scale dynamics features such as tidal mixing. Physical-biological modelling is in actuality a fruitful approach promoted by international programs as GLOBEC.

In this framework, a 2-layer 1-D model was developed by Macias et al. (2007) by extracting an along-strait W-E section (red line in Figure 3a) from the advanced hydrodynamic model of Izquierdo et al. (2001). This 1D model was created to improve the representation of vertical mixing and a biogeochemical model was coupled to the hydrodynamic one to examine the effect of the mixing and advection processes in the main channel of the Strait on the distribution patterns of biogeochemical fields in the region.

In this section of the chapter the predictions of the model developed by Macías et al. (2007) will be used to explore the consequences of the tidal dynamics on the biogeochemical budget through the Strait. The validity of this 1D approach will be checked by comparing the model predictions with the field observations taken on the diel cycles at a fixed position in the main channel of the Strait.

However, these two approaches focused on the temporal evolution of the hydrological and biogeochemical signatures in the Strait, sacrificing the vertical resolution of such patterns to get the adequate temporal resolution. Nevertheless, it is expected that those processes also have consequences on the vertical distribution of both water masses and pelagic biomass within the Strait.

It is widely recognized, for example, that one of the most distinct and ubiquitous biological features in the world's ocean is the presence of a Deep (or sub-surface) Chlorophyll Maximum (DCM) associated with discontinuities in the water column, such as the seasonal thermocline or the presence of haloclines, particularly where different water masses meet. The existence of a DCM, in conjunction with seasonal variations in the depth of the mixed layer, is of crucial importance to explain the annual cycle of primary production in the open ocean (Gran, 1931; Sverdrup, 1953). Hence, coupled biological-physical studies in regions with marked vertical gradients (such as the Strait itself) should focus on the vertical dimension, since it is presumably in these interfaces characterized by marked gradients, where phytoplankton cells tend to accumulate (Mann and Lazier, 1991; Rodriguez et al., 1998).

The relationship between DCM, water masses and hydrodynamic conditions has been extensively studied both in the Mediterranean basin and in the Gulf of Cadiz, the two oceanographic regions connected by the Strait of Gibraltar (Estrada et al., 1993; Morán et al., 2001; Navarro et al., 2006); its distribution within the Strait itself should be far more complex because of the simultaneous presence of the different typical water masses (SAW, NACW and MOW) leading to the presence of multiple density interfaces. This complex dynamic will be described in detail in the final part of this section. All of this information will serve to propose a conceptual 2D model which represents patterns of water mass circulation and biological variables distribution in relation to the tidal cycle within the Strait, expanding, thereby, the conclusion obtained with the more simplistic 1D approach.

A brief description of the used numerical model and the field data samplings is presented in subsection 3.1; subsection 3.2 focuses on describing the 1D-along-strait dynamic based on the diel cycle observations and the model simulations while in subsection 3.3 the conceptual 2D (vertically) model of water circulation and DCM dynamics is presented.

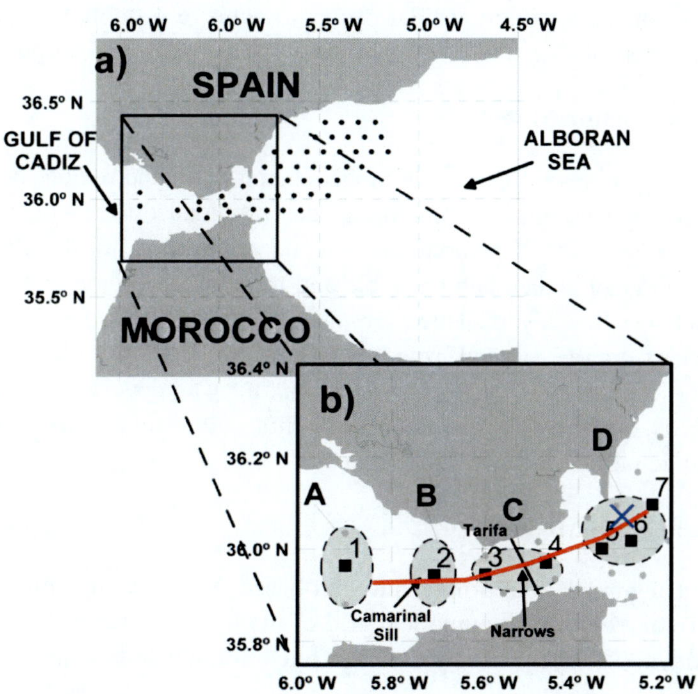

Figure 3. a) Grid of stations sampled in the Strait and NW Alboran Sea. b) Stations selected within the main channel of the Strait to create the conceptual model of water masses circulation and DCM position. Blue line marks the section simulated with the biological-hydrological coupled model and the blue cross indicates the position of the diel sampling used to compare with the model.

3.1. Models and Field Data Sampling Descriptions

3.1.1. Hydrodynamic Model

The physical part of the physical-biological model is a 2D, nonlinear, two-layer, free-surface, hydrostatic model with boundary-fitted curvilinear coordinates. Sea-water density is uniform and prescribed in each layer. A complete model description, including governing equations and parameter values used can be found in Izquierdo et al. (2001). From the complete grid of the hydrodynamic model, an along-strait section was selected (red line in Figure 3a) situated in the center of the main channel of the Strait.

The model is forced at the open boundaries with radiation-type boundary conditions ensuring that when short-wavelength disturbances in the fields of variables are generated they all propagate away from the region of interest.

At the coastal boundaries a condition of null normal flow is applied. In order to reduce the influence of any inaccuracies in boundary forcing upon the sought-for solution the waves produced within the strait are allowed to propagate freely through its open boundaries. The M_2, S_2, K_1 and O_1 surface tidal elevation amplitudes and phases used to set the tidal forcing at the open boundary grid points were derived by interpolating the relevant values from a 0.5° gridded version of the FES95.2 global tidal solutions of Le Provost et al. (1998). The model was run for 30 identical semidiurnal tidal cycles to achieve a stable time-periodic solution. After establishing this solution, the model run was continued for a 13-month period.

3.1.1.1. Mixing-Advection Model

The physical processes influence the biogeochemistry of the region through mixing and advection. Advection velocities for all tracers are provided directly by the model output. Mixing, which is responsible for variations of concentration, is not directly computed from the hydrodynamic model, which is immiscible and therefore does not allow for any exchange of properties between the layers. However, taking into account that interfacial mixing is strongly dependent on the vertical velocity shear across the interface (Briscoe, 1994) a parameterisation scheme to estimate interfacial mixing was developed. The success of this parameterisation is supported by the good reliability of the tidal current predictions of the used hydrodynamic model (Brandt et al., 1996).

3.1.1.2. Biogeochemical Model

The biological model is a simple nitrogen-based Nutrient-Phytoplankton-Zooplankton (NPZ) model. To agree with the numerical grid of the hydro-dynamical model for the along-strait transect, the upper layer was divided into fixed volume cells with the same along-strait length.

Boundary and initial conditions for the concentration of nutrients, phytoplankton and zooplankton come from the analysis of more than 150 field data taken all over the Strait at five different depths during four different cruises carried out in several years being in good agreement with previous observations in the region (Gómez et al., 2000b; Minas et al., 2001; Dafner et al., 2003).

An NPZ model is a very simple description of the pelagic plankton ecosystem. A more complex model, like that described by Fasham et al. (1990) with seven different compartments including bacteria and detritus, was also tested but the differences in model performance were very small for all N, P and Z.

3.1.2. Diel Cycle Sampling

Data were collected during four different cruises carried out from November 2002 until November 2003 in the Strait of Gibraltar. In each cruise one or two fixed stations were sampled in the Eastern side of the Strait (blue cross in Figure 3b).

In each fixed station (3 to 26 h total observation times) several CTD profiles were made following an interval from 0.25 hours to 1 hour. In each profile, salinity, temperature and fluorescence distribution were sampled from surface to 300 meters depth by using a combined CTD probe. This limit of depth was chosen to ensure observation of the Atlantic-Mediterranean-Interface (AMI) and MOW but increasing time resolution. The dates of the fixed stations were selected to include different tidal amplitude cycles, from spring to neap tides, with associated along-strait current velocities ranging from 2.5 m/s to 0.7 m/s.

3.1.3. DCM Sampling

Data were obtained during two simultaneous cruises in November 2003 performed in the area of the Strait of Gibraltar and western Alboran Sea on board two research vessels, BIO *Hespérides* and BO *Mytilus*. A grid that covered the studied area (Figure 3a) was surveyed twice at different tidal amplitudes (spring and neap tides) and wind regimes (westerlies and easterlies). From the complete grid, a set of seven stations that characterised the main hydrographic features of the Strait was selected (Figure 3b). Station 1 was used to illustrate the characteristics of the water column on the western side of the Strait (region A, Figure 3b). Station 2 described the water column on the main sill of the Strait (region B, Figure 3b). The vertical distribution of variables in the central section of the channel (region C, Figure 3b) was examined using the casts performed in stations 3 and 4. The description and monitoring of the Eastern sector of the Strait (region D, Figure 3b) was based on data acquired at stations 5, 6 and 7. Each station was sampled several times making a total of 32 observations within the main channel of the Strait.

Basic information on physical structure of the water column at each station was obtained from CTD casts while discrete water samples were also collected at different depths to analyse several variables related to the dynamics of the pelagic ecosystem. Each observation was classified according to the specific period of the tidal cycle to generate the conceptual model of vertical distribution of water masses and biogeochemical properties along the main channel of the Strait.

3.2. 1D (Horizontally) along-Strait Dynamics

3.2.1. Field Data

Tidal induced dynamics can be studied by analysing the time evolution of basic descriptive variables (temperature, salinity, chlorophyll fluorescence and tidal current speed) recorded during the seven samplings events (Figure 4). The details about the prediction of the tidal current shown in plate d could be revised in Alonso et al. (2003). This figure is but an example of the seven sampling performed on the Eastern entrance of the main channel of the Strait (see Figure 3a) and it could be used to track the presence and vertical position of the different water masses that appears in the Strait.

Each CTD cast was classified into two different types; when NACW was clearly present it was assumed that no significant interfacial mixing (between Atlantic and Mediterranean waters) have taken place, these casts were, thereby, classified as "not mixed". On the other hand, if no NACW was detected the TS diagram (see details in Macías et al., 2006) depicted a progressive gradient from surface SAW to deep MOW, indicating some degree of interfacial mixing, being then classified as "mixed". These two types of water masses appear clearly separated in time during each cycle at the fixed stations. The presence of mixed waters is usually detected around HW-4 (i.e. 4 hours before high-water), when the current velocity over the Sill changes from inflowing (positive values in plate d Figure 4) to outflowing (negative values in plate d Figure 4).

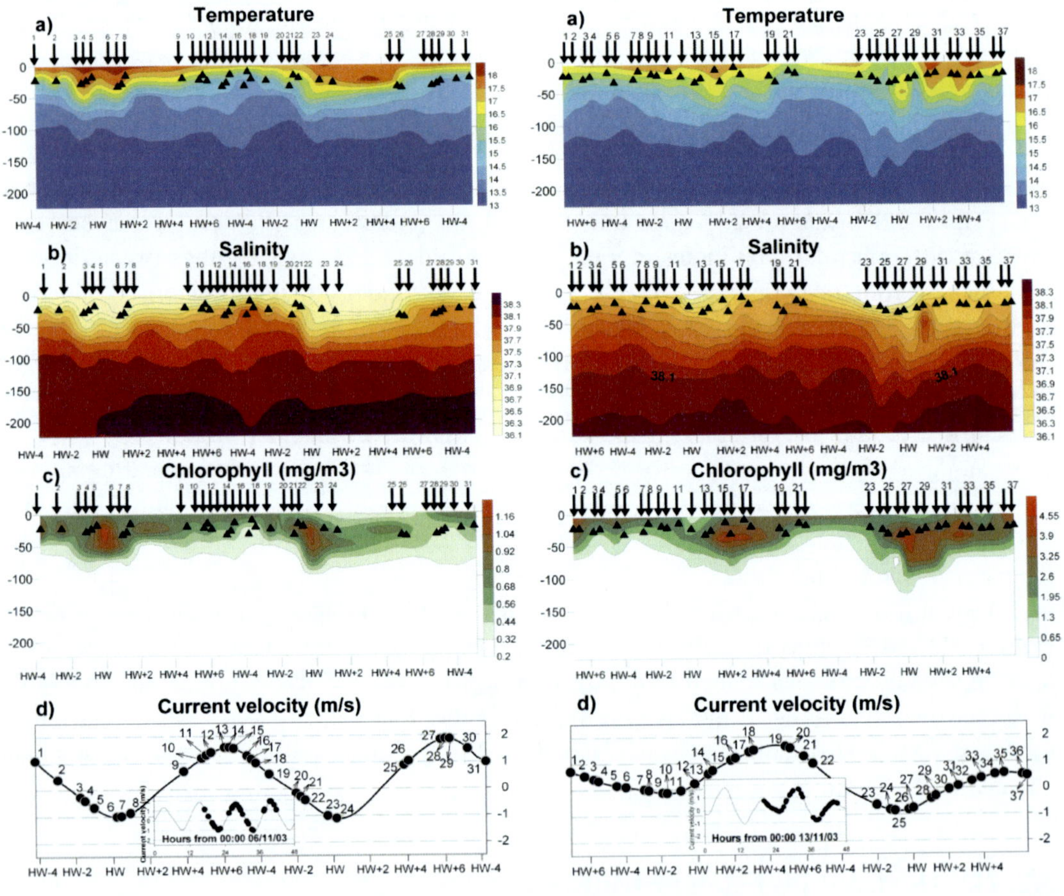

Figure 4. Temporal evolution of salinity (p.s.u.), temperature (ºC), Chlorophyll (mg/m^3) and current velocity prediction over the Camarinal Sill (m/s) during the diel cycle sampling. Squares showed the presence of mixing of AMW and the triangles the presence of MMW. The arrows and circles mark the moments when CTD profiles were made.

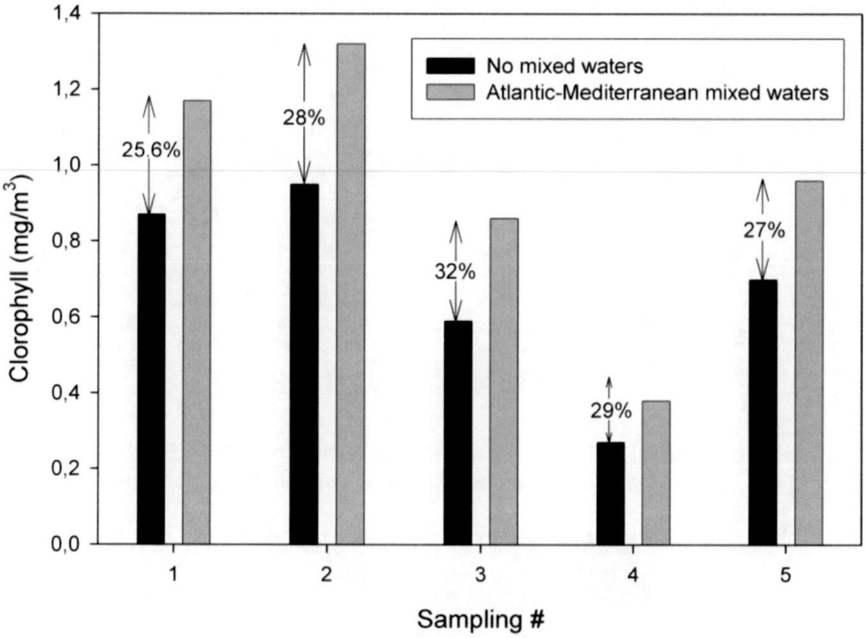

Figure 5. Mean chlorophyll concentration in the upper 75 meters of the water column during the different sampling in the fixed station. Black bars correspond to moments of no mixed water column and grey bars to mixed waters.

Several undulatory phenomena were detected during the different sampling events, generally associated with the presence of such mixed waters (Figure 4). These disturbances are marked by vertical displacement of isotherms and isohalines and can be related with the internal waves generated over the Camarinal Sill, travelling within the main flux towards the Mediterranean.

Another common pattern observed is that the highest levels of chlorophyll coincide with the presence of these mixed waters and internal waves (Figure 4), suggesting a strong coupling of the physical forcing and the biological patterns that will be discussed later. Actually, mean chlorophyll concentrations in the mixed waters were in average 28% higher than in the Atlantic waters (Figure 5) independently of the sampling date and of the absolute value of chlorophyll concentration.

3.2.2. Model Simulations – Field Data Comparison

In order to evaluate the validity of the mixing and advection schemes adopted in the model a non-reactive tracer such as salinity was firstly chosen to be compared with the field data, as the temporal evolution of such variable should not be affected by the biogeochemical simplifications assumed by the model. Figure 6 shows the salinity in the upper layer predicted by the model (blue line) and the actual data taken from the field data sampling presented above (black dots) along 24 hours during a spring-tide period. The observed and predicted values were in the same range and the temporal dynamics very well reproduced by the model (see Macías et al., 2007 and 2007a).

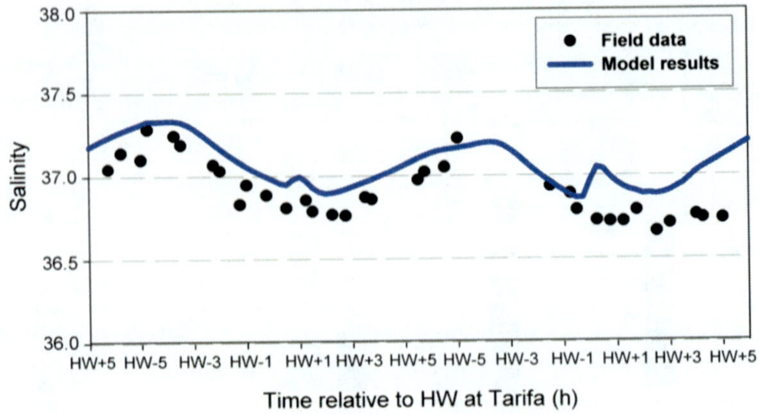

Figure 6. Measured (black dots) and modelled (blue line) upper layer salinity along 24 h. at the Eulerian station using the coupled model. Horizontal axis is time referred to the High Water (HW) at Tarifa.

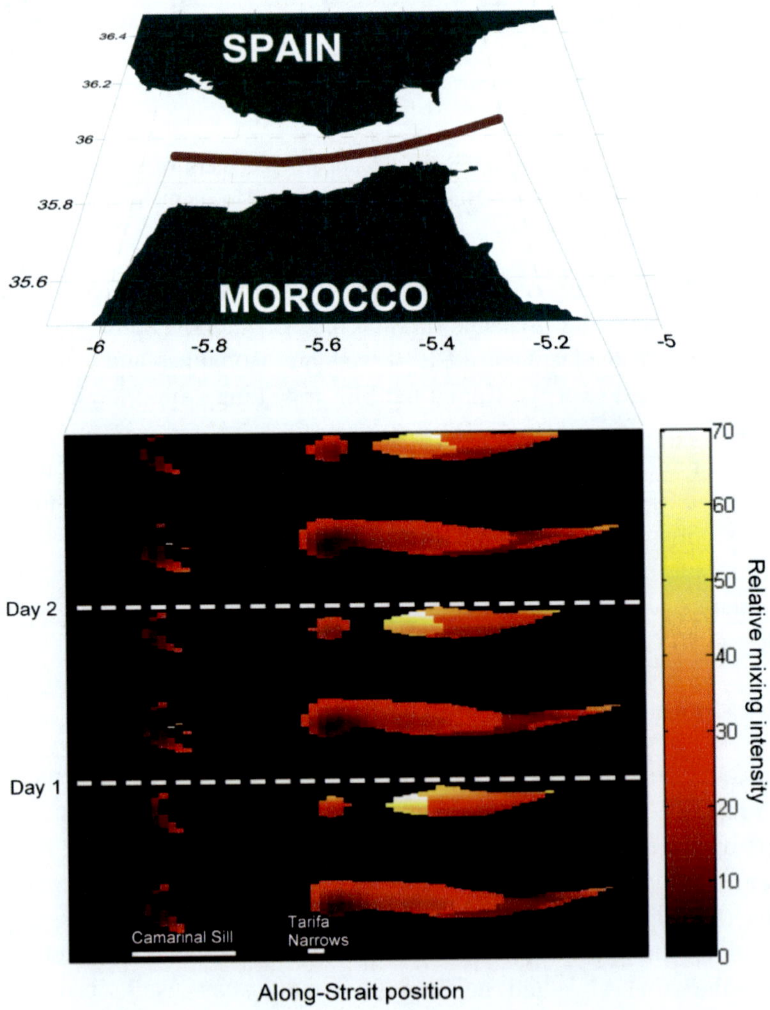

Figure 7. Mixing intensity (in relative units) between upper and intermediate layers during three days of simulations calculated according to the mixing-advection model.

This model could also be used to evaluate the temporal and spatial (in the along-strait direction) distribution of the interfacial mixing intensity (Figure 7). Unsurprisingly, there is intense and pulsating mixing over the Camarinal Sill although the most intense interfacial mixing occurs eastwards of Tarifa Narrows (see label in the bottom panel of Figure 7). This happens because in this part of the Strait the interface tends to be shallower (Bray et al., 1995) and thicker (García-Lafuente et al., 2002) than elsewhere. The upper layer becomes thinner gradually and, thereby, its velocity increases to satisfy mass conservation. Therefore, the velocity shear is enhanced in this region and the interfacial mixing increases. A similar result was suggested by Sannino et al. (2004) when analysing water entrainment/detrainment forced by tides. As expected, two mixing-enhanced events happen every day, indicating the tidal periodicity of the phenomenon. They correspond to the increased shear that takes place during flood tide (García-Lafuente et al., 2000; Izquierdo et al., 2001; García-Lafuente et al., 2002).

However, there is some uncertainty in the estimates of the amount of interchange between layers driven by vertical mixing processes in the Strait of Gibraltar based on field measurements (Minas and Minas, 1993) due to the high variability of the physical environment. Estimates of nutrient fluxes given in the introduction for the case of no-mixing and steady exchange are not realistic due to the role that mixing plays in the inter-layer exchange of properties. Furthermore, the comparison exercise (between field data and model simulation) carried out for salinity demonstrates the necessity to include mixing in any model of the Strait in order to obtain realistic representation of the biogeochemical signatures in the area.

Thereby, the coupled hydrodynamic-biogeochemical model could be used to estimate the amount of nitrogen in the outflowing layer that is introduced in the upper layer by mixing and hence, recirculates to the Mediterranean Sea (Figure 8). The concentration of nutrients in the upper layer of the model is a function of the tidal amplitude (figure 8b) and the percentage of the outflowing nitrate recirculating back into the Mediterranean due to interfacial mixing is significantly correlated with tidal amplitude ($r^2=0.7$; $p<0.01$), varying between 4% and 35% with an average of 16.3% (Macías et al., 2007).

This mean value of recirculation is in good agreement with previous estimates based on field measurements (Wesson and Gregg, 1994; Gómez et al., 2000b; Dafner et al., 2003), although it is in the lower end of the measured range. One hypothesis for the slightly low value predicted by this model is its 1-D character, which neglects 2-D effects. Flow-topography interactions favour mixing and upwelling processes in the shallower marginal areas to the sides of the modelled transects. These processes, in turn inject nutrients laterally into the upper layer in a manner that the 1-D model is unable to reproduce.

Also, the only source of nutrients to the upper layer in the model is the mixing between the surface Atlantic layer and the deeper Mediterranean waters (see Macías et al., 2007). The contribution of the relatively nutrient-rich North Atlantic Central Water (Packard et al., 1988) is not considered, though it is known that this water could be upwelled during some phases of the tidal cycle, being incorporated into the main along strait circulation (e.g. Gómez et al., 2001; Macías et al., 2006).

These two processes not considered in the model will lead to surface nutrient concentration underestimation but the good agreement with the previously reported data based on field sampling indicates the likely underestimation is not significant.

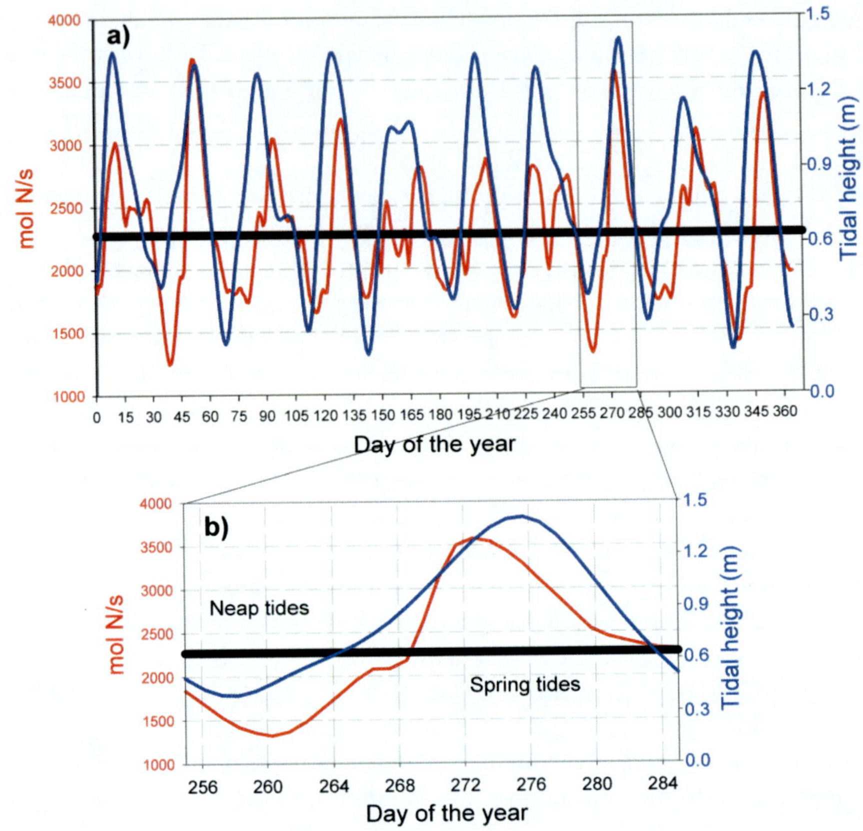

Figure 8. Mean daily flux of nutrient in the upper layer towards the Mediterranean (red line) and tidal height (blue line). a) During the whole simulation. b) During 30 days. Thick solid black line indicates the mean carbon nitrogen flux over all the period.

The year-averaged amount of nutrients introduced into the Mediterranean by the Atlantic inflow computed from the model is 2272.8 mol N/s (Figure 8), but it fluctuates between a minimum of around 1200 mol N/s and a maximum greater than 3500 mol N/s. The average value must be compared with the mean amount of nutrients advected into the Mediterranean Sea in the case of no-mixing, steady exchange as discussed in the introduction. The value of 972 mol N/s computed for that case increases to 2272.8 mol N/s if interfacial mixing is taken into account, that is, an increase of 130%. Obviously, mixing driven by tides cannot be neglected when computing nutrient fluxes. A consequence of this is that the estimates made with field data collected over a limited period of the year will differ from the real annual mean values by a factor dependent on the amplitude of the tide during the sampling period.

One of the most interesting effects of the tidally-induced mixing from a biological point of view is the fertilisation of illuminated surface waters. Tidally-induced mixing in the Strait supplies nutrients to the phytoplankton in the photic layer, which might be thought to result in a positive relation between tidal amplitude and phytoplankton abundance as described above for the field data (see Figure 5). However, a negative correlation ($r^2=0.2$; $p<0.01$) between tidal amplitude and phytoplankton concentration in the upper layer was predicted by the model, indicating that the phytoplankton is unable to utilise fully the amount of nutrient available through mixing.

An explanation for this behaviour is provided by the model simulations, which show that interfacial mixing produces strong dilution of the phytoplankton in the upper layer because of the low concentration of cells in the lower layer. The population growth (triggered by the nutrient injection) is not large enough to compensate losses due to mixing-related dilution and therefore the net concentration of phytoplankton in the upper layer decreases towards the east. The main reason of the low increase of phytoplankton concentration in the model along the Strait is the short residence time of cells within the model domain. An average phytoplankton cell only takes 25 hours to cross the Strait from West to East in the upper layer and, as the maximum growth rate of the modelled phytoplankton is 3.0 d^{-1} and, considering the small initial concentrations of phytoplankton, there is no time for sufficient growth of the population regardless of the amount of available nutrients.

When comparing the temporal dynamics of measured and modelled phytoplankton concentration at the Eastern side of the Strait (Figure 9) appreciable differences can be observed. In this figure it is clear that observations exhibit two peaks of high concentration around HW-4 simultaneous to the peaks of salinity shown in Figure 6, although the distribution of phytoplankton has higher dispersion than that of salinity. The peaks are twelve hours apart, indicating tidal periodicity in the forcing of the phytoplankton peaks. Although the model output shows similar behaviour, it has two important drawbacks. Firstly, it must be noticed that scales for modelled (left) and observed (right) concentrations in Figure 9 are different, so the observed concentration of phytoplankton is approximately twice the concentration predicted by the model. Secondly, there is also a clear difference in the time when predicted and observed maxima appear, the maximum observed concentration coinciding with a minimum of predicted abundance and vice versa (correlation coefficient r^2= 0.3, $p<0.01$). So, although the model predicts the spatial distribution of salinity reasonably well, it can not predict phytoplankton abundance satisfactorily (more details about this discussion can be obtained in Macías et al., 2007).

With the residence times predicted by the model, it is clear that the high levels of chlorophyll measured in the eastern section of the Strait must be the result of something other than local growth. Another source of chlorophyll is necessary in order to explain the high concentration observed in the Eastern section.

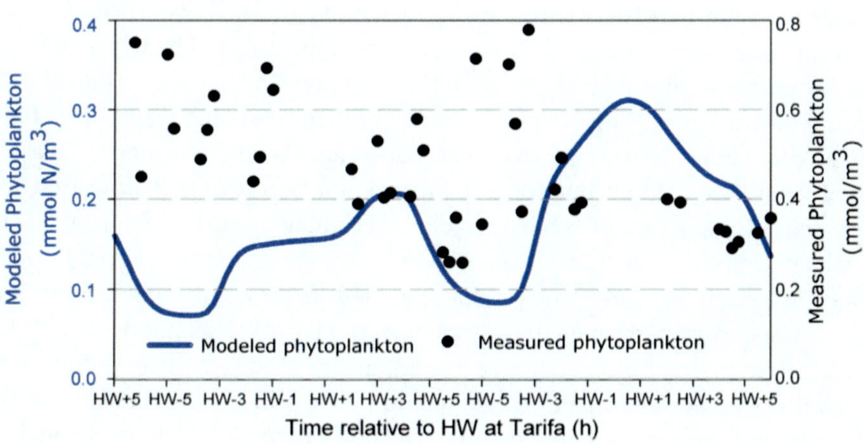

Figure 9. Measured (black dots) and modelled (blue line) Phytoplankton concentration (mmol N/m^3) in the upper layer along 24 h. at the Eulerian station.

3.2.3. Some Conclusions about the 1D Dynamics

From the comparison between field data and modelling results shown above some conclusions emerge. On the one hand, the 1D simplification of the Strait seems to be good enough to represent the general dynamics of some tracers such as salinity or nutrients as both the temporal evolution and absolute values coincides with the reported values from field sampling.

However, important and large differences between observation and predictions were found when phytoplankton concentrations are compared. This seems to be incongruent with the previous paragraph and could lead one to suspect a possible misrepresentation of the biogeochemical compartment by the simplified NPZ model used in the simulations. However, changing the complexity of the biogeochemical model (see Macias et al., 2007) does not lead to an improvement of the fitting between observations and predictions.

The high chlorophyll concentration in the Eastern side of the Strait is recurrently shown by field data (e.g. Minas et al., 1991; Gómez et al., 2001; Echevarría et al., 2002; Macias et al., 2006). Local growth of phytoplankton populations while crossing the Strait in an environment enriched by the mixing-driven injection of nutrients in the Camarinal Sill has previously been put forward to explain these observations (Reul et al., 2002). Other works (Packard et al., 1988; Ruiz et al., 2001) propose that noticeable increase of phytoplankton concentrations in the Atlantic layer away from the Strait can be better explained by the influence of mixing in the Strait than by local enrichment in the Alboran basin, due to the usual large advection velocities of the Atlantic jet which would imply the observation of these effects in the North-Western Alborán Sea or, even, much further west, in the Almería-Orán front, at the east of the Alborán Sea (Arnone et al., 1990). These latter results would agree with the predictions of the model, indicating a residence time in the main channel too short to allow for a significant growth of phytoplankton during its transit through the Strait, that is much more probable to register noticeable results of growth within the Alborán basin itself.

An alternative hypothesis explaining the lack of concordance between observed and modelled phytoplankton in the Eastern section of the Strait is related with tidally-induced dynamics of the surface layer in the western part of the Strait. In this region the currents in the surface layer reverse during part of the semidiurnal tidal cycle (Béthoux and Copin-Montégut, 1986; Candela, 1990) while they hardly ever reverse east of Tarifa Narrows (García-Lafuente et al., 2000), giving rise to intense divergences in the upper layer. The physical model used to force the biogeochemistry reproduces this phenomenon and indicates that the divergence takes place between Tarifa Narrows and Camarinal Sill around high water time (Figure 10).

Internal oscillations of the interface can partly account for the upper layer divergence (García-Lafuente et al., 2000) but some compensation by horizontal flow from both North and South coastal areas cannot be disregarded. The input from the chlorophyll-rich coastal surface waters into the central channel would increase the chlorophyll concentration periodically by means of a mechanism not specified in the used model. Some evidence of suction of coastal waters through the Strait is provided by Van Geen et al. (1988), who suggest that high concentrations of trace metals in the Mediterranean surface waters when compared with open ocean Atlantic waters could be explained by the advection through the Strait of Gibraltar of waters from the continental shelf of the Gulf of Cadiz, which are mainly enriched by the Tinto and Odiel rivers discharges (Elbaz-Poulichet et al., 2001).

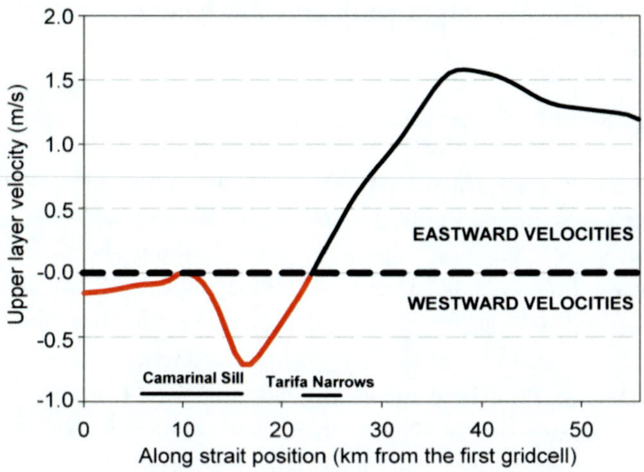

Figure 10. Upper layer velocity (m/s) along the central section near the High Water time.

Horizontal advection from the North and South is also supported by the following discussion. If there is no horizontal advection, salinity and phytoplankton concentration must be negatively correlated because salinity increases when interfacial mixing is enhanced but phytoplankton concentration diminishes by dilution. The model predicts this behaviour but observations behave in the opposite manner. The contradiction can be overcome if simultaneously with intense interfacial mixing there is advection of coastal chlorophyll-rich waters to the interior of the channel. Since interfacial mixing is favoured by the undulatory processes in Camarinal (Wesson and Gregg, 1994) and divergence usually happens when undulatory processes are active (Izquierdo et al., 2001), it follows that interfacial mixing and divergences are simultaneous and, hence, horizontal advection and intense mixing occurs at the same time. Notice that horizontal advection has no impact on salinity since coastal and open channel waters have similar values (Navarro, 2004; García-Lafuente and Ruiz, 2007).

The coincidence of the highest concentrations of chlorophyll with the presence of undulations would lead to a packaging of the cells in patches as such undulations are able to create convergence and divergence alternating areas. Those concentration mechanisms could, thereby, explain the high-chlorophyll patches observed in the diel cycles carried out in the Eastern sector of the channel of the Strait.

The mechanisms, hypothesis and proofs related with this 2D horizontal interaction between the coastal areas and the main channel of the Strait will be further discussed in the last section of this chapter.

3.3. 2D-Tidally-Induced Vertical Pattern of Water Masses Circulation and Biogeochemical Signatures

What is presented above in this section of the chapter is but a simplification of the biogeochemical signatures associated with the tidal dynamics in the Strait of Gibraltar as this physical-biological coupling should also have consequences on the vertical distribution of the planktonic biomass. In this sub-section we will focus on describing the different types of planktonic biomass accumulation that can be observed in the Strait as well as its tidally-forced dynamics while travelling between basins following the main results presented by Macías et al. (2008).

3.3.1. Identified Chlorophyll Maxima

From the observations performed along the main channel of the Strait (see Figure 3a) at least three different types of chlorophyll maxima could be identified, each with different absolute chlorophyll values, position within the water column and physiological properties. The three types of maxima were: (i) those associated with the SAW-NACW interface, (ii) those situated between SAW and MOW, and (iii) those found in the transition between NACW and MOW (Macías et al., 2008). The mean characteristics of each type of maximum are summarised in Figure 11, where the confidence intervals calculated using the t-student analysis (Zar, 1984) are also shown. There was a clear relationship between chlorophyll and carbon concentration for all the identified maxima, with correlation coefficients of > 0.9 ($p < 0.001$) in all cases. Thus, photoadaptation did not seem to be a key factor determining the presence of the DCM suggesting that the occurrence of the maxima could be attributable to an increase in phytoplankton biomass due to accumulation of cells at the water interfaces.

The interface between SAW and NACW is usually characterised by a strong thermocline between the surface and the central waters (Gascard and Richez, 1985; Ochoa and Bray, 1991). A chlorophyll maximum linked to that shallow thermocline has been previously registered in the Western section of the Strait (Gómez et al., 2000; Echevarría et al., 2002) and in the Gulf of Cadiz (Navarro et al., 2006). Because of this, the denomination of 'Atlantic maxima' (**AM**) was used to define this SAW-NACW chlorophyll maximum.

Navarro et al. (2006) suggested the chlorophyll maxima associated with that interface was attributed to a functional response of the picoplankton to the existence of a nutricline between the nutrient-rich NACW and the impoverished SAW (Herbland and Voituriez, 1979). This hypothesis was supported by the observations made within the Strait as mean Nitrate + Nitrite (NN) concentration in the overlying waters was 0.6 μM ($n = 26$; CV = 70 %), a concentration below the threshold of 1 μM accepted as limiting for phytoplankton proliferation (Eppley et al., 1969; Wash, 1988), whereas NN concentration in the deeper water layer (between the 26.35 and 26.5 isopicnals) was a mean value of 2.0 μM ($n = 21$; CV = 76 %), confirming the existence of a marked nutricline in this region of the water column.

The predominance of small cells in the phytoplankton assemblage of the AM (Figure 11) is typical of oligotrophic environments (Li, 2002) were maxima are usually associated with gradients in nutrients distribution (Moran et al., 2001). Picoplankton composition also support this hypothesis, as *Prochlorococcus* (the more abundant picoplankter in AM) has been reported to be more abundant in offshore waters (Goericke et al., 2000) and is prevalent in the open region of the Gulf of Cadiz (Reul et al., 2006; Echevarría et al., 2009). The low percentage of active chlorophyll and the high TEP/Chla ratio (Figure 11) of these maxima are indicative of a senescent phytoplankton population; mucilage production is known to be a good indicator of declining plankton communities (Prieto et al., 2006). This extent is confirmed by production measurements using C^{14} incorporation ratios by Macias et al. (2009) as a low maximum production (Pm) and initial slope of the production-irradiance curve (α) was found in those AM.

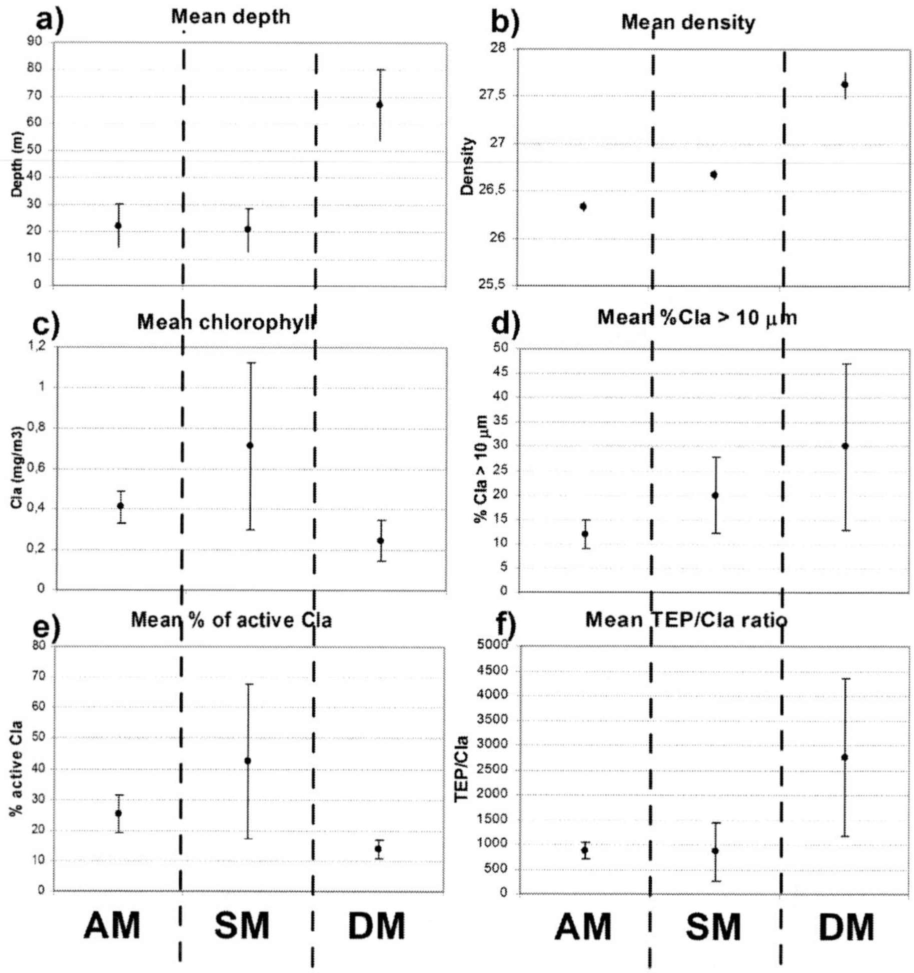

Figure 11. Mean characteristics and 95% confidence intervals (t-student analysis) of the identified maxima.

The larger chlorophyll maxima (Figure 11) were those found associated with the SAW-MOW interface being classified as "Suction Maxima" (**SM**). These maxima correspond with the observed previously in the diel cycle sampling (see above) and do not seem to be linked to the presence of a nutricline, as NN concentration in the overlaying waters was around 2.2 μM ($n = 9$, CV = 53 %), well above the threshold limiting value that limits phytoplankton growth. Therefore, the generation of these chlorophyll patches is likely to be related to physical processes, rather than *in-situ* growth.

The higher mean concentration of chlorophyll, percentage of larger cells and active chlorophyll in these maxima give further support to the coastal origin' hypothesis presented above, as larger cells are generally found in coastal environments where nutrient inputs to the surface layer are more frequent (Malone, 1980). Moreover, the increase in the quantity of *Synechococcus* and the decrease of *Prochlorococcus* (see Macías et al., 2008) also supports the coastal origin of these chlorophyll patches as *Synechococcus* occurs mainly in shallow inshore areas (Chisholm, 1992; Goericke et al., 2000; Pan et al., 2004) and it is usually found in the coastal region of the Gulf of Cadiz (Huertas et al., 2005; Reul et al., 2006; Echevarría et al., 2009). Also, the relative low TEP/Chla ratio in SM is an indicator of active population growth as it has been documented that in early bloom stages the production of TEP is minor (Corzo et al., 2000). These SM also presented higher Pm and α in the C^{14} incubation experiments performed by Macías et al. (2009).

The origin of the last chlorophyll maximum (the Deep Maxima (**DM**)) is not clear, but examination of their characteristics suggests they are generated by sedimentation of large aggregates (high percentage of chlorophyll in particles larger than 10 μm) formed by senescent cells (low quantity of active chlorophyll, high TEP/chla ratio and extremely low values of Pm and α in the C^{14} incorporation measurements). These accumulate at the interface due to a reduction in their settling velocity caused by either the density gradient created by the strong halocline (Longhurst and Harrison, 1989; Margalef, 1989) or because of modifications in turbulence levels (Ruiz et al., 2004). The accumulation of larger particles at this interface, observed using an Underwater Video Profiler (UVP), has been previously described by Gómez et al. (2001).

3.3.2. Tidal Dependence of Water Masses Circulation and DCM Positions

As circulation patterns of water masses through the Strait are expected to influence the position and, even, the composition of the chlorophyll maxima, the conceptual diagrams of along-strait water masses and DCM position during different phases of the tidal cycle created by Macías et al. (2008) are presented in Figure 12. Vertical scale maintains the original values recorded in the CTD casts (position and thickness of the interfaces and DCM) while horizontal scale has been modified to include the entire longitude of the Strait.

As already pointed out, shortly before the High Water time (Figure 11, upper plate) there is an enhanced outflow (westwards) of deeper water over the Camarinal Sill towards Atlantic (García-Lafuente and Vargas, 2003). When the tidal amplitude is high enough, even the upper layer can flow towards the West (LaViolette and Arnone, 1988; Candela, 1990) creating a surface divergence zone over the Camarinal Sill (see Figure 10), which is thought to be responsible for the entrainment of coastal waters into the central region of the Strait. At this stage of the tidal cycle internal waves have already formed (if the tidal amplitude is sufficient), but remain arrested over the Camarinal Sill because propagation against the main outflowing flow is impeded.

West of the Sill (position **A,** Figure 12) SAW is situated in the upper 100 m of the water column, followed by NACW between 100 and 150 m and MOW at depths of < 150 m. Maximum chlorophyll is observed at the SAW–NACW interface situated at 75 m and corresponds to the AM. During this tidal stage, only SAW and MOW are detected over the Camarinal Sill (position B, Figure 12), possibly due to the current regime, which prevents NACW entering this area. Surface chlorophyll maximum observed in this area are associated with the SAW-MOW interface, with the characteristics of the SM (i.e. with a coastal origin). Towards the central region of the Strait (position C, Figure 12), the three water masses are again detected in the water column. In this situation, NACW forms a "tongue" lying over MOW and below SAW, which is located only in the central area of the Strait; no NACW signal is present at positions "B" or "D" (Figure 12). The NACW tongue found at intermediate depths at station "C" during this tidal stage, was originated in the previous tidal cycle when, in a situation similar to that shown in the lower plate of Figure 12 (LW time), the

NACW is able to overcome the sill after the flow Eastward is re-established being dragged by the main surface current towards the Mediterranean side. At this position "C" two chlorophyll maxima are commonly observed, an AM at the upper interface and a DM at the lower. On the Eastern side of the Strait (position D, Figure 12), only SAW and MOW are commonly observed. In this area, the chlorophyll maximum linked with the interface appears to be very similar to those found over the Sill (position B), i.e., a coastally-originated SM.

Four hours later (i.e two hours after the high water time) the inflow of Atlantic water is reinforced and a weakening of the Mediterranean outflow takes place (Figure 12, intermediate plate) (Armi and Farmer, 1988). The internal waves have been already released during HW and therefore, the NACW can cross the sill and flow towards the Mediterranean Sea.

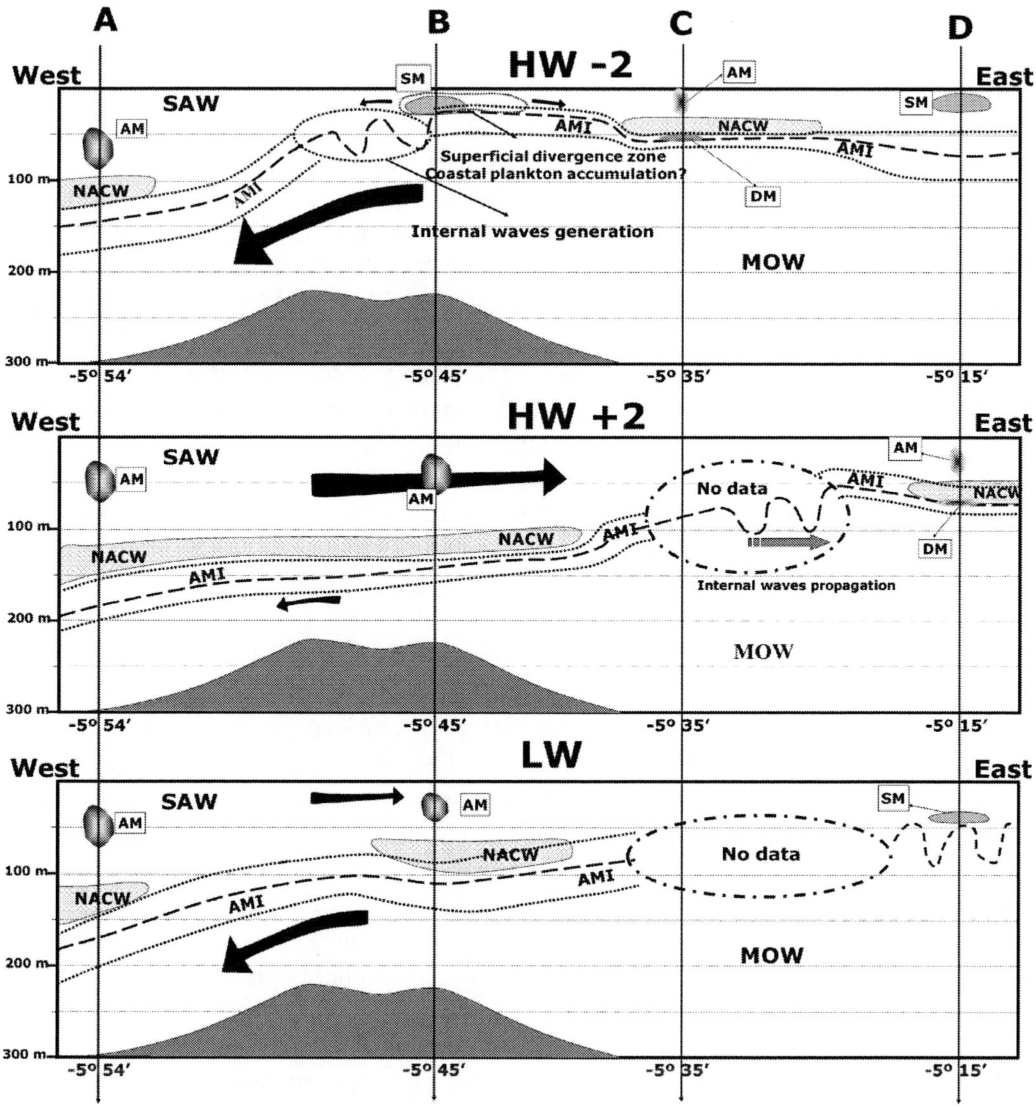

Figure 12. Conceptual diagram showing interface and DCM positions and water mass circulation within the Strait during different stages of the tidal cycle. Areas encircled by a dashed-dotted line represent values obtained from the literature (no data available).

West of the Camarinal Sill (position A, Figure 12), the three water masses (SAW, NACW and MOW) are detected and an AM is observed at the SAW-NACW interface. Above the Camarinal Sill (position B) the three water masses are still present, a clear NACW signal is observed between 120 and 130 meters and an AM is still associated with the SAW-NACW interface. During this stage of the tidal cycle the internal waves are propagating within the Strait, and are expected to be around position C, probably with the SM described in position "B" of Figure 12, upper plate (HW-2). However, there was no data available for this section of the channel during this tidal stage to confirm this assumption. At position "D", the tongue of NACW described in region "C" of Figure 12, upper plate was also detected. Furthermore, the two maxima linked to the interfaces SAW-NACW and NACW-MOW were again clearly observed in the CTD profile.

Finally, four hours later coinciding with the low water time the internal waves are expected to be close to the Eastern side of the Strait (Farmer and Armi, 1986; Macías et al., 2006). At this stage, the current velocity of the surface layer over the Sill progressively decreases and the deeper outflow is reinforced. In this situation a shallowing of the AMI takes place, obstructing the Eastward flow of NACW (Figure 12, lower plate).

In position "A" (west of the Sill), the composition of the water column is very similar to that described in the two previous diagrams i.e., the presence of all three water masses and an AM at the upper interface. The signature of NACW was still detectable in the profiles performed over the Camarinal Sill (position B). Four hours later, NACW will form the tongue described in position "C" at HW-2 (Figure 12, upper plate). An AM was registered between the SAW and the NACW. During this phase of the tidal cycle the internal waves generated in the previous outflowing event are expected to be approaching the Eastern side of the Strait (position D). In the cast performed at this position there was no NACW signal and a shallow SM connected to the SAW-MOW interface was observed. This maximum was very similar to the DCM associated with the generation of internal waves located over the Camarinal Sill at HW-2 (position B, Figure 12, upper plate). Completing the tidal cycle and four hours after the situation above, the scenario was again comparable to the one described in the upper plate of Figure 12.

These results clearly indicated that the nature of the DCM observed in the Strait of Gibraltar is strongly influenced by tidal amplitude, so a discontinuous entrance of different phytoplankton assemblages into the Alboran Sea would be expected (Macías et al., 2008). This discontinuous entrance of biogeochemical material through the Strait is important for the pelagic ecosystem of the Alboran Sea, as a pulsating input of both plankton and nutrients could drive the behaviour of the system. This extent has been recently addressed by Macías et al. (*in press*) using numerical models to explore how changing from a continuous to a pulsating nutrient input (while maintaining constant the total amount supplied) affects the dynamics and structure of the pelagic community. These last authors showed that a discontinuous pattern of nutrient input to the ecosystem similar to the one happening at the Strait induce larger biomass production in the system and also differentially favours some plankton groups. For example, under fast and repetitive nutrient pulses (like the ones associated with the tidal mixing at the Strait) favours fast-growing and big cells when compared to the continuous supply case. This has consequences on trophic web dynamics as energy flow would be more or less efficient depending on the biomass distribution between groups (Smetacek, 1999) and, thereby, should have consequences on the higher trophic levels of the Alboran basin. So, the nature of this input should be taken into account when explaining the spatial and temporal patterns of biological constituents of the Alboran ecosystem.

In summary, tidal induced variability within the Strait of Gibraltar not only affects the transport and circulation of the water masses that meet in the area, but also the distribution and characteristics of chlorophyll patches. This fact corroborates the importance of tidal induced phenomena in explaining physical and biological structures at the regional scale.

4. Hypothesis, Evidences and Proofs of the Tidally-Induced Coastal-Channel Interactions in The Strait

Across the entire chapter it has been stressed that the hydrodynamic and associated biogeochemical pattern within the Strait of Gibraltar is mainly controlled by the tidal dynamic through various and complicated processes including changing advection directions, vertical interfacial mixing and (probably) complex coastal-channel interactions.

This last process is particularly relevant for the pelagic ecosystem of the adjacent Alboran Sea as it implies the injection of coastal elements into the open sea at the same time that the surface layer (i.e. the lighted one) is being fertilized by the interfacial mixing associated to the internal wave's generation. The coincidence of high nutrient concentration with high plankton biomass should affect the structure and dynamic of the entire pelagic ecosystem not only within the Strait itself but also on the nearby North-Western Alboran Sea.

This mechanism of coastal-channel interaction had been unnoticed until the performance at the beginning of the last decade of several exhaustive samplings, some of them already commented in the previous sections of the chapter. In this last section we will focus on direct evidences and proofs of the exact mechanism of such interactions based on both field data samplings, tracers release experiences and modelling simulations.

In the first part of this last section we will focus in describing some results on how plankton distribution and community composition is affected by the generation and development of the surface divergence and internal waves, based on observations made on the Camarinal Sill area and presented in Macías et al. (2010).

The second part will be a description of an *ad-hoc* experiment designed to confirm the existence of these interactions by combining a water-tracer release experience with computing simulation using one of the latest most advanced 3D hydrodynamic models of the Strait available nowadays.

4.1. Plankton Distribution Pattern and Physical Forcing at the Camarinal Sill

During a cruise carried out in November 2003 on board BIO Hespérides, a specific zooplankton sampling was designed over the Camarinal Sill trying to study the effect of the internal wave's generation and development on the planktonic community distribution and composition (see details in Macías et al., 2010).

Zooplankton was sampled using a Longhurst-Hardy Plankton Recorder (LHPR) net, which is a multiple sampling net designed for plankton capture taking several samples at each cast (Wiebe and Benfield, 2003). Collected samples were analysed using a software for plankton determination known as PVA (Plankton Virtual Analyzer) which counts and measures planktonic organisms (Boyra et al., 2005).

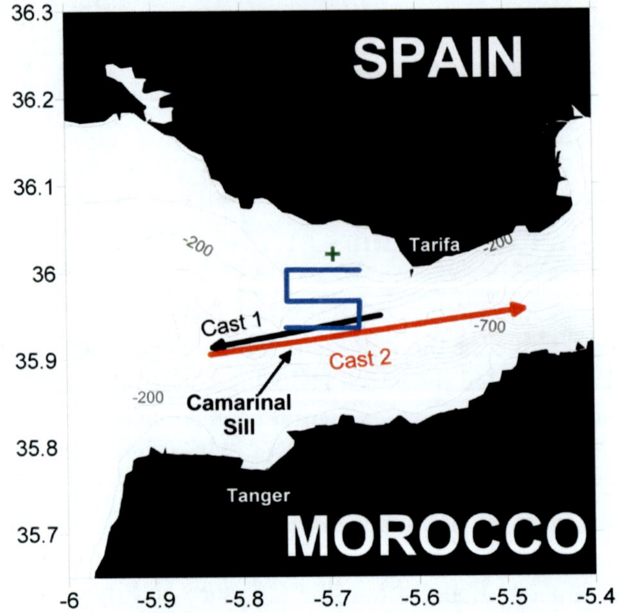

Figure 13. Position of the Rhodamine initial release (green cross). Path followed by the vessel during the Rhodamine sampling (blue line). Westward (black line) and eastward (red line) casts of the LHPR sampling over the Camarinal Sill.

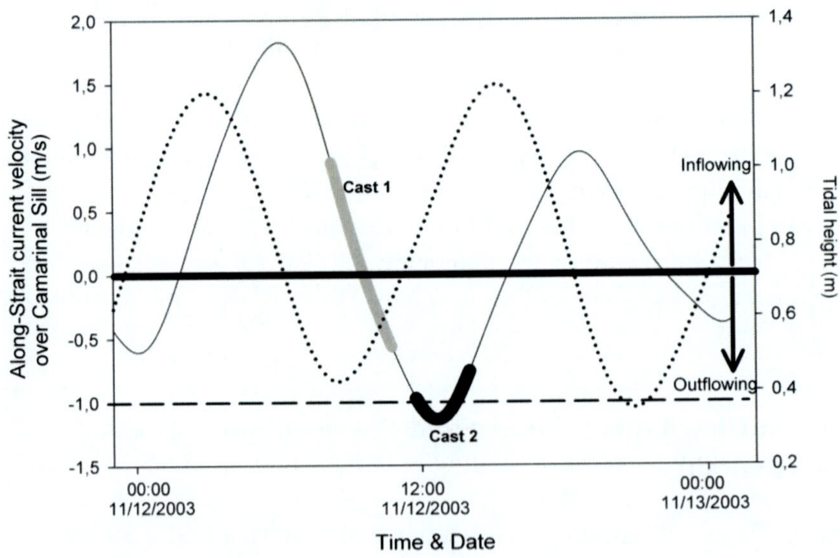

Figure 14. Tidal conditions over the Camarinal Sill during the sampling period. Solid line represents current velocity over the sill (positive are Westward). Dotted line marks the tidal height at Tarifa harbor. Dashed line indicates minimum outflowing velocity for the generation and liberation of internal waves (according to Vazquez et al., 2008).

The LHPR was also equipped with a CTD SBE-19 and fluorometer sensor that registered pressure, temperature, salinity and fluorescence of the water masses as the LHPR passed. This design allows a large number of samples to be obtained in a single haul (between 30and 60), and is able to resolve partially the variability of zooplankton distribution at short scale and the aggregation phenomena that were expected to occur in the Strait.

Two casts were performed over the Camarinal Sill (Figure 13) just before (cast 1) and after (cast 2) the generation of internal waves (Figure 14). During cast 1 the tidal current changes from Eastward to Westward and the cast finished shortly after the low-water time. Tidal currents were towards the West during the entire period of cast 2, since it coincided with the tidal phase when the internal waves are being developed over the Camarinal Sill. This cast ended just before the high-water time, when these waves are typically released (Figure 14).

From the collected samples total plankton biomass was determined by using a wet weight measurement while community composition was assessed by using the PVA software and a discriminant analysis based on the Fisher's functions (Krzanowski, 1998).

4.1.1. Cast 1 (Westward Hauling- Eastward Tidal Flow)

This entire cast has been divided into five sectors (designated A to E in Figure 15) based on the spatio-temporal distribution of temperature, salinity and chlorophyll.

In sector **A** (East of the sill), NACW was located around 100 meters depth marked by a salinity minimum in Figure 15. The transition between sectors A and B is indicated by a strong horizontal salinity gradient (Figure 15), and a loss of the NACW signal. Sector **B** is characterized by a temperature gradient from surface to deeper waters while salinity is very constant from the surface down to 100 meters. Between sectors B and **C**, a new horizontal gradient in salinity values is present (Figure 15) with increasing values towards the West. The location of sector **D** was directly above the Camarinal Sill, and presents a marked horizontal salinity gradient (Figure 15) increasing from 36.5 up to 37.5, and no vertical structure. Sector **E** starts with a sharp change in salinity conditions of the upper water layer, from a uniform value of 37.5 to a steep vertical gradient (Figure 15). A salinity minimum of 35.5 is observed at 50 meters depth, indicating the presence of NACW and absence of MOW.

Chlorophyll distribution (Figure 15, lower plate) shows a DCM located around 50 meters depth and extending from sectors C to E. In general chlorophyll concentration is relatively low (average 0.9 mg m^{-3}) and even the DCM is not very marked, with values of 1 - 1.3 mg m^{-3}.

Zooplankton biomass along this cast does not seem to be correlated with any of the hydrological (temperature or salinity) or biogeochemical (chlorophyll) variables (Figure 16, upper plate) being the average biomass particularly low for all this cast. Composition of the zooplankton community during this sampling was analysed in 36 samples taken in sectors B to E. Very little difference is found between samples, indicating a very similar composition of the zooplankton community throughout the sampling path.

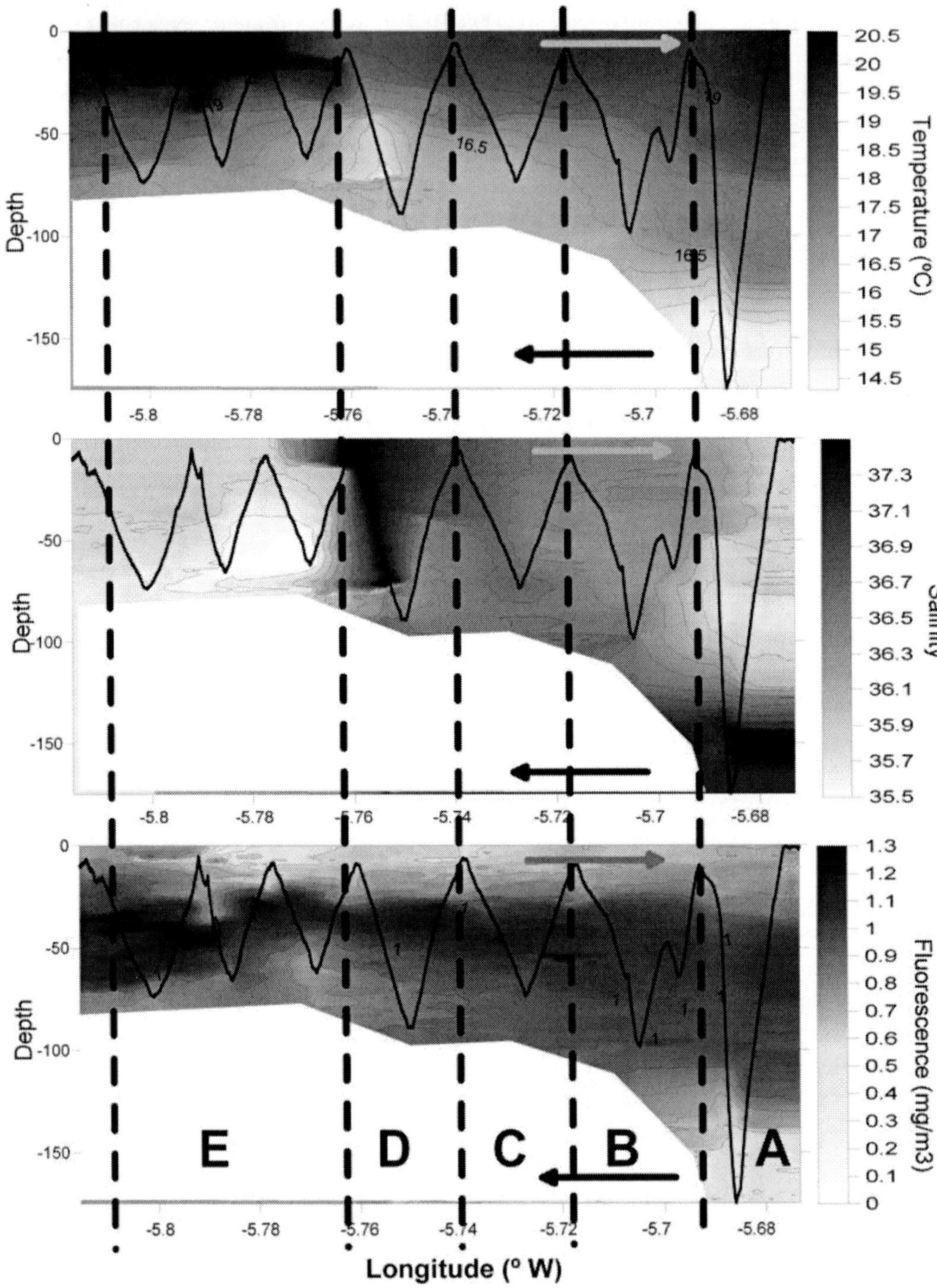

Figure 15. Temperature (°C), salinity and chlorophyll (mg/m^3) distributions during cast 1. Grey arrows show water current direction and black arrows show sampling direction. In sector E water flow velocity decreased and even changed its direction (i.e. flowing eastward) as indicated in figure 14.

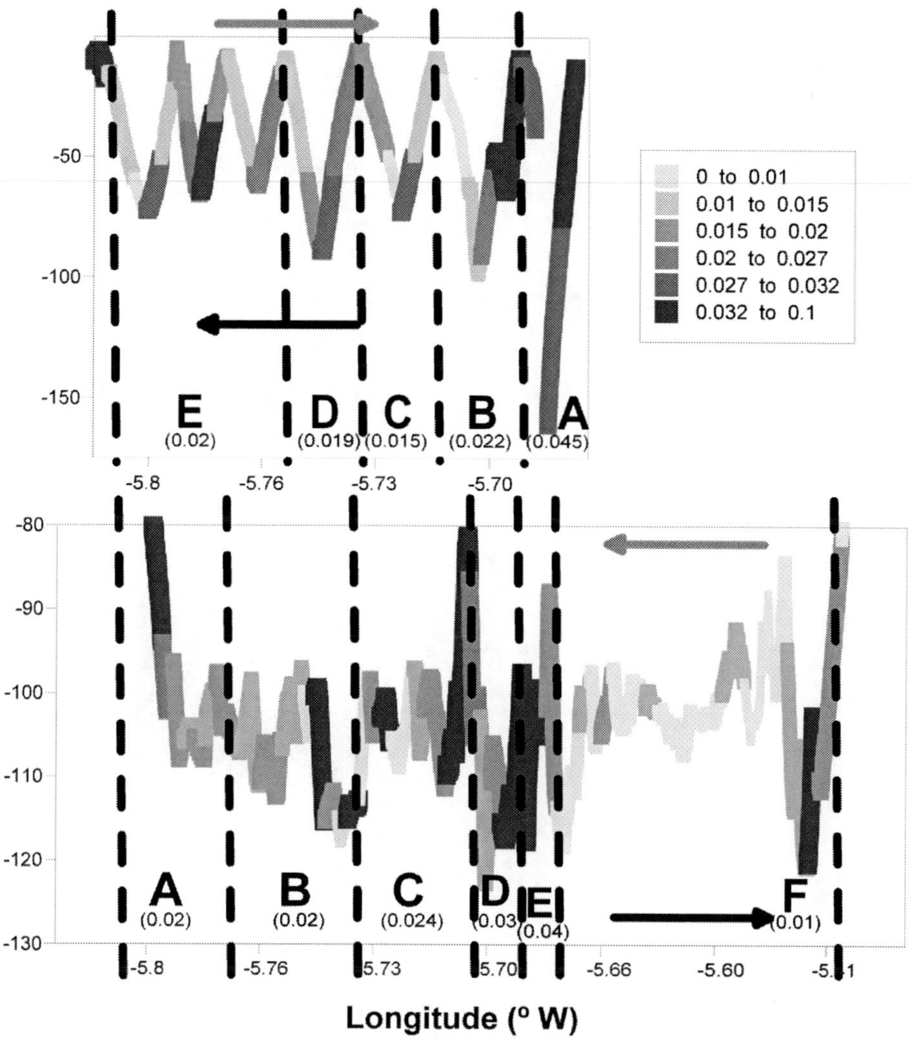

Figure 16. Zooplankton biomass (g/m^3) during cast 1 (upper plate) and cast 2 (lower plate). Values in parenthesis are mean biomass concentration in each sector. Grey arrows show water current direction and black arrows show sampling direction.

4.1.2. Cast 2 (Eastward Hauling- Westward Tidal Flow)

This second cast was performed entirely during the west-flowing phase of the tidal cycle (Figure 14) and a different strategy was used: the LHPR was hauled from east to west (Figure 13) at a constant depth (nominal depth 100 m).

As in the previous casts, it has been divided into 6 sectors taking into account the hydrological conditions of the water column (mainly water masses, Figure 17). Sector **A**, situated west of the Camarinal Sill is characterised by a constant temperature and salinity (Figure 17) with the presence of NACW and low values of chlorophyll throughout the whole sector (Figure 17, lower plate). Further east, in sector **B**, a gradient of both temperature and salinity is recorded with an increase in salinity and a decrease in temperature towards the eastern sectors (Figure 17).The NACW signal disappears, and even lower chlorophyll values are found (Figure 17, lower plate).

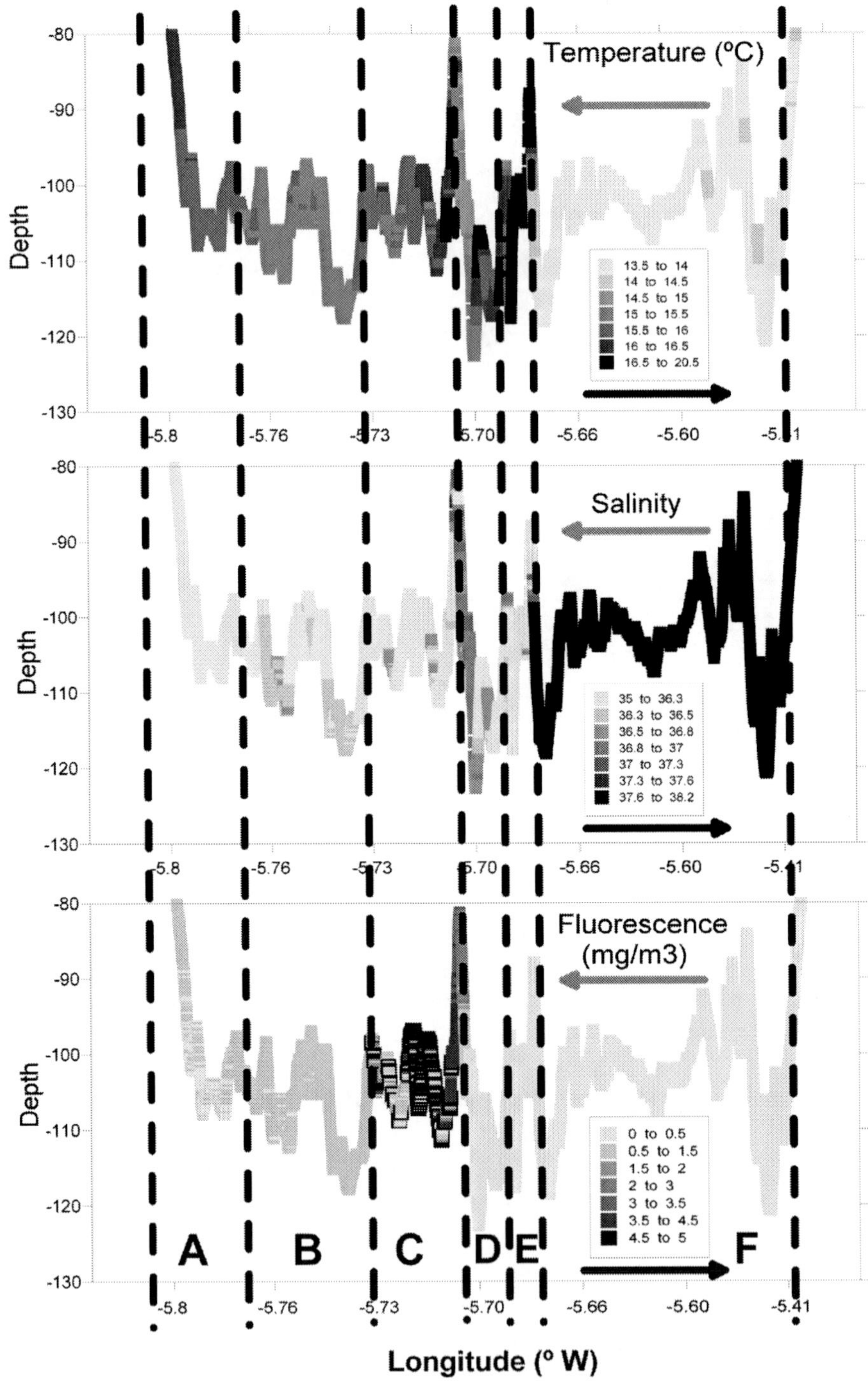

Figure 17. Temperature (ºC), salinity and chlorophyll (mg/m^3) distributions during cast 2. Grey arrows show water current direction and black arrows show sampling direction.

The boundary between sectors B and **C** is located directly over the Camarinal Sill and abrupt changes in the depth of the net occur in this transition zone, from 120 m at the end of sector B and just 100 meters in the beginning of sector C (Figure 17). An approximate vertical velocity of around 0.3 m/s can be calculated from this vertical displacement of the net. There is a slight increase in salinity values and a decrease in temperature with respect to the preceding sector (Figure 17). The most dramatic change corresponds to the chlorophyll concentration, which is 10 times greater than in the preceding sectors (Figure 17). The chlorophyll concentration recorded in sector C is the highest found during all of the casts carried out in the Strait.

Between sectors C and **D** strong vertical oscillations of the LHPR net are again observed (Figure 17), changing from 80 to 120 meters very rapidly, with associated vertical velocities of around 0.4 m/s. Corresponding with this oscillation an abrupt change to colder and more saline waters is recorded (Figure 17). At the same time, chlorophyll concentration decreases very rapidly (Figure 17). At the beginning of sector **E** the LHPR again undergoes sharp vertical movements and salinity values decrease whereas temperature increases (Figure 17). As for the preceding sector, chlorophyll concentration remains very close to zero (Figure 17). Finally, in sector **F** (which starts also with a severe oscillation of the LHPR) water composition changes. The salinity increases and the temperature decreases (Figure 17), indicating that only MOW is present in this zone. No changes in chlorophyll concentration are detected, and the values remain very low.

During this cast, zooplankton biomass tends to accumulate in the transition zones between the different sectors defined previously (Figure 16, lower plate).The density is greater in those transition zones characterised by marked horizontal gradients in salinity and/or temperature (such as the C-D and D-E boundaries). Taxonomic composition of the community during this last cast shows considerable difference in the various sectors, and is unlike the composition found in the previous sampling. Although copepods are still the most important group in the smaller fraction (<1000 μm), there are a number of different taxa showing increased abundance, such as ostracods, decapods larvae, etc. Within this smaller size fraction the second most abundant group (ostracods) show large difference, accounting for 20% in sectors A and F (the first and last sampled) but decreasing to less than 2% in sector E. In the vicinity of the Camarinal Sill (sections C and D), groups like siphonophores, salps and larvaceans (appendicularia) are more abundant than in the rest of the cast, in both the larger and smaller fractions. Finally, in sectors C, D and E mucilaginous material was present in the samples.

4.1.3. Evidence of Quick Changes and Coastal-Channel Interactions over the Camarinal Sill

For the first time the most dynamic area of the entire Strait (the Camarinal Sill sector) was sampled at two very different phases of the tidal cycle and internal wave development (Macías et al., 2010). So it makes it possible to relate the hydrological environment to the biogeochemical signatures recorded during each cast and confirms the hypothesis of coastal-channel interactions forced by the tidal dynamics.

Phytoplankton distribution during the first cast over the sill could be analysed within the framework of a model of currents-plankton interactions developed by Franks (1992) for frontal areas. Following this model, when a divergent (horizontally) flux is present in a frontal zone the plankton tends to show accumulation in a horizontal band that crosses the pycnoclines. In this case it has been postulated that a surface divergent area could appear between the Camarinal Sill and the Tarifa Narrows (see Figure 13) coinciding with the tidal reversion (Izquierdo et al., 2001, Macías et al., 2007).

Predictions of the Franks' (1992) model agree quite well with the phytoplankton distribution found during this cast, since a high chlorophyll band was situated at about 50 meters depth from sectors B to D (Figure 15, lower plate). At the same time, the average concentration during the entire cast is quite low, corresponding to a situation of divergence. This same hydrological condition could be responsible for the generalised decrease in the zooplankton biomass recorded during this cast (Figure 16, upper panel). As these organisms possess swimming capabilities, the Franks' (1992) model predicts an effect different from the one observed for phytoplankton. In this case, a progressive reduction of the population density from the point of divergence is expected, with the faster swimmers located further from that point. This is observed in our data, as mean zooplankton concentration is very low over the Sill (sector C) but is higher to the West and East (Figure 16a). In almost all the samples from this cast the dominant organisms are copepods of the genus *Oithona* which has a relatively slow swimming speed (0.4 mm s^{-1}, Yamasaki and Squires, 1996). Larger and faster swimmers, like chaetognaths (swimming speeds of 1-2 cm s^{-1}, Mutlu, 2006), big euphausiids and siphonophores (30 cm s^{-1}; Bone, 2005) were scarce and only found in the samples from sectors B and E (the furthest from the area of divergence).

On the other hand, hydrological conditions in the Camarinal Sill area were substantially different during the second sampling (see Figure 14). In such tidal conditions and with enough tidal amplitude (tidal current over the Sill above -1m/s) internal waves are created along the AMI by the interaction of tidal currents and sill topography (Armi and Farmer, 1988, Bruno et al., 2002). Such undulatory processes remain arrested over the Sill until the outflowing velocity falls below -1 m/s (Vázquez et al., 2008) which happens at the end of this sampling (Figure 14).

The presence of the internal waves throughout this entire cast is revealed by the oscillation of the LHPR at the sector boundaries (Figure 17), since the sampling strategy in this cast was to maintain the net at a constant depth of around 100 meters. A schematic representation of the crest and trough of the internal waves during this cast is presented in Figure 18, constructed from the differential vertical movements of the LHPR. In this scheme the maximum vertical velocities (around 0.4 m/s as stated above) are associated with transitions between trough and crest and coincide with the sectors boundaries defined in Figure 17. The distribution expected was a symmetrical accumulation of zooplankton biomass above the troughs and below the crests of the internal waves. Since the convergence-divergence processes were actually taking place during the sampling, the accumulation recorded was higher towards the end of the cast (darker accumulation areas on the right-hand side of Figure 17).

Sector C of this final cast coincides with sector C of the previous one (see Figure 16) and is located directly over the sill (Figure 13) or slightly westward. If the divergent processes do occur, as proposed above, it is precisely in this area where they will be most likely (see Macias et al., 2007; 2008). As commented above, during the final cast a high chlorophyll concentration (Figure 17) was found in this sector, while during the previous cast this concentration was extremely low (Figure 15). The short time between one cast and the next (insufficient time for *in situ* growth), and the presence of macroalgae debris (gen. *Plocamium*) in the samples of this sector in the last cast, are two factors suggesting the hypothesis of lateral advection of chlorophyll-rich coastal areas to be very likely. This is also supported by the intermediate values of salinity and temperature recorded in this sector C (Figure 17), which match well with the presence of partially-mixed waters from the coastal and shallower sectors (Navarro et al., 2006, Macías et al., 2007). The steep horizontal gradients of temperature and salinity between sectors C and D impede the propagation of this chlorophyll patch eastward. The effect is to pack individuals together and concentrate the phytoplankton in a reduced area, making the chlorophyll concentration in the other sectors virtually zero (Figure 17).

Within this last cast sharp differences in zooplankton biomass distributions are found between sectors (Figure 16, lower panel). Maximum concentrations appear situated over the wave trough (see Figure 18) and below the wave crest (sectors C, D and E; Figure 16, lower panel) as predicted by the model of Lernnet-Cody and Franks (1999). However, not only biomass distribution is affected by the internal waves but also taxonomic composition (Munk et al., 2003). In this study, bigger (and likely better) swimming organisms appear in the samples from sectors D and E, induced by the presence of the arrested internal waves in this region, while ostracods with lower swimming velocities decrease in abundance from sectors A to E, but increase again in sector F where the effects of internal wave generation are no longer felt.

The reported data present evidence of intense coastal-channel interactions in the Camarinal Sill region, and help to explain how the combined presence of such processes with the internal waves created over the sill could create the spatio-temporal pattern of plankton biomass and its distribution (Macías et al., 2010). This is the first time that direct evidences (i.e. quick planktonic biomass accumulation) of such interactions are obtained and give a strong support to the hypothesis presented in this chapter about the generation mechanisms of those processes as the timings and places of the aggregations match perfectly with the theoretical interpretation exposed elsewhere.

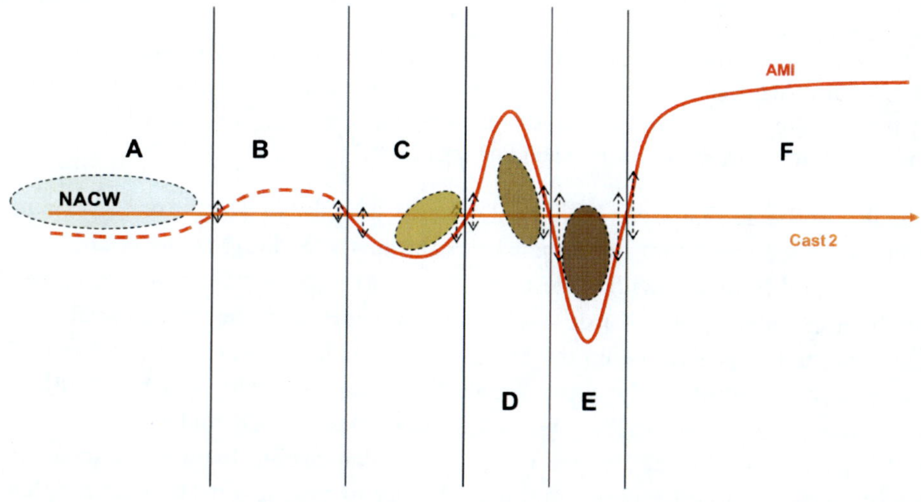

Figure 18. Schematic representation of internal wave's develop during cast 2 inferred from the vertical movement of the LHPR. Small arrows represent zones of stronger vertical movement which usually happens between crest and trough and coincides with sector limits (from figure 17).Shaded areas are preferential accumulation zones for plankton biomass.

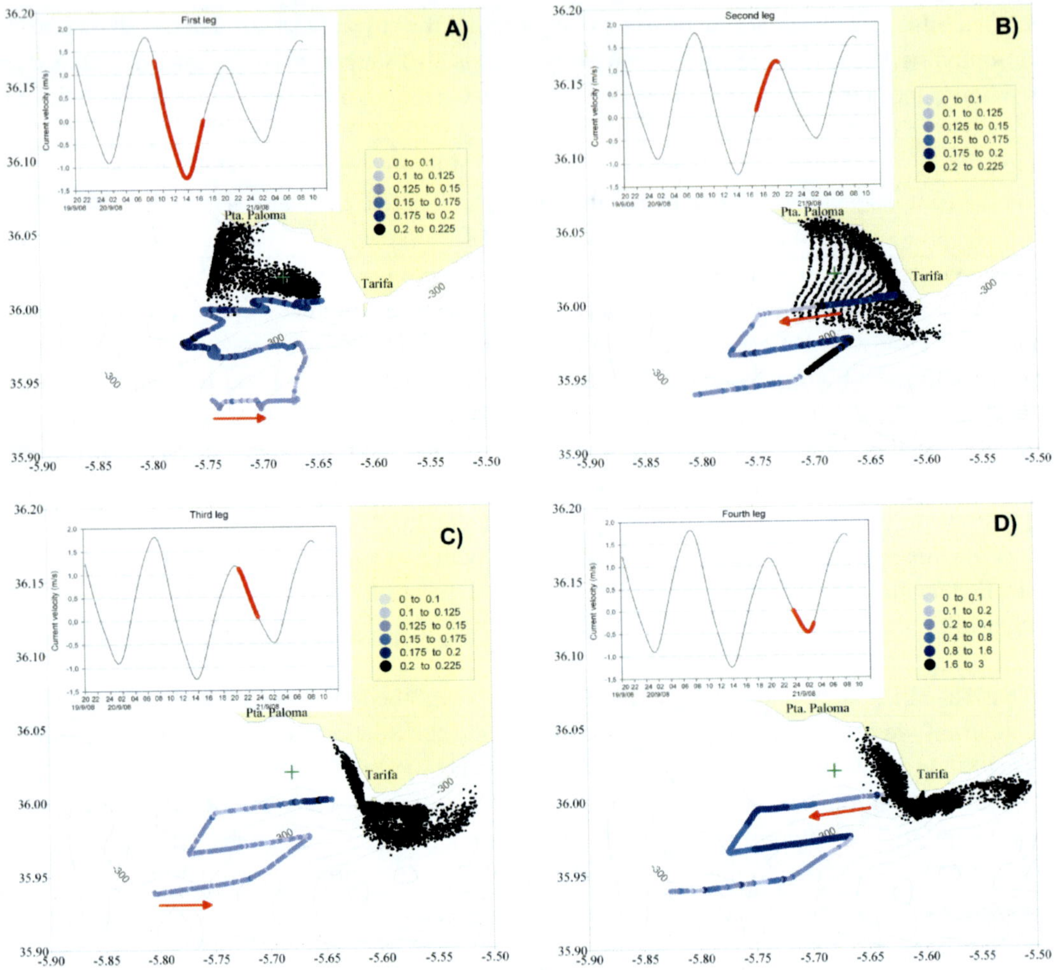

Figure 19. Rhodamine concentration in the surface waters during the different sampling legs (blue dots). Predicted dispersion of the virtual tracer by the hydrological model (black dots). Red arrow marks the travel direction of the vessel when following the sampling path showed in figure 13. Inlets show the tidal current during each of the sampling legs.

It is also probable that these biological patterns could also be part of the explanation of the special food web structure in the pelagic realm of the Strait (Macías et al., 2010) which presents notable differences with respect to other nearby basins (Gulf of Cádiz and Alboran Sea), as high abundance of seabirds, high-level predators and marine mammals have being reported to permanently live within the Strait itself. This high abundance has been associated with the strategic position of the Strait between two marine basins and two continents (as a natural pathway for large-scale migration). However, the characteristics of its marine environment may also partially explain those particular trophic features as accumulation of biomass of a size suitable for consumption by higher trophic levels (such as zooplankton) is of vital importance for the dynamics and structure of the marine food web, as seabirds and top-predator fish (i.e tuna, swordfish,..) have been reported to accumulate on oceanic fronts (Russell et al., 1999) following higher zooplankton abundance in such areas. Furthermore, marine mammals are usually found in systems where physical forcing permanently reinforces primary and secondary production (Fielder, 2002). As both these processes take place in the Strait, they could partially explain the abundance of all these high-trophic-level organisms.

4.2. Direct Evidences of Coastal-Channel-Tidally-Induced Interactions

With the accumulation of evidences about the actual mechanisms of coastal-channel interactions and taking into account its potential effects on the dynamic of the marine pelagic ecosystem of the Strait and nearby adjacent basins (see discussion above in the chapter) a specifically designed observation was carried out in September 2008 during a cruise on board BO Sarmiento de Gamboa.

This experiment consisted on a water-tracer release in a coastal situation (green cross in Figure 13 and 19) from a small boat during the shift from inflowing to outflowing current at a spring tide period. The used tracer was Rhodamine-WT which is a fluorescent material not found naturally in marine environments being, thereby, easily detected by using a conventional fluorometer (Turner TD-10) equipped with specific excitation/emission filters.

The surface Rhodamine concentration was recorded from BO Sarmiento de Gamboa while following the track presented with a blue line in Figure 13. The ship moved along this zigzag line during 18 hours, i.e. covering an entire tidal cycle. Water was collected by a surface intake (nominal depth 5 meters) and conducted to the registration chamber of the fluorometer which recorded Rhodamine-related fluorescence at one hertz rate (i.e one record per second). Data were stored at the fluorometer-physical memory and later downloaded to a compatible PC computer.

The moment of the tracer release was selected to be before the more suitable moment for the horizontal divergences appearance to allow the initial patch to be expanded a little bit and being, thereby, suitable to be suctioned towards the central channel of the Strait. The results of the water samplings were later divided into "legs", each of the leg being a complete track of the blue line showed in Figure 13.

On the other hand, numerical simulations of the dispersion of a virtual tracer were carried out using a simplified version of the recently-developed UCA-3D numerical model which is nonlinear and weakly non-hydrostatic, has a spatial high-resolution and uses finite-difference schemes and sigma-coordinate in the vertical.

Current fields provided by the model for the same dates that the experiment was conducted were used to feed-in the advection module of the modelling tool Ichthyop (Lett et al., 2008) which simulate the horizontal advection and dispersion of an inert tracer with neutral buoyancy (i.e. assimilated to a Rhodamine-marked water mass). The release of the virtual tracer was simulated at the same position and time (relative to the tidal cycle) as the real one so the results of the simulation could be directly compared to the Rhodamine data collected from the boat.

4.2.1. Field Data and Modelling Results

As stated above, the tracer was released shortly before the tidal current reversion over the Camarinal Sill (see inlet in Figure 19a). Coinciding with the release, the vessel started registering the surface Rhodamine-fluorescence along the sampling path beginning by the south-west corner (red arrow in Figure 19a). As can be observed, at the south of the sampling area no Rhodamine was detected at this stage as the recorded concentration were very close to the analytical determination threshold (situated at 0.1 ppb). While the vessel navigated on-shore, Rhodamine concentration increases progressively particularly at the western side of the medium line and in the closest region to the release area.

Simulations of the advection-diffusion model for the inert tracer shows a quite good agreement with the observations. The initial patch is moving Westward when the surface current reinforces its outflowing characteristics (from 1 hour after the release). At "Punta Paloma" cape (see Figure 19) the patch is stretched in a North-South direction coinciding with the curvature of the bottom-ridge which constitutes the Camarinal Sill. This region of southward transport matches exactly with the area of higher Rhodamine concentration found in the vessel-based sample.

The second sampling leg started when the currents over the Sill changed again to East flowing (i.e. after the HW time) and begins in the North-East tip of the sampling path (see Figure 19b). Higher Rhodamine concentrations were found in the eastern side of the sampled area, corresponding to a quick advection of the marked water masses with the increasing inflowing flux. At the same time quite high concentrations were detected in the middle of the channel of the Strait (at about 35.9ºN) confirming the suction of coastal waters during this tidal cycle.

The advection pattern predicted by the numerical model also agrees well with the reported data as the virtual markers were quickly moving towards the East with the increased inflowing current although the path followed by the virtual particles seems to be closer to the shore that the Rhodamine data. This second sampling leg finished almost 12 hours after the Rhodamine release so nearly one tidal cycle was already sampled so far.

Immediately after the end of this sampling, the third leg started (Figure 19c), which only lasted three hours. It was conducted from the South (i.e. from the Camarinal Sill area) towards the shore and in this case only significant Rhodamine fluorescence was found at the easternmost side of the inner line, almost in front of the town of Tarifa. The model also predicts accumulation of the virtual tracer in this same region though slightly displaced towards the East.

The final sampling leg was done from inshore and lasted also three hours (Figure 19d). In this case two separate patches of high Rhodamine concentration were detected in the inner and middle lines which could indicate that some of the water which was moving eastward during the previous leg was able to reverse its movement and come back forced by the new change on the tidal current direction over the Camarinal Sill (see inlet in Figure 19d). This effect is not well reproduced by the model which predicts that the virtual tracer keep travelling Eastward during this period.

This lack of agreement in the last sampling leg could be attributable to the slightly quicker Eastward advection of the marked water masses predicted by the model. As already pointed out, during the third leg almost all the virtual particles were found to the east of Tarifa point while some of the Rhodamine still remains westwards. It is already known that surface-tidally-induced reversion only happens westwards of Tarifa (see section 3 of the chapter) so this could explain why the model does not reproduce the reversion of the marked water masses observed in the field data.

However, and in spite of this slight incongruence, both model and field data clearly show that coastal waters are effectively transported towards the main channel of the Strait during the east flowing phase of the tidal cycle. Both dataset also points to the region South of "Punta Paloma" cape as the most likely to the offshore suction of the water masses.

This same region has been identified as very relevant for these processes by Vázquez et al. (2009) in a very recent study combining Advanced Synthetic Aperture Radar (ASAR Image Mode) images and surface color imagery (MODIS and MERIS sensor). These authors use the remote information from the different sensors to study the generation (and propagation) of several internal waves episodes over the Camarinal Sill and associated 2D surface structures and relate them to the tidal conditions within the Strait. These are the first direct evidences of such interactions which confirm the proposed hypothesis and mechanisms already exposed above.

As a repetitive pattern in all the internal waves generation processes studied, a tongue of high chlorophyll waters appears extending Southward in the region of the "Punta Paloma" cape, thereby, confirming the findings of the tracer-release experiment and the model simulations showed here.

5. Conclusion

The main conclusion which could be outdrawn of this entire chapter is that tidal dynamics control and shapes as the main agent the general hydrodynamics of the Strait of Gibraltar. In this sense it is necessary to acknowledge that other factors such as winds (e.g. Macías et al., 2008b) or atmospheric pressure distributions (e.g. Vázquez et al., 2008) could also play a role in controlling the hydrography of the Strait and nearby areas but at a much lower level than that of the tide.

As usually happens in the marine environment, a strong coupling between physical forcing and biogeochemical patterns is found in the Strait with the main biological processes being shaped and controlled by hydrological factors. The interaction mechanisms found in the area include:

i) A strong interfacial mixing (between Mediterranean and Atlantic waters) taking place along the main channel of the Strait giving rise to the development of a intermittent upwelling with a tidal-related periodicity. This process injects nutrients in the upper layer of the incoming Atlantic Jet and not only alters the biogeochemical budget between both basins but also should have consequences on the dynamic of the pelagic ecosystem of the nearby Alboran Sea.

ii) The strong advection associated to the baroclinic and barotropic tidally-related along-strait currents prevents the effect of this upper-layer fertilization to take place within the Strait itself but rather makes it more suitable to be used (the injected nutrients) in the pelagic ecosystem of the Eastern Alboran Sea.

iii) Complex water-masses circulation pattern through the Strait are induced by the tidal dynamic and associated current fields. This creates a discontinuous presence of different biomass accumulations with specific characteristics and particular origins.

iv) There are coastal-channel pulsating interactions driven by the horizontal surface divergence induced by the tidal-currents reversion. This makes coastal material to be injected within the main along-strait flux into open regions of the Alboran Sea.

References

Alonso del Rosario, J.J., Bruno Mejías, M., Vázquez-Escobar, A. (2003). The influence of tidal hydrodinamic conditions on the generation of lee waves at the main sill of the Strait of Gibraltar. *Deep-Sea Research I,* 50, 1005-1021.

Armi, L., Farmer, D. (1988). The flow of Mediterranean Water through the Strait of Gibraltar. *Prog. Oceanog.*, 21, 41-82.

Arnone, R.A., Wiesenburg, D.A., Saunders, K.D. (1990). The Origin and Characteristics of the Algerian Current. *Journal of Geophysical Research*, 95, 1587-1598.

Basheck, B., Send, U., García Lafuente, J., Candela, J. (2001). Transport estimates in the Strait of Gibraltar with a tidal inverse model. *Journal of Geophysical Research*, 106, 31033-31044.

Béthoux, J.P., Copin-Montégut, G. (1986). Biological fixation of atmospheric nitrogen in the Mediterranean Sea. *Limnol. Oceanogr.,* 31(6): 1353-1358.

Béthoux, J.P., Morin, P., Chaumery, C., Connan, O., Gentili, B., Ruiz-Pino, D. (1998). Nutrients in the Mediterranean Sea, mass balance and statistical analysis of concentrations with respect to environmental change. *Marine Chemistry*, 63, 155-169.

Bone, Q. (2005). Gelatinous animals and physiology. *Journal of Marine Biology Association of U.K.,* 85, 641-653.

Boyce, F.M. (1975). Internal Waves in the Strait of Gibraltar. *Deep-Sea Research*, 22, 597-610.

Boyra, G., Irigoien, X. and Arregi, I. (2005). *Plankton visual analyzer*. Globec International Newsletter, April.

Brandt, P., Alpers, W., Backhaus, J.O. (1996). Study of the generation and propagation of internal waves in the Strait of Gibraltar using a numerical model and synthetic aperture radar images of the European ERS 1 satellite. *Journal of Geophysical Research*, 101, 14237-14252.

Bray, N.A., Ochoa, J., Kinder, T.H. (1995). The role of interface in exchange through the Strait of Gibraltar. *Journal of Geophysical Research*, 100, 10755-10776.

Briscoe, M.G. (1984). Tides, solitons and nutrients. *Nature*, 312, 15-17.

Bruno, M., Alonso, J.J., Cózar, A., Vidal, J., Ruiz-Cañavate A., Echevarría, F., Ruiz, J. (2002). The boiling-water phenomena at Camarinal Sill, the strait of Gibraltar. *Deep-Sea Research II*, 49, 4097-4113.

Bryden, H.L., Kinder, T.H. (1988). Gibraltar experiment: a plan for dynamic and kinematic investigation of strait mixing, exchange and turbulence. Oceanologica Acta, N_SP.

Bryden, H.L., Kinder, T.H. (1991). Steady two-layer exchange through the Strait of Gibraltar. *Deep-Sea Research*, 38, 445-463.

Bryden, H.L., Candela, J., Kinder, T.H. (1994). Exchange through the Strait of Gibraltar. *Progress in Oceanography*, 33, 201-248.

Candela, J., Winant, C., Bryden, H. (1989). Meteorologically Forced Subinertial Flows Through the Strait of Gibraltar. *Journal of Geophysical Research*, 94, 12667-12679.

Candela, J. (1990). The barotrophic tide in the Strait of Gibraltar. In: K.A. Publisher (Editor), *The Physical Oceanography of Sea Straits,* pp. 457-475.

Candela, J., Winant, C., Ruiz, A. (1990). Tides in the Strait of Gibraltar. *Journal of Geophyisical Research- Oceans*, 95, 7313-7335.

Candela, J. (1991). The Strait of Gibraltar and its role in the dynamics of the Mediterran Sea. *Dynamics of Atmospheres and Oceans*, 15, 267-299.

Castro, M.J., García-Rodriguez, J.A., González-Vida, J.M., Macías,J., Parés, C., Vázquez-Cendón, M.E. (2004). Numerical simulation of two-layer shallow water flows through channels with irregular geometry. *Journal of Computational Physics*, 195, 202-235.

Chisholm, S.W. (1992). Phytoplankton size. In: P.G. Falkowski, Woodhead, A.D. (Editor), *Primary productivity and Biogeochemical cycles in the Sea*. Plennun Press, New York, pp. 213-237.

Corzo, A., Morillo, J.A., Rodríguez, S. (2000). Production of transparent exopolymer particles (TEP) in cultures of Chaetoceros calcitrans under nitrogen limitation. *Aquatic Microbial Ecology,* 23, 63-72.

Crepon, M. (1965). Influence de la pression atmospherique sur le niveau moyen de la Mediterranee Occidentale et sur le flux a travers le detroit de Gibraltar. *Cahiers Oceanographiques*, 1, 15-32.

Dafner, E.V., Boscolo, R., Bryden, H.L. (2003). The N:Si:P molar ratio in the Strait of Gibraltar. *Geophysical Research Letters,* 30, 13.1-13.4.

Echevarria, F., García-Lafuente, J.,Bruno, M.,Gorsky, G, Goutx, M.,González, N.,García, C.M.,Gómez, F.,Vargas, J.M.,Picheral, M.,Striby, L.,Varela, M.,Alonso, J.J.,Reul, A., Cózar, A.,Prieto, L.,Sarhan, T.,Plaza, F.,Jiménez-Gómez, F.(2002). Physical-biological coupling in the Strait of Gibraltar. *Deep-Sea Research II,* 49, 4115-4130.

Echevarria, F., Zabala, L., Corzo, A., Navarro, G., Macías, D., Prieto, L. (2009). Spatial distribution of autotrophic picoplankton in relation to physical forcings: the Gulf of Cádiz, Strait of Gibraltar and Alborán Sea case study. *Journal of Plankton Research*, 33, 1339-1351.

Elbaz-Poulichet, F., Morley, N.H., Beckers, J.M., Nomerange, P. (2001). Metal fluxes through the Strait of Gibraltar: the influence of the Tinto and Odiel rivers (SW Spain). *Marine Chemistry*, 73, 193-213

Eppley, R.W., Rogers, J.N., McCarthy, J.J. (1969). Half saturation constrats for uptake of nitrate and ammonium by marine phytoplankton. *Limnology and Oceanography*, 14, 912-920.

Estrada, M., Marrase, C., Latasa, M., Berdalet, E., Delgado, M., Riera, T. (1993). Variability of the Deep Chlorophyll Maximum characteristics in the northwestern Mediterranean. *Marine Ecology Progress Series*, 92, 289-300.

Farmer, D., Armi, L. (1986). Maximal two-layer exchange over a sill and through the combination of a sill and contraction with barotropic flow. *Journal of Fluid Mechanic*, 164, 53-76.

Fasham, M.J.R., Ducklow, H.W., McKelvie, S.M. (1990). A nitrogen-based model of plankton dynamics in the oceanic mixed layer. Journal of Marine Research, 48, 591-639.

Fielder, P.C. (2002). The annual cycle and biological effects of the Costa Rica Dome. *Deep-Sea Research I*, 49, 321-338.

Franks, P. (1992). Sink or swim: accumulation of biomass at fronts. *Marine Ecology Progress Series*, 82, 1-12.

García-Lafuente, J., Vargas, J.M., Plaza, F., Sarham, T., Candela, J., Basheck, B. (2000). Tide at the eastern section of the Strait of Gibraltar. *Journal of Geophysical Research*, 105, 14197-14213.

García-Lafuente, J., Criado, F. (2001). La climatología y la topografía del Estrecho de Gibraltar determinantes de las propiedades termohalinas del agua del Mar Mediterráneo., *Física de la Tierra*,13, pp. 43-54 (in Spanish).

García-Lafuente, J., Delgado, J., Vargas, J.M., Vargas, M., Plaza, F., Sarhan, T. (2002). Low frequency variability of the exchanged flows through the Strait of Gibraltar during CANIGO. *Deep-Sea Research II*, 49, 4051-4067.

García-Lafuente, J., Vargas Domínguez, J.M. (2003). Recent observations of the exchanged flows through the Strait of Gibraltar and their fluctuations at different time scales. *Recent Research Development in Geophysics*, 5, 73-84.

García-Lafuente, J., Ruiz, J. (2007). The Gulf of Cádiz pelagic ecosystem. *Progress in Oceanography*, 74, 228-251.

García-Lafuente, J., Sánchez Román, A., Díaz del Río, G., Sannino, G., Sánchez Garrido, J.C. (2007). Recent observations of seasonal variability of the Mediterranean outflow in the Strait of Gibraltar. *Journal of Geophysical Research*, 112, C10005.

García-Lafuente, J. (2008). *Gulf of Cádiz and Strait of Gibraltar in The Seas of Spain*. Ministerio de Medio Ambiente y Medio Rural y Marino, 508 pp.

Gascard, J.C., Richez, C. (1985). Water Masses and Circulation in the Western Alboran Sea and in the Straits of Gibraltar. *Progress in Oceanography*, 15, 157-216.

Goericke, R., Olson, R.J., Shalapyonok, A. (2000). A novel niche for Proclorococcus sp. in low ligth suboxic environments in the Arabian Sea and the Eastern Tropical North Atlantic. *Deep-Sea Research I*, 47, 1183-1205.

Gómez, F., Echevarría, F., García, C.M., Prieto, L., Ruíz, J., Reul, A., Jiménez-Gómez, F., Varela, M. (2000a). Microplankton distribution in the Strait of Gibraltar: coupling between organism and hydrodinamics structures. *Journal of Plankton Research,* 22, 603-617.

Gómez, F., González, N., Echevarría, F., García, C.M. (2000b). Distribution and fluxes of dissolved nutrients in the Strait of Gibraltar and its relationships to microphytoplankton biomass. *Estuarine, Coastal and Shelf Science*, 51, 439-449.

Gómez, F., Gorsky, G.,Striby, L.,Vargas, J.M.,Gónzalez, N.,Picheral, M.,García-Lafuente, J.,Varela, M.,Goutx, M. (2001). Small-scale temporal variations in biogeochemical features in the Strait of Gibraltar, Mediterranean side- the role of NACW and the interface oscillation. *Journal of Marine Systems*, 30, 207-220.

Gómez, F., Gorsky, G.,García-Górriz, E., Picheral, M. (2004). Control of the phytoplankton distribution in the Strait of Gibraltar by wind and fortnightly tides. *Estuarine*, *Coastal and Shelf Science*, 59: 485-497.

Gran, H.H., 1931. On the conditions for the production of plankton in the sea. Raap. P.-v *Réun. Cons. int. Explor. Mer.*, 75, 37-46.

Herbland, A., Voituriez,B. (1979). Hydrological structure analysis for estimating the primary production in the tropical Atlantic Ocean. *Journal of Marine Research*, 37: 87-101.

Hopkins, T.S. (1999). The thermohaline forcing of the Gibraltar exchange. *Journal of Marine Systems*, 20, 1-31.

Huertas, E., Navarro, G., Rodríguez-Gálvez, S., Prieto, L. (2005). The influence of phytoplankton biomass on the spatial distribution of carbon dioxide in surface sea water of a coastal area of the Gulf of Cádiz (southwestern Spain). *Canadian Journal of Botany*, 83, 929-940.

Izquierdo, A., Tejedor, L., Sein, D.V., Backhaus, J.O., Brandt, P., Rubino, A., Kagan, B.A. (2001). Control Variability and Internal Bore Evolution in the Strait of Gibraltar: A 2-D Two-Layer Model Study. *Estuarine, Coastal and Shelf Science*, 53, 637-651.

Krzanowski, W.J. (1998). *Principles of Multivariate Análisis*. A user´s Perspective. Oxford Press.

La Violette, P.E., Arnone, R.A. (1988). A tide-generated internal waveform in the western approaches to the Strait of Gibraltar. *Journal of Geophysical Research*, 93, 15653-15667.

Lacombe, H., Richez, C. (1982). The regime of the Strait of Gibraltar. In: J.C.J. Nilhoul (Editor), *Elsevier Oceanography Series*. 34.

Le Provost, C., Lyard, F., Molines, J. M., Genko, M. L. & Rabillloud, F. (1998). A hydrodynamic ocean tide model improved by assimilating a satellite altimeter-derived data set. *Journal of Geophyisical Research*, 103, 5513-5529.

Leet, C., Verley, P. , Mullon, C., Parada, C., Brochier, T., Penven, P., Blanke, B. (2008). A lagrangian tool for modelling ichthyoplankton dynamics. *Environmental Modelling & Software*, 23, 1210-1214.

Lennert-Cody, C.E., Franks, P.J.S. (1999). Plankton patchiness in high-frequency internal waves. *Marine Ecology Progress Series*, 186, 59-66.

Li, W.K.W. (2002). Macroecological patterns of phytoplankton in the northwestern North Atlantic Ocean. *Nature*, 419, 154-157.

Longhurst, A., Harrison, A. (1989). The biological pump: profiles of plankton production and consumption in the upper ocean. *Progress in Oceanography*, 22, 47-123.

Macías, D., García, C.M., Echevarría, F., Vázquez-Escobar, A., Bruno, M. (2006). Tidal induced variability of mixing processes on Camarinal Sill (Strait of Gibraltar).A pulsating event. *Journal of Marine Systems*, 60, 177-192.

Macías, D., Martin, A.P., García Lafuente, J., García, C.M.,Yool, A., Bruno, M.,Vázquez, A., Izquierdo, A., Sein, D., Echevarría, F. (2007). Mixing and biogeochemical effects induced by tides on the Atlantic-Mediterranean flow in the Strait of Gibraltar. An analysis through a physical-biological coupled model. *Progress in Oceanography*, 74, 252-272.

Macías, D., Martin, A.P., García Lafuente, J., García, C.M.,Yool, A. Bruno, M.,Vázquez, A., Izquierdo, A., Sein, D., Echevarría, F. (2007[a]). Mixing and biogeochemical effects induced by tides at the Strait of Gibraltar: a modelling study. *Globec International Newsletter*, 13(2), 25-27.

Macías, D., Lubian, L.M., Echevarría, F., Huertas, E., García, C.M. (2008). Chlorophyll maxima and water mass interfaces: tidally induced dynamics in the Strait of Gibraltar. *Deep-Sea Research I*, 55, 832-846.

Macías, D., NAvarro, G., Bartual, A., Echevarría, F., Huertas, I.E. (2009). Primary production in the Strait of Gibraltar: carbon fixation rates in relation to hydrodynamic and phytoplankton dynamics. *Estuarine Coastal and Shelf Science*, 83, 197-210.

Macías, D., Somavilla, R., González-Gordillo, I., Echevarría, F. (2010). Physical control on zooplankton distribution pattern at the Strait of Gibraltar during an episode of internal wave generation. *Marine Ecology Progress Series*, in press.

Macías, D., Ramírez, E., García, C.M. (submitted). Effect of nutrient input frequency on the structure and dynamics of the marine pelagic community. A modelling approach. *Journal of Marine Research.*

Malone, T.C. (1980). Algal size. In: I. Morris (Editor), *The Physological Ecology of Phytoplankton*. Blackwell Scientific Publications, Oxford, pp. 133-463.

Mann, K.H., Lazier, J.R.N. (1991). Dynamics of marine ecosystems. *Biological-Physical Interactions in the Oceans*. Blackwell Scientific Publications.

Margalef, R. (1989). *Ecología. Ediciones Omega* (in Spanish).

Mercado, J.M., Ramírez, T., Cortés, D., Sebastián, M., Vargas-Yáñez, M. (2005). Seasonal and inter-annual variability of the phytoplankton communities in an upwelling area of the Alborán Sea (SW Mediterranean Sea). *Scientia Marina*, 69, 451-465.

Minas, H.J., Coste, B., Le Corre, P., Minas, M., Raimbault, P. (1991). Biological and Geochemical Signatures Associated With the Water Circulation Through the Strait of Gibraltar and in the Western Alboran Sea. *Journal of Geophyisical Research*, 96, 8755-8771.

Minas, H.J., Minas, M. (1993). Influence du Détroit de Gibraltar sue la biogéochime de la Méditerranée et du proceh Atlantique. *Annales de L'Institute Océanographique*, 69, 203-214.

Morán, X.A.G., Taupier-Letage, I., Vázquez-Dominguez, E., Ruiz, S., Arin, L., Raimbault, P., Estrada, M. (2001). Physical-biological coupling in Algerian Basin (SW Mediterranean): Influence of mesoscales instabilities on the biomass and production of phytoplankton and bacterioplankton. *Deep-Sea Research I*, 48, 405-437.

Munk, P., Hansen, B.W., Nielsen, T.G., Thomsen, H.A. (2003). Changes in plankton and fish larvae communities across hydrographic fronts off West Greenland. *Journal of Plankton Research*, 25, 815-830.

Mutlu, E. (2006). Diel vertical migration of Sagitta setosa as inferred acoustically in the Black Sea. *Marine Biology,* 149, 573-584.

Navarro, G. (2004). Escalas de variación espacio-temporal de procesos pelágicos en el Golfo de Cádiz, PhD Thesis.Universidad de Cádiz, 207 pp (in Spanish).

Navarro, G., Ruiz, J., Huertas, I.E., García, C.M., Criado-Aldeanueva, F., Echevarría, F. (2006). Basin scale structures governing the position of the subsurface chlorophyll maximum in the gulf of Cádiz. *Deep-Sea Research II,* 53, 1261-1281.

Ochoa, J., Bray, N.A. (1991). *Water mass exchange in the Gulf of Cadiz. Deep Sea Research*, 38, S465-S503.

Packard, T.T., Minas, H.J., Coste, B., Martinez, R., Bonin, M.C., Gostan, J., Garfield, P., Christensen, J., Dortch, Q., Minas, M., Copin-Montegut, G., Copin-Montegut, C. (1988). Formation of the Alboran oxygen minimun zone. *Deep-Sea Research*, 35, 1111-1118.

Pan, L.A., Zhang, L.H., Zhang, J., Gasol, J.M., Chao, M. (2004). On board flow cytometry observations of picoplankton comunity structure in the East China Sea during the fall of different years. *FEMS Microbiology Ecology*,52, 243-253.

Pettigrew, N.R. (1989). Direct measurements of the Flow of Western Mediterranean Deep Water Over the Gibraltar Sill. *Journal of Geophysical Research*, 94, 18089-18093.

Prieto, L., G. Navarro, A. Cózar, F. Echevarría, C.M. García (2006). Distribution of TEP in the euphotic and upper mesopelagic zones of the southern Iberian coasts. *Deep-Sea Research II*, 53, 1314-1328.

Reul, A., Vargas, J.M., Jiménez-Gómez, F., Echevarría, F., García-Lafuente, J., Rodriguez,J. (2002). Exchange of planktonic biomass through the Strait of Gibraltar in late summer conditions. *Deep-Sea Research II*, 49, 4131-4144.

Reul, A., Muñoz, M., Criado-Aldeanueva, F., Rodriguez, V. (2006). Spatial distribution of phytoplankton<13 □m in the Gulf of Cadiz in relation to water masses and circulation pattern under westerlies and easterlies wind regimes. *Deep-Sea Research II*, 53, 1294-1313.

Richez, C. (1994). Airbone synthetic-aperture radar tracking of internal waves in the Strait of Gibraltar. *Progress in Oceanography,* 33, 93-101.

Rodríguez, J., Blanco, J.M. , Jiménez-Gómez, F., Echevarría, F., Gil, J., Rodríguez, V., Ruiz, J., Bautista, B., Guerrero, F. (1998). Patterns in the size structure of the phytoplankton community in the deep fluorescence maximum of the Alboran Sea (southwestern Mediterranean). *Deep-Sea Research I,* 45, 1577-1593.

Ruiz, J., Echevarría, F., Font, J., Ruiz, S., García, E., Blanco, J.M., Jiménez-Gómez, F., Prieto, L., González-Alaminos, A., García, C.M., Cipollini, P., Snaith, H., Bartual, A., Reul, A., Rodríguez, V. (2001). Surface distribution of chlorophyll, particles and gelbstoff in the Atlantic jet of the Alborán Sea: from submesoscale to subinertial scales of variability. *Journal of Marine Systems,* 29, 277-292.

Ruiz, J., Macías, D., Peters, F. (2004). Turbulence increases the average settling velocity of phytoplankton cells. *PNAS,* 101, 17720-17724.

Russell, R.W., Harrison,N. M. and Hunt Jr.,G. L. (1999). Foraging at a front: Hydrography, zooplankton, and avian planktivory in the northern Bering Sea. *Marine Ecology Progress Series*, 182, 77-93.

Sannino, G., Bargagli, A., Artale, V. (2002). Numerical model of the mean exchange through the Strait of Gibraltar. *Journal of Geophysical Research*, 107, 3044.

Sannino, G., A. Bargagli, Artale, V. (2004). Numerical modeling of the semidiurnal tidal exchange through the Strait of Gibraltar. *Journal of Geophyisical Research*, 109, C05011.

Sánchez-Garrido, J.C., García-Lafuente, J., Criado Aldeanueva, F., Baquerizo, A., Sannino, G.(2008). Time-spatial variability observed in velocity of propagation of the internal bore in the Strait of Gibraltar. *Journal of Geophyisical Research*, 113, C07034.

Sein, D.V., Backhaus, J.O., Brandt, P., Izquierdo, A., Kagan, B.A., Rubino, A., Tejedor, L. (1998). *Flow exchange and tidally induced dynamics in the Strait of Gibraltar as derived from a two-layer, boundary-fitted coordinated model.* Unesco Workshop on Ocenic Fronts and Related Phenomena.

Smetacek, V., Passow, U. (1990). Spring bloom initiation and Sverdrup´s critical-depth model. *Limnology and Oceanography*, 35, 228-234.

Sverdrup, H.U. (1953). On conditions for the vernal blooming of phytoplankton. *J. Cons. Perm. Int. Exp. Mer.,* 18, 287-295.

Tsimplis, M.N., Bryden, H.L. (2000). Estimation of the transports through the Strait of Gibraltar. *Deep-Sea Research I,* 47, 2219-2242.

van Genn, A., Rosener, P., Boyle, E. (1988). Entrainment of trace-metal-enriched-Atlantic-shelf water in the inflow to the Mediterranean Sea. *Nature*, 331, 423-426.

Vargas, J.M., García-Lafuente,J., Candela, J., Sánchez, A.J. (2006). Fortnightly and monthly variability of the exchange through the Strait of Gibraltar. *Prog. Oceanog*., 70, 466-485.

Vázquez, A., Stashchuk, N., Vlasenko, V., Bruno, M., Izquierdo, A., Gallacher, P.C. (2006). Evidence of multimodal structure of the baroclinic tide in the Strait of Gibraltar. *Geophysical Research Letters*, 33, L17605.

Vázquez, A., Bruno, M., Izquierdo, Macías, D. (2008). The effect of meteorologically forced subinertial flows on internal waves generation at the main sill of the Strait of Gibraltar. *Deep-Sea Research I,* 57, 1277-1283.

Vázquez, A., Flecha, S., Bruno, M., Macías, D., Navarro, G. (2009). Internal waves and short-scale distribution patterns of chlorophyll in the Strait of Gibraltar and Alborán Sea. *Geophysical Research Letters*, 36, L23601.

Wang, D.P. (1989). Model of mean and tidal flows in the Strait of Gibraltar. *Deep-Sea Research*, 36, 1535-1548.

Wang, D.P. (1993). The Strait of Gibraltar Model: Internal tide, diurnal inequality and fortnightly modulations. *Deep-Sea Research I,* 40A, 1187-1203.

Wash, J.J. (1988). On the nature of continental shelves. Academic Press. New York, 126.

Wesson, J.C., Gregg, M.C. (1994). Mixing at Camarinal Sill in the Strait of Gibraltar. *Journal of Geophysical Research*, 99, 9847-9878.

Wiebe, P.H., Benfield, M.C. (2003). From the Hensen net toward a four-dimensional biological oceanography. *Progress in Oceanography*, 56, 7-136.

Wu, P., Haines, K. (1996). Modelling the dispersal of Levantine Intermediate Water and its role in Mediterranean deep water formation. *Journal of Geophysical Research*, 101, 6591-6607.

Yamasaki, H., Squires, K.D. (1996). Comparison of oceanic turbulence and copepod swimming. *Marine Ecology Progress Series,* 144, 299-301.

Zar, J.H. (1974). *Biostatistical analysis*. Prentice Hall International Editions.

In: Sea Level Rise, Coastal Engineering, Shorelines and Tides ISBN: 978-1-61728-655-1
Editor: Linda L. Wright, pp. 47-82 © 2011 Nova Science Publishers, Inc.

Chapter 2

TIDAL ENERGY-POTENTIAL ENERGY IN FUTURE

Liu Li-qun*

Taiyuan University of Science & Technology, Taiyuan, China

Abstract

Energy provides an essential impetus to support the development of countries in the whole world, with the increasing demand for energy, the traditional fossil energy sources will eventually be depleted in the foresee future, such as coal, oil, natural gas, and a large number consumption of fossil energy leads the enormous environmental issues, such as acid rain, greenhouse effect and pollution of river and soil, in order to the world's sustainable development, in order to future generations of breeding, renewable energy research has been regarded by governments and the general public, such as wind, solar, ocean energy, biomass, etc., renewable and sustainable energy has been applied to the daily life of ordinary people. But for now, the application of ocean energy is still relatively backward compare with the application of wind energy and solar power, and the fundamental reason is that the application of ocean energies need huge investment and the technologies either are not mature enough or are too expensive, and the gains of project are limited, and the investment has great risk, so the normal companies are unwilling to invest in this area, and the support of government is necessary. With the improved of ocean energy developed technology and the increase of traditional energy prices and ordinary people aware the important of ocean energy application, the ocean energy in the future can enter a new era. Tide is due to the earth and the celestial movement and the interaction, which lead the seawater fluctuations. In hundreds years ago, the tide is used to grind corn by the ancient human in order to reduce the labor intensity of people and livestock, and the modern use of tide is tidal energy generation. According to the statistics data, the total theoretical potential of tidal energy in the whole world is more than 3 billion kW, and the potential development of tidal energy is about 200TWh. At the same time, the total installed capacity of electric power around the world in 2004 is about 4 billion kilowatts, so the development of tidal energy has great potential and bright prospects in future. This chapter analyses the formation of tide, distribution, power generation principle and application examples in the world, and forecast the prospects of tidal energy power generation and the various problems for tidal energy development.

* E-mail address: llqd2004@163.com

1. Introduction

There are more than 6 billion mankind and incomputable life-form which live in Earth, and the Earth is just the drop in the infinite universe ocean, but it is still more than $12700\,km$ in diameter, and the surface area is about $5.1\times10^{8}\,km^{2}$, in which the land surface area has accounted for 29.2%; and the marine area has accounted for 70.8%. The average land elevation is about 840 meters to sea-level terms, and the average ocean depth is about 380 meters, the volume of sea water reach $1.37\times10^{9}\,km^{3}$, and it brings together 97% of the water on the planet, and only 3% of fresh water can be direct use by humans or animals or plant. A vast expanse of ocean is the cradle of pregnant life, and billions of mankind and life-form birth from here, due to it can offers a rich variety of bio-food and nutrients since time immemorial. While in the modern society, *ocean* is still to provide mankind with a huge amount of aquatic products, and rich in mineral, and convenient shipping. The living of billions of people around the world is closely and inseparably related to the ocean, so it can say that the ocean is the mother of mankind, and she gave birth to humans, and raising human. In addition to the ocean is also a vast reservoir of energy, and the ocean surface is rich in solar energy and wind energy, and the sea has abundant sea energy, such as the theoretical potential of global ocean wave energy reserves is about 70 billion kW, and the amount is about 3 billion kW which can be developed and utilized, so the entire ocean will make a greater contribution to mankind in the foresee future [1-8].

Of human development from primitive to modern humans, the whole of human history is only about three million years, compared with the Earth billions of years of history, humanity history is nothing which is only a waves in a long history river, and humanity history after the industrial revolution is only a few hundred years, but the industrial revolution change the entire planet in the past few hundred years is more than the changes of the past hundreds of thousands of years. From the ancient times of the hunting society to the agricultural society until the modern industrial society, then the information society, with the raising of the human's quality of life and level, and the scale of ocean and Earth is snatched and destroyed by human beings is huge, such as there are several billions of tons of mineral resources per year is exploited by human being, and several billions of tons of waste water every year is discharged into the ocean, and thousands of species of plants and animals become extinct every year, and every year hundreds of millions of tons of greenhouse gases into the atmosphere, and hundreds of km^{2} of polar glacier and mountain glacier are melting, and the earth's temperature rise, and some small island countries will disappeared in foreseen future. These are warning we have to change the current way of life, we only have a common earth, but unfortunately not all were aware of this, the result of just concluded in Copenhagen to bring all mankind is a regrettable. human has time? Only the future can answer this question.

Of course right now in many countries around the world have begun to realize the importance of environmental issues, the development and application of renewable energy sources in the whole world in recent years like a raging fire, such as wind energy, solar energy, hydroenergy, ocean energy, bioenergy, geothermal and nuclear energy. Thousands of experts and scholars have been pay attention to renewable energy domain, especially in wind and solar power is now the hot spots of the international development, for example, as of 2008, the global cumulative installed capacity of wind power has more than 120 million kw [9]. As of 2006, the world's annual production and cumulative installed capacity of photovoltaic (PV) are 2500MW and 8.59GW, respectively [10]. In addition, development and application of hydropower and nuclear energy are very broad, but for now, the development and application of ocean energy is still at a low stage due to the development cost and technical reasons, and not only the capacity is small, and related technology is high-performance. At present, because the developed technology of wind energy is relatively mature, and the prices is relatively low, so which has become the world focus on new energy development over the last decade, and the annual growth rate maintained at about 30%, and the new installed capacity is 27261 MW in 2008, and total cumulative installed capacity of wind power is 121188 MW, while wind power development costs are coming down, and more and more developing countries began to regard the development of wind power. Let us take China as an example, the current market costs of wind power in China is about 5000RMB/kW, and wind farm costs is about 8000RMB/kW, and China ranks fourth in the world, as of 2008, the world's top four wind energy development country were the United States, Germany, Spain and China, the global wind energy resources in total of about 20 billion kW, but it should be noted that wind power and solar power generation is big affected by the weather compare with the ocean energy [9] [11]. While the utilization of wind and solar energy is rapid developed, the development and utilization of marine energy is relatively backward, for now, the capacity is smaller, the technology is immature, and the cost of development is high. For example, Chinese wind power development scale is big, the planned installed capacity of wind farms will be more than 10000MW in Jiuquan, Gansu Province and Yancheng, Jiangsu Province, and some secondary wind farm have been built in other areas and the installed capacity is over hundreds of MW-class [12-13] . But the world's largest tidal power plant is the La Rance tidal power plant in French and which have been completed about 40 years, and the installed capacity is 240MW. The SIHWA tidal power plant built in South Korea, but the installed 260MW. While the world's major maritime countries also planned large-scale tidal power plant, such as the installed capacity of Meijin and Tuguer tidal power plant in Russia is also exceeded the 10000MW, but when can be completed is still unknown, while the other ocean energy installed capacity even smaller, such as ocean wave energy, ocean salinity energy, ocean thermal energy and ocean current energy.

Ocean energy, usually refer to renewable energy of ocean, including tidal energy, wave energy, ocean current energy, ocean salinity energy and ocean thermal energy, and more broadly over the ocean energy, including ocean wind energy and solar energy on sea surface, and marine biomass and so on. In addition to tidal energy and tidal current energy can be derived from the changes of place of the sun and the moon, as well as the Earth's gravitational force. All other energy are derived from solar radiation. According to estimates by United Nations Educational Scientific and Cultural Organization (UNESCO) in 1981, the total global theory reservoir of ocean energy resources is about 76.6 billion kilowatts, and of which ocean thermal energy can be about 40 billion kW, and the ocean salinity energy can be about 30 billion kW, and the tidal energy is greater than 3 billion kW, and the ocean wave energy is about 3 billion kW, and ocean currents for 600 million kW. But also should be noted that although the reserves of ocean energy are large, but it is difficult to use all of its, only the high energy density region has the development value, so there are a higher development value of the ocean energy, and the total amount is about 6.4 billion kW, of which ocean salinity energy is about 3 billion kW, and ocean thermal energy can be 2 billion kW, and ocean wave energy is about 1 billion kW, and ocean current energy can be 300 million kW, and tidal energy more than 100 million kW [2]. Now look at the development of ocean energy costs are relatively high, the La Rance tidal energy power plant built in 1966, for example, its total investment is more than 480 million francs, and the total installed capacity is 240MW, and the installed cost per kilowatt is 2000 francs, about the time Hydropower 2.5 times the cost, so the development of ocean energy are not currently a national focus on renewable energy domain, but in the future, as advances in development technology and people to deepen understanding of ocean energy, and ocean energy in the whole world will occupy the important place in entire energy structure [6].

In general, ocean energy is a renewable and clean energy, but also vast amounts in global ocean, but it should see that the ocean energy have some shortcomings, such as the uneven distribution of the energy density, low energy density, low efficiency, poor economy, and the energy instability. The tidal energy are currently the most technologically mature and a better economy among a variety of marine energy development, so this chapter will first introduce the cause of tide, and tide distribution, and then describes the current tidal power technologies and application examples, and at the last, the future development of the tide is forecasted and some suggestions is recommended.

2. Tidal Energy

2.1. Tide

Tidal phenomenon is a natural phenomenon, which is due to the sun and the moon at different locations on the planet and the Earth's rotation in different gravity generated, and the joint action of the centrifugal force generated the Earth's tidal force, and the tidal force caused the regular and periodic fluctuation phenomenon of ocean surface, known as the tides. From a global point of view, the phenomenon of tidal water is actually a kind of fluctuations, and tidal energy in the vertical direction refer to the tide rise and tide fall, that is, the potential energy of tides. And the tidal energy in the horizontal direction expressed as the trend of tidal advance and tidal retreat, that is, the kinetic energy of tides. Tidal phenomenon from a location view, sea level rise is the flood tide, and sea level drop is ebb tide, and the flood tide and ebb tide replace in a certain period of time. When the tide changed from flood tide to ebb tide, and the sea level is the highest, known as the high tide, and the tide changed from ebb tide to flood tide, and the water level is the lowest, known as the low tide, and the high tide and low tide replacement in a certain period of time.. Tidal current, from one high tide to the next one high tide or from one low tide to the next one low tide as the average time interval between the average tidal cycle, generally 12h25m, and a lunar day for astronomy will be 24h50m. Each tidal wave tidal peak and valley level difference is called tidal range, while the tidal water level changes over time during the process of line known as the tide line. Each time tidal fluctuation, not a simple repetition of the last time, which has not the same height of tide, and low tide is not equally low, and high tide is not equally high. Tides are also cyclical changes. In a cycle, tidal range is changing from big to small, and then from small to big, the largest tidal range is called spring tide, and when the tidal range is smallest which is called the neap tide. From the spring tide or a neap tide over until the next spring tide or neap tide of time interval is called the Semi-Lunar month, and the cycle is 14.77 days, so every month have two spring tides and neap tides.

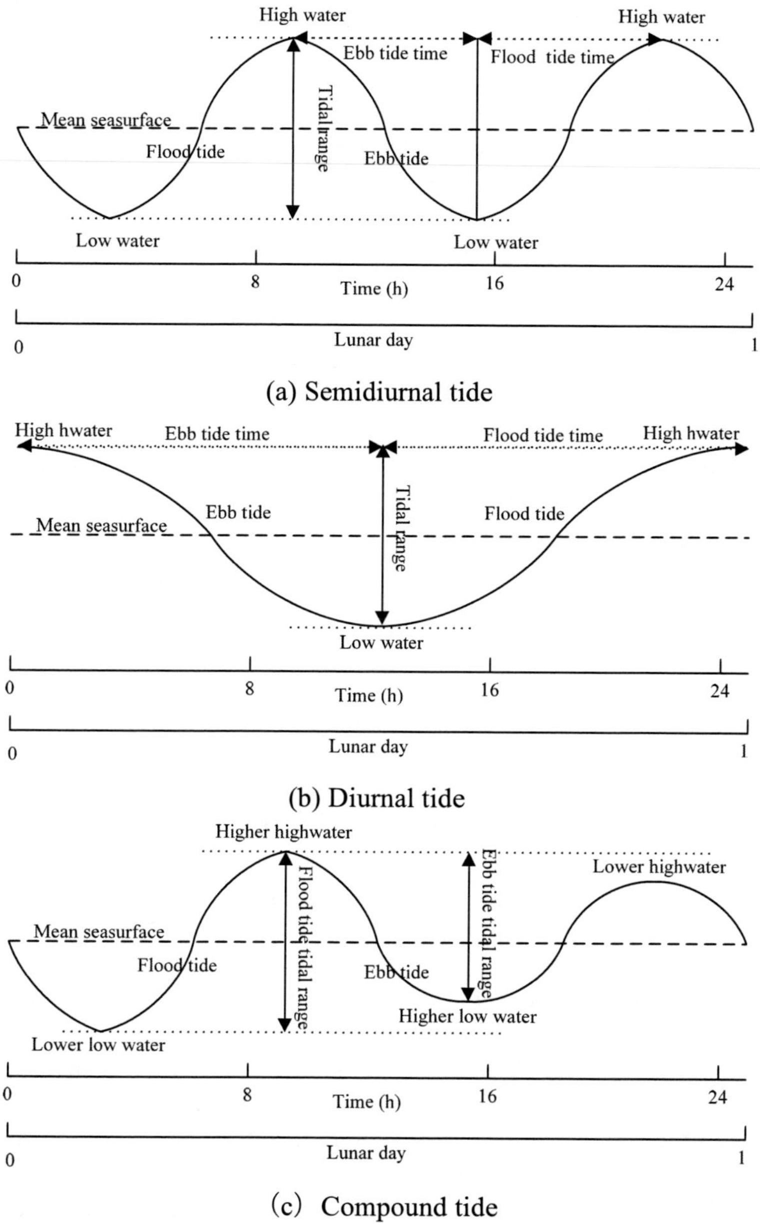

Figure 1. Three types of tides [6].

Tidal fluctuation cycles vary due to time and location, according to the Sun, Moon, Earth placement can be classified as "spring tide" and "neap tide"; according to distance from the Moon and the Earth can be classified as perigean tide and apogean tide; under the direction of tidal force can be divided into fair tide and counter tide and so on. From the fluctuation of the cycle can be divided into three types of tide: semidiurnal tide, diurnal tide, and compound tide. Semidiurnal tide is defined as a lunar day, there were two high tides and two low tides, that is, tidal fluctuation twice in a lunar day and a period of 12h25m, it is characterized by two adjacent ebb tides and high tides have almost equal height of tide, and duration of tide is almost equal. Diurnal tide is defined as a lunar day tide, which occur only one low tide and high tide, tidal fluctuation time has a period of 24h50m. Compound tide is between semidiurnal tide and diurnal tide, and the tidal wave form is defined as within a month on certain days, there were two high tides and two low tides, and some days there are one high tide and one low tide, and it is can be divided into two types of tides: irregular semidiurnal tide and irregular diurnal tide. The irregular semidiurnal tide have twice high tide and two low in a lunar day, but the adjacent height of tide of high tide and low tide height are different, and duration of tide is different. Irregular diurnal tide is defined as the majority lunar days of Semi-Lunar month have the irregular semidiurnal tide, but sometimes also taken place the diurnal tide phenomena, namely, one low tide and high tide phenomena in a lunar day, and three types of tides are shown in Figure 1 [6]. Here, the Figure 1 (a) shows the semidiurnal tide, and Figure 1 (b) shows the diurnal tide, and Figure 1 (c) shows the compound tide, which also known as mixed tide.

So tidal cycle is a very complex issue, including the half-day cycle, diurnal cycle and lunar cycle changes, the Earth and the Moon and other celestial bodies rotation also produced a number of other periodic cycles, and the cycle from a few days to several years. In addition to the Moon and the Sun have big effect to the tides of Earth, all other celestial bodies on the Earth's tides also affect the size and periodicity, although the effect is very small and can be negligible because which is away from the Earth or the quality is small, so the tide force is small. With the gravity calculation, the tide force of Moon on the tides of the sea surface of Earth can increase 0.563m, the tide force of Sun on the tides of sea level of Earth can rise 0.246m, because although the mass of Sun is much greater than the mass of Moon, but the distance from Sun to Earth than the distance between the Moon and the Earth is much greater, so the tidal force of the Sun is less than half of the tidal force of the Moon. Compared with the results and actual observations, the actual tidal range is much greater than the above values, so the above values can be applied the open ocean and far from the coast ocean. In addition, the actual height of tide is related with the shape of the coast and the seabed. Such as the shore was trumpet shape, and it will inevitably lead to accumulation of more and more sea water, so the height of tide is raised due to narrow channel. Such as China, Hangzhou Bay, the largest tidal range up to 8.93m., Canada in North America, Fundy Bay, the largest tidal range up to 19.6m. This difference between the actual and calculated there were no precise explanation. In generally believed that when the tidal current hit the continental shelf and the coastline, and the tidal range will increases through raised and pinching and sympathetic vibration. Energy of tidal current is proportional to the tide volume and tide range, or, with the square of tide range and the area of reservoir. And hydroelectric power compared to the energy density of tidal energy is very low, equivalent to the level of micro-head power generation. The largest the value of the world's tidal range is about 13 ~ 15m, but in general, if the average tidal range is more than 3m which have practical value [1].

Tide is varying with time and location, and different regions often have different tidal systems, which obtain energy from tidal waves of deep sea, but it has its own unique characteristics. Although the change of tide is very complex, but the tides of any place can be accurate forecasted. Ocean tides relative to the Earth's solid surface constantly on the movement, it get energy from the rotation of the earth, at the same time, friction with the Earth's solid surface to consume the absorbed energy, that the Earth's rotational energy, the course of consumed energy will result in slowing down the Earth's rotation and rotation speed decreases. It is calculated that every day will extend 28 seconds after each million years, there is evidence that, in more than 300 million years ago, the Earth is about 400 days a year, but now only about 365 days a year. But mankind is almost not aware the slow down of the Earth's rotation speed in the person's life, but also the slow speed will not speed up due to the development and application of tidal energy. The Earth's rotational energy is dissipated through the friction with shallow sea area and coastal area, and the rate of dissipation is about 1.7TW.

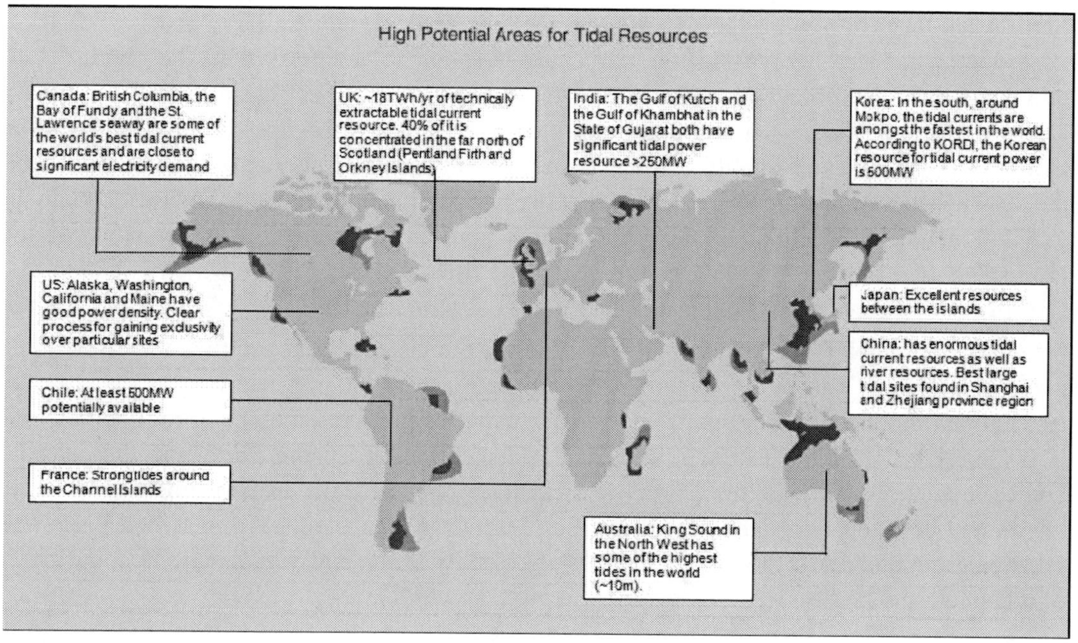

Figure 2. High potential areas of tide energy resources in the whole world [20].

Not all the parts of the world are suitable for construction of tidal power plant, the suitable location should be only appears high tide and energy concentration and excellent geographical conditions for the construction, where extracting energy from the tides is possible. Although these place is not everywhere, but countries around the world have been selected for a considerable number of suitable sites for tidal energy development. According to the latest estimates, the development potential of tidal energy is about annually 200TW•h [1] [17]. Around the world, the phenomenon of tides in coastal areas will occur, but the main tides and tidal currents occurs in the Arctic Ocean, English Channel, Irish Sea, Skagerrak, Kattegat, the Hebrides, , Mexico, and St. Lawrence Bay, Fundy Bay, Magellan Strait, the Strait of Gibraltar, the Strait of Messina, Sicily Island, the Bosphorus, near Taiwan, the Kuril Islands, north-west and west of Australia, as well as the Amazon and La Plata, as well as many other locations, such as Yellow sea and Bohai sea and Donghai sea of China, as shown in Figure 2 [19] [20].

2.2. Tides Cause

Tidal energy is an important form of ocean energy, which includes the potential energy due to the tidal changes in sea levels and the kinetic energy due to the flow of sea water, namely, the flood tide and ebb tide sea water produced the potential energy and kinetic energy of sea water. The energy of tidal fluctuations of sea water get from the movement phenomenon between Earth and the celestial, and caused by the interaction between them, and simply say that only a small part of the tidal energy get from the Earth's rotation, and most of the tidal energy get from the gravitational interaction between the Earth and Moon and Sun, namely, the gravitational interaction generated the kinetic energy and potential energy of tides. Moon's gravity on Earth's sea water generated lunar tide, which raise the Earth's sea level of toward the Moon's surface and back Moon's surface, and the sea level raise due to the Sun's gravity on the earth tide is called solar tide. Where the gravity is an interaction between object and other object due to every one object has a mass. The gravity value is related with the size and the quality of objects, as well as the distance between the two objects. The greater quality of objects has the greater gravitational force between them, the distance between objects farther, the more gravitational force between them is small, gravitational force is equal to the gravitational constant multiplied by the quality of the product of two objects, divided by the square of their distance, as shown in equation (1). Here, G is the gravitational constant, its value is about $6.67\times10^{-11}Nm^2/kg^2$. m_1 and m_2 are the quality of the two objects, respectively. And r is the distance between two objects [6]. The gravitational force can be applied to between the Earth and other celestial bodies, and the gravitational interaction is proportional to the quality of gravitation and celestial bodies, and is inversely proportional to the square of the distance between the Earth and the celestial bodies, and the quality of the Earth as a constant [14-18]. It is usually, the gravitational interaction between Earth and Moon and Sun is greater than the gravity between Earth and other celestial bodies, so the tide on Earth is affected by Moon and Sun is more than the effect of other celestial bodies.

$$F = \frac{Gm_1m_2}{r^2} \tag{1}$$

The tidal forces generate tide from which get from Moon and Sun, because the tide processes produce by Sun is similar with Moon, so the moon's tidal force is only discussed in this paper. On the Earth, the force acting on the earth there are two: one is the moon's gravitational interaction, and the other is the inertia centrifugal force which generated by the movement of the earth and the moon, due to it is generally believed that the Earth's movement around the Sun, while the Moon as the Earth's satellites around the earth rotation, in fact, the movement orbit is elliptical, and the conventional center of gravity of Earth on elliptical orbit around the Sun in accordance with the movement is not the center of the Earth, actually, a public mass of Earth and Moon is produced in rotation process because of the Moon gravity on the Earth and Earth has gravitational interaction on Moon, a mutual attraction of gravity system constitutes between the Earth and Moon, so have a common-centroid in the gravity system, and common-centroid mean the quality of the gravity system. The common-centroid calculation of Earth and Moon gravity system is as follows: E and M are the quality of the Earth and the Moon, respectively, and r' is the distance between them. The Earth's distance from the common-centroid is r_1, the moon's distance from the common-centroid is r_2, and $r_1 + r_2 = r'$, due to gravitational interaction between the Earth and Moon are equal, so the common-centroid is formulated the formula (2), r_1 and r_2 can be calculated as equation (3) and

(4), respectively. At the same time, because the quality of the Earth is 81.5 times of the Moon's quality, and the average distance between the Earth and moon is $r^{'}=384404$ kilometers. The Earth's average radius is about $R=6371$ kilometers, so the common-centroid at 4659.44 kilometers in the Earth's interior, that is 0.73 times the Earth's radius.

$$\mathrm{r_1 \times E = r_2 \times M} \tag{2}$$

$$\mathrm{r_1 = r^{'} \times [M/(E+M)]} \tag{3}$$

$$\mathrm{r_2 = r^{'} \times [E/(E+M)]} \tag{4}$$

$$\mathrm{F} = \frac{\mathrm{GEM}}{r^{'2}} \tag{5}$$

$$\mathrm{F_c} = -\frac{\mathrm{GEM}}{r^{'2}} \tag{6}$$

$$\mathrm{F_B} = \frac{\mathrm{G\Delta mM}}{(r^{'}-R)^2} \tag{7}$$

$$\mathrm{F_{Bc}} = -\frac{\mathrm{G\Delta mM}}{r^{'2}} \tag{8}$$

$$\mathrm{F_{Bi}} = \mathrm{F}_B + F_{Bc} = \frac{\mathrm{G\Delta mM}}{(r^{'}-R)^2} - \frac{G\Delta mM}{r^{'2}} = G\Delta mM\frac{2Rr^{'}-R^2}{r^{'4}-2r^{'3}R+r^{'2}R^2} \tag{9}$$

$$\frac{r^{'}}{R} = 60.3 \tag{10}$$

$$\mathrm{F_{Bi}} = G\Delta mM\frac{2R-\frac{R^2}{r^{'}}}{r^{'3}-2r^{'2}R+r^{'}R^2} \approx G\Delta mM\frac{2R}{r^{'3}} \tag{11}$$

$$\mathrm{F_A} = \frac{\mathrm{G\Delta mM}}{(r^{'}+R)^2} \tag{12}$$

$$\mathrm{F_{Ac}} = -\frac{\mathrm{G\Delta mM}}{r^{'2}} \tag{13}$$

$$\mathrm{F_{Ai}} = \mathrm{F}_A + F_{Ac} = \frac{\mathrm{G\Delta mM}}{(r^{'}+R)^2} - \frac{G\Delta mM}{r^{'2}} = G\Delta mM\frac{-2Rr^{'}-R^2}{r^{'4}+2r^{'3}R+r^{'2}R^2} \tag{14}$$

$$F_{Ai} = G\Delta mM \frac{-2R - \frac{R^2}{r'}}{r'^3 + 2r'^2 R + r' R^2} \approx -G\Delta mM \frac{2R}{r'^3} \quad (15)$$

The Moon around the center of common-centroid and the movement include the revolution and rotation, and the revolution cycle is about one month, while the Earth's movement around the common-centroid is a kind of translational motion. In such circumstances, the Earth's inertial centrifugal force generated by all the points on Earth are equal to the size, and the orientation is the same, and with the Moon in the opposite direction of gravity, so to be able to maintain on the average same distance between Earth and Moon, and the vector of force acting on the Earth must be separate zero. As the inertial centrifugal force on the planet are same direction, but the lunar gravity of different point of Earth is different from the size of the force vector and direction, so the synthesis force vector of all the points are not the same, so the tidal force of different points of Earth is different, the tidal forces of Moon on the Earth in different points are shown in Figure 3. The Moon's gravity on Earth is expressed as equation (5). Any object in the geocentric reference frame suffered inside the centrifugal force of inertia is equal to the negative gravity of the Moon which is put on the center of the earth. So the inertia of centrifugal force of all objects on Earth can be expressed as equation (6), the gravity of Moon of an object on Earth B point is expressed as equation (7), and the quality of object is Δm . The lunar gravity and the centrifugal force of Earth are two opposing forces, and the two forces have combined to produce the force vector, which produce the tide phenomenon of seawater, known as the Moon's tidal force. The lunar tidal force of an object on Earth B point is expressed as equation (8). Then the object on B point were affected by an unbalanced force such as the equation (9) as shown, while the equation (10) shows the relationship between the Earth-Moon distance and the radius of the Earth, then equation (9) is simplified the equation (11), the tidal force of B point can be expressed as equation (11). The gravity of Moon of an object on Earth A point is expressed as equation (12), and inertial centrifugal force of A point can be expressed as equation (13), then the unbalanced force of A point is expressed as equation (14). Then the equation (14) is simplified the equation (15) by using the equation (10), namely, the tidal force of A point can be expressed as equation (15). Therefore, the tidal force of A point and the lunar gravity are opposite direction, while the tidal force of B point and the lunar gravity in the same direction [17-18].

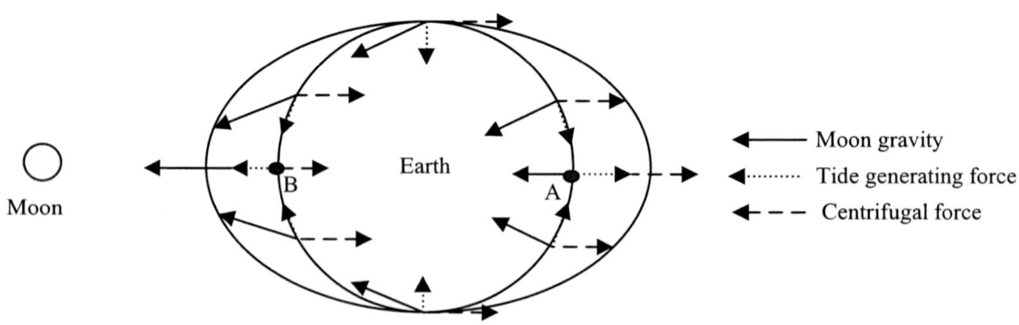

Figure 3. Tide generating force sketch map [6].

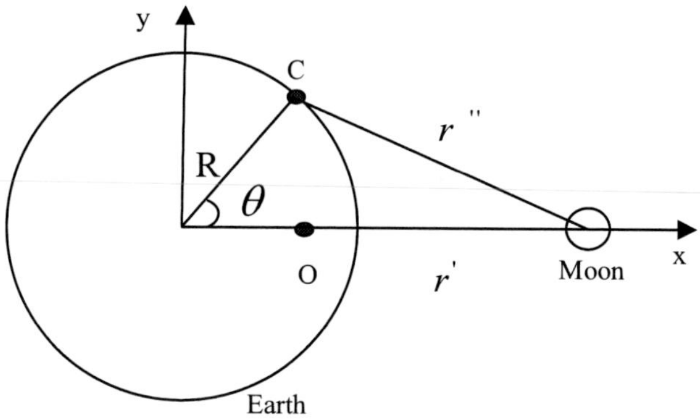

Figure 4. Tide generating force analysis diagram [18].

If the Earth as a centroid, all mass concentrated in the center of the Earth, then the Earth's centroid is affected by two forces, and the combined force vector can be expressed as (16). As the tidal force of the Moon in different parts on Earth is not the same, so the size and direction of the combined force vector in all the points on Earth are not same, as shown in Figure 4, and O points is the common-centroid. We can see from the diagram: $r^{"} = |\vec{r}^{'} - \vec{R}| = \sqrt{r^{'2} + R^2 - 2r^{'}R\cos\theta}$, $\vec{r}^{"}_x = (\vec{r}^{'} - \vec{R})_x = r^{'} - R\cos\theta$, $\vec{r}^{"}_y = (\vec{r}^{'} - \vec{R})_y = -R\sin\theta$, $\vec{r}^{'}_x = r^{'}$, $\vec{r}^{'}_y \approx 0$, then the inertial centrifugal force in the x-axis of C point object can be expressed as equation (17), and the quality of object is Δm. And the lunar gravity of C point object is expressed as equation (18), and then the combined force of two forces is expressed as equation (19), so the equation (19) can be converted into the equation (20), and to simplify the equation (21). The inertial centrifugal force in the y-axis of C point object is expressed as equation (22), and the lunar gravity is expressed as equation (23), the combined force can be expressed as (24), which can be transformed into the equation (25), and simplify the equation (26), that is, the tidal forces of different points on the planet have different size and direction. In $\theta = 0$ and $\theta = \pi$ place, that is, the most recent and most distant from the Moon, the tidal force is a departure from the center of the Earth and form the peak of sea water. In $\theta = \pi/2$ place, where point to the center of the Earth and form the sea water minimum. As the Earth's rotation, there are about two peaks and two troughs swept every place every day, forming two peaks and two low tide, where f is the inertial centrifugal force, f_c is the inertial centrifugal force of C point, and I is the unit vector of the inertial centrifugal force, and j is the unit vector of the lunar gravity [17-18].

$$f\vec{i} + \frac{\mathrm{GEM}}{r^{'2}}\vec{\mathrm{j}} = 0 \tag{16}$$

$$\mathrm{f_{cx}} = -\frac{\mathrm{G\Delta mM}}{r^{'2}} = -\frac{\mathrm{G\Delta mM}}{r^{'3}}\vec{r}^{'}_x \tag{17}$$

$$\mathrm{F_{Cx}} = \frac{\mathrm{G\Delta mM}}{r^{"2}} = \frac{\mathrm{G\Delta mM}}{r^{"3}}\vec{r}^{"}_x \tag{18}$$

$$F_{\mathrm{Cix}} = F_{Cx} + f_c = \mathrm{G\Delta mM}(\frac{\vec{r}''_x}{r''^3} - \frac{\vec{r}'_x}{r'^3}) \quad (19)$$

$$F_{\mathrm{Cix}} = F_{Cx} + f_c = \mathrm{G\Delta mM}(\frac{\vec{r}''_x}{|\vec{r}' - \vec{R}|^3} - \frac{\vec{r}'_x}{r'^3}) = \mathrm{G\Delta mM}\frac{2r'R\cos\theta - R^2}{(r'^2 + R^2 - 2r'R\cos\theta)r'^2} \quad (20)$$

$$F_{\mathrm{Cix}} \approx \frac{2RG\Delta mM\cos\theta}{r'^3} \quad (21)$$

$$\mathrm{f}_{\mathrm{cx}} = -\frac{\mathrm{G\Delta mM}}{r'^3}\vec{r}'_y \approx 0 \quad (22)$$

$$F_{\mathrm{Cy}} = \mathrm{G\Delta mM}\frac{\vec{r}''_y}{r''^3} \quad (23)$$

$$F_{\mathrm{Ciy}} = F_{Cy} + f_c = \mathrm{G\Delta mM}(\frac{\vec{r}''_y}{r''^3} - \frac{\vec{r}'_y}{r'^3}) = \mathrm{G\Delta mM}\frac{\vec{r}''_y}{r''^3} \quad (24)$$

$$F_{\mathrm{Ciy}} = \mathrm{G\Delta mM}\frac{\vec{r}''_y}{|\vec{r}' - \vec{R}|^3} = \frac{\mathrm{G\Delta mM}}{(r'^2 + R^2 - 2r'R\cos\theta)^{\frac{3}{2}}}R\sin\theta \approx \frac{\mathrm{G\Delta mM}}{r'^3}R\sin\theta \quad (25)$$

$$F_{\mathrm{Ciy}} \approx \frac{\mathrm{G\Delta mM}}{r'^3}R\sin\theta \quad (26)$$

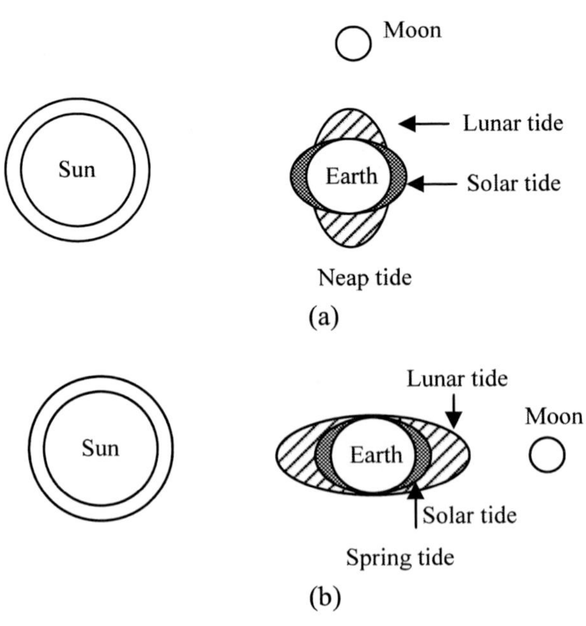

Figure 5. Explanatory drawing for neap tide and spring tide [3].

The derivation also applies to the tidal force of Sun, only the M and r' should be replaced by the Sun's mass M_s and the distance between Sun and Earth R', respectively. By equation (20) can be seen that the tidal force is proportional to the quality of celestial bodies and is inversely proportional to the cube of the distance. The quality of the sun is the moon's 27.1 million times and the distance between Sun and Earth is about 389 times the distance from Earth and Moon, so the tidal force on Earth of Sun's quality is less than the tidal force on Earth of Moon's distance, the ratio is about 1:2.17. As the Earth's rotation, the tide level of lunar tide spread from east to west, and the period is 12h25min and the amplitude is less than 1m deep-sea wave. The role of solar gravity is similar, but because of the distance between the Sun and the Earth is more than the distance from Earth to Moon, so the solar tidal force is smaller than the lunar tidal force, its period of 12h. When the Sun and Moon and Earth at right angles, the combined tidal force creates a small tide, namely, Neap tide. When they are in a straight line, the combined tidal force creates a big tide, namely, Spring tide, as shown in Figure 5 [1].

2.3. The Calculation of the High Tide

The cyclical fluctuations of sea level is affected by the function of tidal force, and the height of the rising tide can be ascertained by using the Newton's mechanics, within x and y direction of the Earth dug two silos and the silos are directly communicate with geocentric, and the x silo point to the direction of the Sun. The silo depths of x and y direction are h_1 and h_2, respectively, cross-sectional area is ds, and wells filled with water [17-18].

The resulting water pressure of h_1 wells in the center of the Earth is p_1, and the density of water is expressed as ρ, as a constant. The water quality of per unit length dr is $dm = \rho ds dr$, and the water received by the Earth gravity is $dmg(r) = \rho g(r) ds dr$, here $g(r)$ is the acceleration due to gravity at r location, so the water pressure of dr length by using the tidal force equation is expressed as equation (27).

$$dp_{h_1} = \frac{\left[\rho g(r) ds dr - \dfrac{2GM\rho r dr ds}{r'^3}\right]}{ds} = \rho[g(r) - \frac{2GrM}{r'^3}]dr \tag{27}$$

Integral to the entire depth h_1, the pressure of h_1 well bottom is expressed as equation (28).

$$p_{h_1} = \rho \int_0^{h_1} [g(r) - \frac{2GMr}{r'^3}]dr \tag{28}$$

By the same token, the pressure of h_2 well bottom is expressed as equation (29).

$$p_{h_2} = \rho \int_0^{h_2} [g(r) + \frac{GMr}{r'^3}]dr \tag{29}$$

In stable situations, $p_{h_1} = p_{h_2}$, such as the equation (30), and the equation (30) can be transformed into equation (31).

$$\int_0^{h_1}[g(r)-\frac{2GMr}{r'^3}]dr=\int_0^{h_2}[g(r)+\frac{GMr}{r'^3}]dr \tag{30}$$

$$\int_0^{h_1}g(r)dr-\int_0^{h_2}g(r)dr=\int_0^{h_1}\frac{2GMr}{r'^3}dr+\int_0^{h_2}\frac{GMr}{r'^3}dr \tag{31}$$

Due to $h_1 \approx R$ and $h_2 \approx R$, the value of the Earth's surface acceleration of gravity is $g(r)$, and $g(r)=GE/R^2$, then the equation (31) on the left can be expressed as equation (32).

$$\int_{h_2}^{h_1}g(r)dr=(h_1-h_2)g(R)=\frac{GE}{R^2}\Delta h_m \tag{32}$$

$\Delta h_m=(h_1-h_2)$ can be regarded as a high tide. Equation (31) on the right can be expressed as equation (33), then combined equation (32) and (33) gained the equation (34), then the high tide Δh can be expressed as equation (35).

$$\int_0^R\frac{3GMr}{r'^3}dr=\frac{3GM}{2r'^3}R^2 \tag{33}$$

$$\frac{GE}{R^2}\Delta h_m=\frac{3GM}{2r'^3}R^2 \tag{34}$$

$$\Delta h_m=\frac{3M}{2E}\left(\frac{R}{r'}\right)^3R \tag{35}$$

Then $M=7.35\times10^{22}kg$, $E=5.98\times10^{24}kg$, $r'=38\times10^5km$ and $R=6.4\times10^3km$ are substituted into equation (35), and the height of tide on Earth of the Moon is get, that is, the tide height caused by Moon, and the tide height of lunar tide is $\Delta h_m=0.56m$.

By the same token, the tidal height on Earth caused by Sun, the height tide of solar tide can be expressed as equation (36)

$$\Delta h_s=\frac{3M_s}{2E}\left(\frac{R}{R'}\right)^3R \tag{36}$$

The sun's mass is $M_s=1.99\times10^{30}kg$, and the distance from Sun to Earth is $R'=1.5\times10^8km$, which are substituted into equation (36), then the tide height of solar tide is $\Delta h_s=0.25m$.

In fact, the height tide is the vector superposition of Δh_s and Δh_m. When Moon, Sun and Earth has been almost the same line, the Spring tide is achieved due to the superposition of solar tide and lunar tide, and the height of Spring tide is $0.81m$, as shown in Figure 5 (b). When Moon, Sun and Earth has been almost vertical orientation, part tidal height is offset, and the

Neap tide is achieved due to the reduce of solar tide and lunar tide, as shown in Figure 5 (a). So there are take place two Spring tide and Neap tide within one month [17-18].

3. Tidal Power Generations

Human uses the tide energy to improve the standard of living since ancient times, such as tide mill have been used to grind flour and rice, as early as a thousand years ago, the tidal power is use to grind millet by China's coastal residents, and in currently China's Shandong province still preserved the ancient tidal mill. More than 900 years ago, Luoyang Bridge in Quanzhou was installed in Luoyang River by the Chinese residents, and the building stone is conveyed by using the tide power. 11th century, the tide mills of west coast of Europe make the early industrial country on the riches road, and Suffolk County in England still has a one in the 12th century tide mill for visitors, then immigrants bring it to the Americas in the New World, such as the French set up the first tide mill in the east coast of Canada in the Americas [1] [4].

The use of modern tidal energy was mainly used to tidal power generations, the principle of tidal power are similar to the principle of hydroelectric power. The potential energy of tide can be used to generate electricity, which uses the tidal range between high tide and ebb tide, and the tidal current can generate electricity, that is, the kinetic energy of tide is used to generate electricity, which produces by the seawater flow of high tide and ebb tide. So the potential energy or kinetic energy of tide is converted into mechanical energy, then mechanical energy into electrical energy, the process is the tidal power generation. Traditional tidal power use the tidal potential energy to generate electricity, briefly, a stone check dams is built in the Gulf or the tidal estuary, namely, a reservoirs is built in the bays or estuaries in order to isolate the sea, and then hydro-generating unit is installed in the dam, and then turbine is promoted to generate electricity by using the tidal range of seawater during high tide and ebb tide course. The new tidal power uses the tidal kinetic energy to generate electricity, that is placed hydro-generating unit under the sea level, when the tidal fluctuations, the flow of seawater is used to promote turbine rotation and power generation.

3.1. The Conventional Tidal Power

3.1.1. Power Generation Principle and Equipment Configuration

As the tidewater at the fluctuation when the tidal flow direction is constantly changing, so the tidal power have emerged in different types, such as one-way and single reservoir tidal power station which can power generation at ebb tide, and bi-directional and single reservoir tidal power station which can power generation at ebb tide and flood tide, and dual-reservoir and one-way tidal power station which can be a continuous power generation. In 1912, the world's first small tidal power station was built in Husum, Germany. And the world's first tidal power station with the economic value is La Rance tidal power station in 1966, Brittany, the west coast of France, it uses bi-directional and single reservoir type, its installed capacity is 240 MW, with an annual generating capacity of 544 million degrees, which is still the world's largest one of tidal power station at present.

(1) One-way and single reservoir tidal power station refers to the tidal power station can power generation at ebb tide by using the turbine power generation, and the tide is used to power generation in only one direction. That is, when the flood tide, the strobes will be opened and the sea water into the reservoir, then the strobes are closed during the slack tide course, then the strobes will be opened during ebb tide, and the water of reservoir will be released, and the seawater will promote the hydro-generating unit power generation until the water level of reservoir is same to the sea level. If the pumped-storage is used to increase the water level of reservoir during slack tide course, that is, when the reservoir water surface and sea-level is almost equal, the sea water is pumped into the reservoir by using the water pump in order to improve water quantity and water level of reservoirs, due to the pumping water level difference is small, followed by power generation course is carried out under the high water level difference, so the electricity consumption of pumped storage power is less than the increased power generation of tidal power station, that is, it will increase the tidal power plant generating capacity.

One-way and single reservoir type of tidal power station just need to build a dam and hydroelectric generating set only generate power when the ebb tide, that is, the turbine only work in a single direction of water flow, therefore the equipment has a simple structure and low investment. However, the type of power plant can not be generated power during flood tide course, so the efficiency is low and the power generation capacity is few. The average working time per day is no more than 10-12h. One-way and single reservoir type of tidal power station is shown in Figure 6 (a), which will face four working conditions as shown in Figure 6 (b). ① reservoir water-filled stage during flood tide course, the generator stops turn and strobes open, seawater flows to reservoirs; ② waiting period during slack tide course, generating sets stop, strobes closed, the use of pumped-storage can improve reservoir water level to increase the generating capacity at this time; ③ power generation stage, when the sea level is lower than the reservoir water level to a certain height, the strobe of water turbine engine can be opened, the power generating units is working at the moment, other strobes are closed; ④ waiting period, wait the flood tide, when the water level of reservoir drops to a certain height, the generator shut down again, waiting for the next high tide, the strobes are closed [6].

(2) Bi-directional and single reservoir tidal power station need to build a dam and hydroelectric generating set can generate power at the flood tide and ebb tide by using the water turbine engines, generating sets is stopped only in the inside water level of reservoir is almost equal to the outside water level of reservoir. The bi-directional and single reservoir type of tidal power station is shown in Figure 7 (a), in which the switch of strobe 1 and 3 is synchronized, and the switch of strobe 2 and 4 is synchronized, due to the generating sets generate power during both flood tide and ebb tide, so the building structure is more complex than one-way and single type and the investment should be large, but the daily generation time increases, power generation also increased, on average, every day to work 16-20h.

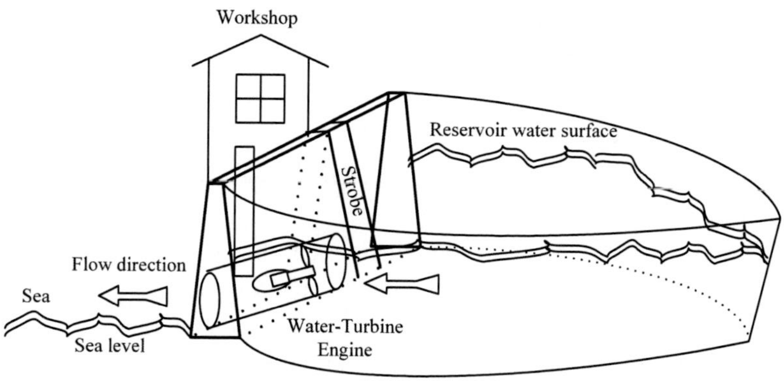

(a) Schematic diagram of a one-way and single reservoir tidal power station

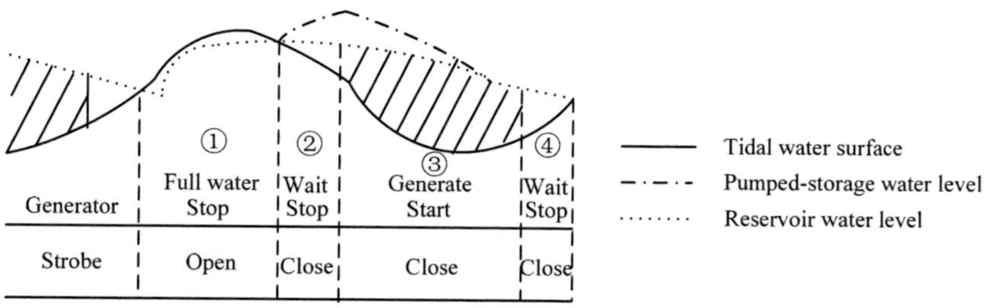

(b) One-way and single reservoir tidal power station operating conditions

Figure 6. One-way and single reservoir tidal power station [6].

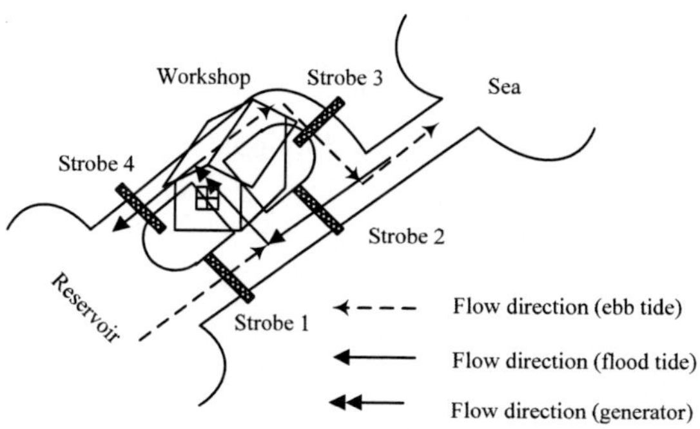

(a) Bi-directional and single reservoir tidal power station diagram

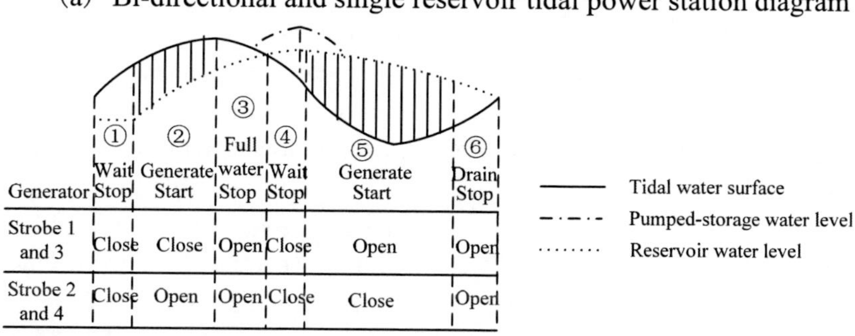

(b) Bi-directional and single reservoir tidal power station operating conditions

Figure 7. Bi-directional and single reservoir tidal power station [6].

The working conditions of this type tidal power station is divided into six kinds, such as waiting, tide power generation, water-filled, waiting, tide power generation, as shown in Figure 7 (b). ① waiting stage at the initial stage of flood tide, in order to wait the appropriate water-head between inside and outside of reservoir, the generator sets stall, strobes all closed; ② generation stage, the generators work, strobe 1 and 3 closed, strobe 2, and 4 open, at the flood tide generation phase, seawater flows from the sea into the reservoir; ③ water-filled stage, power generation unit off, all strobes fully open, the sea water flows to reservoirs; ④ waiting period, generator sets and strobes all closed, waiting for the sea level lower than the reservoir surface to a certain height; ⑤ generation stage, the generators work, strobe 1 and 3 open, strobe 2 and 4 off, at the ebb tide generation stage, the seawater flows from the reservoir into the sea, and the use of pumped-storage can improve reservoir water level to increase the generating capacity at ③ and ④ and initial ⑤ phase; ⑥ drain stage, generating units shut down, all strobes is opened, as much as possible to reduce the water levels of reservoirs. As shown in Fig. 7 (b), the water flow direction is different during flood tide and ebb tide generation phase, but the direction of water flow for the generator room is the same [6].

(3) The dual-reservoirs and one-way tidal power station is the need to build two adjoining reservoirs, namely, high-water reservoir and low-water reservoir. The high-water reservoirs with a water intake sluice has high water level, and the tide water flow to the reservoir only when the water level of sea is higher than the water level of high-water reservoir. The low-water reservoir with a water release gate has a low water level, and the water flows from low-water reservoir to sea only when the water level of low-water reservoir is higher than the water level of sea. The water level of high-water reservoir is always higher than the water level of low-water reservoir, the lake level, so the former is called the previous reservoir or high reservoir, and the latter is called the next reservoir or low reservoir. The water levels difference between high reservoir and low reservoir is always maintained in order to promote the turbine of a hydro-generating unit, the generating unit is placed on the dam between the two dam reservoirs, the water flows through the hydro-generating unit can be all day long, and the power generation process can continuously, so the power output of this type tidal power station is relatively stable.

Dual reservoirs and one-way tidal power station diagram is shown in Figure 8 (a), and this type tidal power station will face a run of four conditions as shown in Figure 8 (b). ① waiting stage, the generator work, all strobes is closed, water flows from high reservoir to low reservoir, the sea is at the flood tide stage, the sea level is lower than the water level of high reservoir and higher than the water level of low reservoir; ② water-filled stage, the generators work, filling water sluices open and drain sluices closed, the sea is at the flood tide stage, the sea level is higher than the water level of high reservoir, sea water flow from sea to the high reservoir; ③ waiting stage, the generating units work, all sluices closed, water flows from high reservoir to low reservoir, the sea is at the ebb tide stage, and the sea level is higher than the water level of low reservoir; ④ Drain stage, the generators work, filling water strobe closed, drain strobe open, in order to make possible to reduce the water levels of low reservoirs, and the water level of low reservoir is higher than the sea level, the sea is at the ebb tide stage [6].

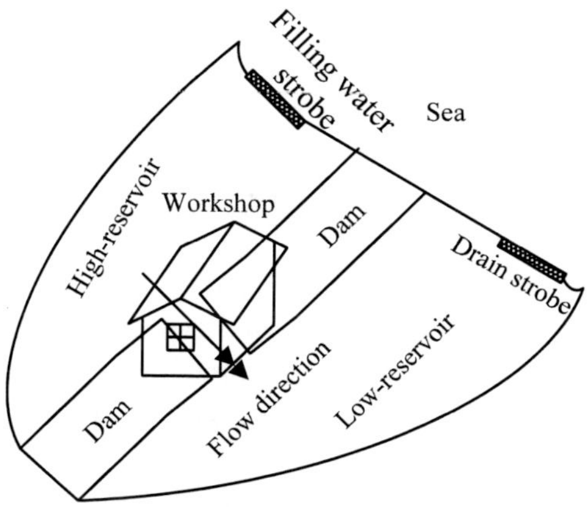

(a) Dual-reservoirs and one-way tidal power station diagram

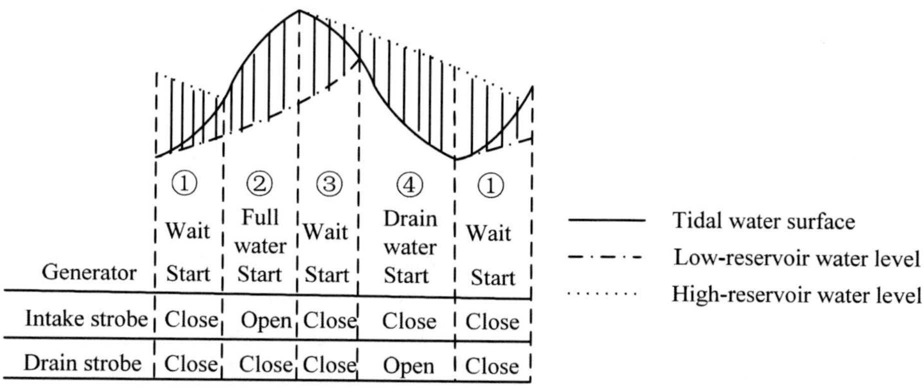

(b) Dual-reservoirs and one-way tidal power station operating conditions

Figure 8. Dual-reservoirs and one-way tidal power station [6].

3.1.2. The Composition of Tidal Power Station

The conventional tidal power station uses the potential energy of tide, which generally includes three parts: dam, sluices, and power plants. In addition, the navigation lock is set in order to ensure the navigation needs, and in the top of the dam can build roads, not only can make tidal power plant into the landscape, but also improve the traffic situation in the Gulf.

(1) Dam, which built in the estuary or harbor areas, the purpose is to estuary or harbor waters and the open sea isolation, and the formation of a reservoir, the inside and outside water levels difference of dam is used to generate electricity. In accordance with the choice of different materials can be divided into earth dam, stone dam, reinforced concrete dam and rubber dam. To note the main problem is to prevent the dam leaking phenomenon, the foundation must be built on a rock foundation on the earth dam and the stone dam if it is needed to build water-resisting walls.

(2) Sluices, used to regulate inflow or outflow reservoir's water. In addition, which be used to control the water flow of generating power unit. The choice of materials can be wood, stone, steel and reinforced concrete.

(3) Power plant, which can transform the tidal energy into electrical energy. The main equipment, including hydro-generating unit, power transmission and distribution equipment, lifting equipment, control room and lower water flow channels, etc. For the hydro-generating unit should meet the low head and large tidal flow characteristics and has a good anti-corrosion, moisture, anti-fouling properties, but also should have good insulation and sealing properties. Hydro-generating unit are generally three kinds of basic structures, i) vertical axis generator unit, that is, vertical axis turbine and generator connected perpendicular to the water surface. This approach requires biggish workshop, bigger investment, while intake pipes and tail pipes is very curving, biggish water energy losses, low efficiency, the structure shown in Figure 9 (a); ii) horizontal-axis unit is turbine and generator shaft parallel to the water surface, the curvature of intake pipes and tail pipes is less than the curvature of vertical axis generator unit, reduce in water energy losses, efficiency increase, but which need the relatively long workshop, the structure shown in Figure 9 (b); iii) through flow turbine (straight-flow turbine), which is based on the horizontal axis generator unit developments as a power generation unit, consists of two types, one is bulb-type units, that is, the turbine, gearbox and generator sets all to do with concrete placed in a into a sealed bulb body, only the exposed turbine is promoted by external water flow, the entire bulb body placed in water flow channels. The merit is that the efficiency is relative high, the curvature is small, and the water energy losses are small. The drawback is that installation is inconvenient, taking up more water channels, the structure shown in Figure 9 (c), the La Rance tidal power station in French use this kinds models, unit capacity of 10MW. The other model is the rim-generator turbine, the stator of generator is installed in the internal wall of water flow channel around the turbine, and generator rotor is installed in the water flow channel in a sealed body, taking up water channel is smaller than the through flow turbine structure, as shown in Figure 9 (d), Annapolis tidal power station in Canada use this model, unit capacity of 20MW [6].

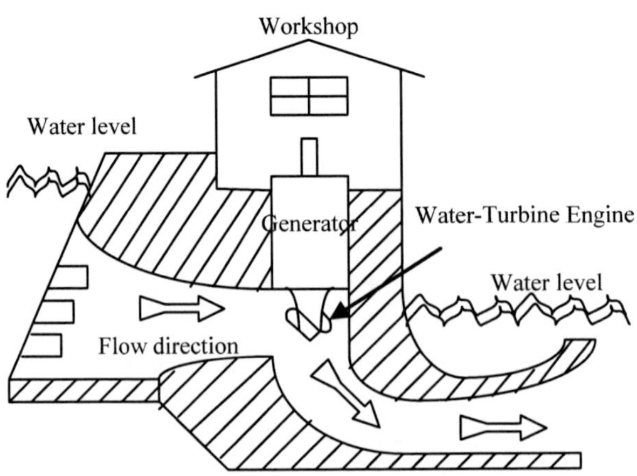

(a) Vertical axis generator unit

Figure 9. Continued on next page

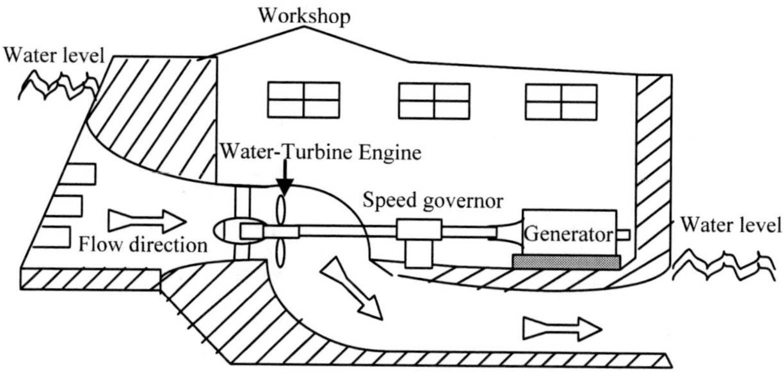

(b) Horizontal-axis generator unit

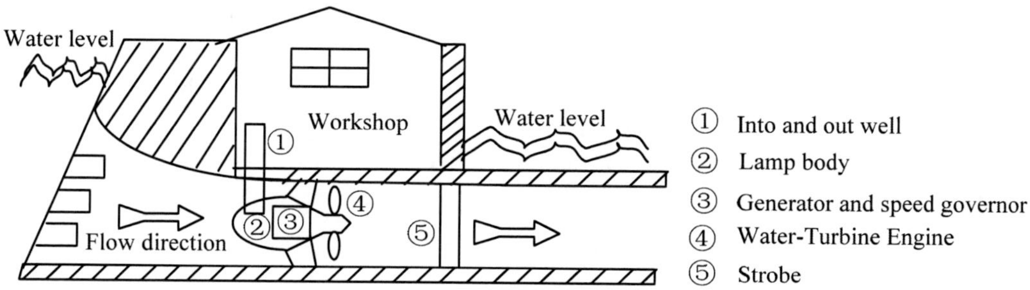

(c) Bulb tubular turbine

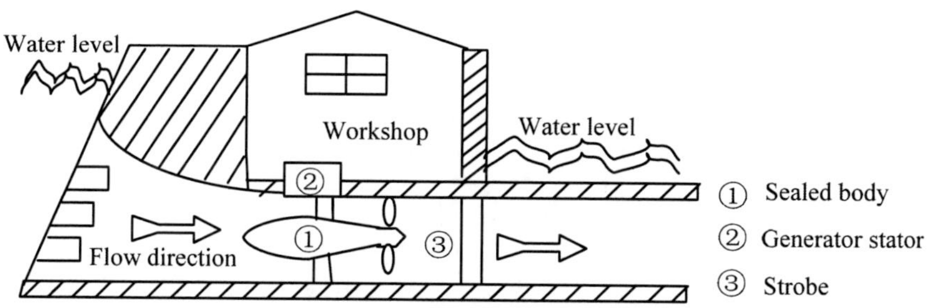

(d) Rim-generator turbine

Figure 9. Types of water-turbine generator set for tidal power station [6] [21].

3.1.3. The Potential Power Generation Capacity of Tidal Power Station

The conventional tidal power plant construct the barrage to isolate the seawater, and the structure is similar with the traditional construction of hydropower stations, and the difference is that the energy of tidal power plant is from tide, while the hydroelectric energy from rivers. In addition, tidal head and direction is constantly changing, but the water head and the direction of hydropower stations is relatively fixed. Tidal power plant is actually a low head and large flow characteristics of the hydropower station. Literature [22-24] have introduced the calculation of tidal energy resources of a tidal reservoir, simple to say that the energy are based on the water level difference of both sides of dam, and assume that the height of reservoir or estuaries is $H_R(t)$, and the sea level height of the dam outside is $H_S(t)$, and water level difference of both sides of dam is $H_b(t)$, in which sea level changing is related to the movement of celestial objects location and geographical environment. For example, the literature [22] described the sea level changing in a given site and all the time, which is defined as the equation (37), where N is harmonic frequency, H_n is the amplitude, ω_n is angular velocity, ϕ_n and V_n are the phase and equilibrium phase, respectively, which are related to astronomical parameters. H_o is the amplitude reference value when $\omega_o = \phi_o = V_o = 0$. $H_b(t)$ can be expressed as equation (38). The size of the water flow $F(t)$ for power generation is related to both the area of reservoir A_R and the water level different of inside and outside of dam $H_b(t)$, which can be expressed as equation (39), where t is the time, water flow unit is m^3/s, the area unit of reservoir is m^2. Is generally believed that the changing of the water level of reservoir is relatively small, that is, the area of the reservoir is fixed, then the total amount of water for power generation can be expressed as equation (40), and the total capacity of the reservoir can be expressed as equation (41), in which h is the depth of the reservoir, T are semi-tidal cycle. In one half tidal cycle, the total amount of water into the reservoir can be approximated as a constant, the total generating power can be expressed as equation (42), in which ρ is the density of sea water [22] [23].

$$H_S(t) = \sum_{n=0}^{N-1} H_n \cos(\omega_n t - \phi_n + V_n) \tag{37}$$

$$H_b(t) = \left| H_R(t) - H_S(t) \right| \tag{38}$$

$$\frac{\partial H_b(t)}{\partial t} A_R(t) = -F(t) \tag{39}$$

$$\int_0^{T/2} F(t)dt = \int_0^{T/2} \frac{dh}{dt} A_R dt \tag{40}$$

$$V_{total} = [h(T/2) - h(0)]A_R \tag{41}$$

$$P = \frac{\rho G \int_0^T F(t) H_b(t) d}{T} \tag{42}$$

The installed capacity of one-way and bi-directional tidal power station based on experience can be expressed as equation (43), and the annual generation capacity of one-way tidal power station based on experience can be expressed as equation (44), and the annual generation capacity of bi-directional tidal power station based on experience can be expressed as a equation (45), here $H_b^{'}$ express the average tidal range [6].

$$N = 200 H_b^{'2} A_R (kW) \tag{43}$$

$$E = 0.4 \times 10^6 H_b^{'2} A_R (kWh) \tag{44}$$

$$E = 0.55 \times 10^6 H_b'^2 A_R (kWh) \quad (45)$$

3.1.4. The Current Issues Involved in Traditional Tidal Power and Major Tidal Power Plant

The traditional tidal power station must construct barrage in order to insulate seawater, and the investment of dam construction accounts for about 45% of the total investment, while the construction of power plants will have an impact on the surrounding environment, the dam construction will affect the sea water temperature and sea currents and salinity impact of neighboring dam, and thus the growth of plankton and fish will be impacted, so it is necessary to study the biological effects of dams, while the engineering structures is selected which have a smaller work amount, such as prefabricated floating reinforced concrete caisson structure. In addition to the cost of hydraulic turbine generating sets accounts for about 50% of the total investment, while the sea water fluctuations brought a lot of sediment, and the sediment will increase the wear and tear of turbine group, while increasing the reservoir sedimentation problem, shorten the life of a reservoir and reduce the generating capacity, so it is necessary to study the movement rule of seawater sediment in local place, there are some question must be thought, such as sediment concentration and sediment size and direction of motion and so on. The dam and generating sets of tidal power stations are immersed in seawater in a long time, and the corrosion of metal components by seawater is a very serious issue and could even cause the entire damage of dam construction and generating sets. At the same time, Sea creatures will attached to the dam, turbine blades, sealed body, and even water channel, and difficult to wash away by water, which will reduce the power capacity of power stations, and even partial blockage of the circulation equipment, and failure of turning facilities, so we must attach importance to the metal structure of the anti-corrosion and marine aquatic life attachment issues. In addition to dual-reservoirs and one-way tidal power station, with tidal fluctuations, the power generation of other types of tidal power station is intermittent, so the tidal power plant must be connected with grid or other types of power plants in order to ensure the quality of the user's power. Of course, we should also see that initial capital outlay of tidal power station is large and unit cost is higher, but the seawater can be stable and automatically supply after completed power plant, and the cost power generation is very low, according to data from have been set up tidal power station, the power generation costs is about 0.05 to 0.08 francs per kWh, and the cost equal to or higher than the hydropower cost, but the cost is lower than the cost of thermal and nuclear power.

At present, a number of tidal power stations have be built in the whole world, such as the most famous La Rance tidal power plant in France, and some are under construction, such as Britain is stepping up the construction of the world's first commercial tidal power station in Strangford Strait and look forward to completion of power generation in 2010, and the installed capacity is one million kW. New Zealand in 2007, started a tidal power plant in the Kaipara river estuary, 200 turbines is installed in the seabed, and the total installed capacity is 2 million kW, the power station will be completed in 2010, will produce enough electricity for 250,000 households. There are trial run tidal power plants in east and west coasts of U.S., respectively: East River tidal power station in New York City and the northern coastal California. Six generating sets is installed in the river on the East River in New York City and the total power is 100,000 kW, which supplied electricity for about 10000 residents of Roosevelt Island in New York City. Furthermore, SIHWA tidal power stations built in South Korean, installed capacity is 260MW. As shown in Table 1, the potential and can be developed locations of traditional tidal power plant in the whole world is described, and the mean range, basin area, potential mean power and potential annual production are described [6,25-29].

Table 1. Major world conventional tidal power plant [6,25-28]

Location	Mean range (m)	Basin area (km^2)	Potential mean power (MW)	Potential annual production (GWh/year)
North America				
Passamaquoddy	5.5	262	1800	15,800
Cobscook	5.5	106	722	6330
Bay of Fundy	6.4	83	765	6710
Minas-Cobequid	10.7	777	19,900	175,000
Amherst Point	10.7	10	256	2250
Shepody	9.8	117	520	22,100
Cumberland	10.1	73	1680	14,700
Petitcodiac	10.7	31	794	6960
Memramcook	10.7	23	590	5170
South America				
San Jose, Argentina	5.9	750	5870	51,500
United Kingdom				
Severn	9.8	70	1680	15,000
Mersey	6.5	7	130	1300
Solway Firth	5.5	60	1200	10,000
Thames	4.2	40	230	1400
France				
Aber-Benoit	5.2	2.9	18	158
Aber-Wrac'h	5	1.1	6	53
Arguenon	8.4	28	446	3910
Frenaye	7.4	12	148	1300
La Rance	8.4	22	349	3060
Rotheneuf	8	1.1	16	140
Mont St Michel	8.4	610	9700	85,100
Somme	6.5	49	466	4090
Ireland				
Strangford Lough	3.6	125	350	3070
Russia				
Kislaya	2.4	2	2	22
Lumbouskii Bay	4.2	70	277	2430
White Sea	5.65	2000	14,400	126,000
Mezen Estuary	6.6	140	370	12,000
Australia				
Kimberley	6.4	600	630	5600
China				
Bayou of Yangtse River	3.04	No data	704	2280
Hangzhou Basin	3.04	480	31600	8700
Leqing Basin	4.51	201	5500	2340
Sanduwo	5.34	233	1293	3560
Xinghua Basin	5.15	451	2397	6592
Fuqing Basin	4.2	326.1	1156	1379
Meizhou Basin	4.55	289.5	1198.7	3297

3.2. The New Tidal Power

The dam of the traditional tidal power plant have a substantial impact on the surrounding environment, such as ocean currents, water temperature and salinity, while a large number of submerged land, not only impact on the survival of humans and animals, but also affect shipping and migration of fish, and the huge amount of investment, all of which limit the traditional tidal power plant development. Therefore, the use of kinetic energy of tidal energy is a current development hotspot, which converts tidal stream energy into electrical energy. Tidal stream power generation devices generally fixed at the sea bed, below the sea surface, which does not affect shipping, sea currents and water temperature and salinity of surrounding region are not a big impact, does not need to flood the land, less impact on people and fish, and because the water turbine slower speed generally will not hurt the fish.

$$P = 1/2\, C_P \rho A V_t^3 \tag{46}$$

$$\lambda = \omega R / V = 2\pi R n / V_t \tag{47}$$

The principle of tidal stream power generator is similar with the principle of wind power generator, which looks like an inverted wind driven generator, so the output power can be expressed as equation (46), the output power affects by V_t, β and λ. Here, V_t is the velocity of seawater flow, β is the pitch angle and λ is the tip speed ratio. The output power is proportional to Cp in different velocity circumstances, and proportional to the cube of wind speed. And the relationship between λ and velocity is expressed as equation (47). Among them, Cp denotes the utilization coefficient of tidal energy. ρ and A are the local seawater density and turbine swept area, respectively, generally which considered to be fixed. ω is the turbine angular velocity of rotation (rad/s), n is the turbine rotational speed (r/s). But it also should be observed that the density of seawater is about 832 times of the air density, and tidal stream speed is lower than the wind speed, so the blade of tidal stream power generation is shorter and squatter than the wind blades, but the direction of tidal stream is more stable than the wind direction, which includes the tidal stream direction of flood tide and ebb tide. The same time, the structure of tidal stream generators is similar with the structure of wind turbines, which include two types of power generation equipment, namely, the horizontal axis and vertical axis generator. The vertical axis tidal stream generator is perpendicular to the water flow direction, as shown in Figure 10, and the figure is the Gorlov helical turbine. The horizontal axis tidal stream generator is parallel to the direction of water flow, as shown in Figure 11, and the figure is the Neptune tidal stream device. Tidal stream power generation devices include several major part, such as the blades which installed in the wheel hub, frame, gear boxes and generators, and the water flows promote the rotate of blades, the gear box increases the rotate speed of blades to satisfy the required speed of generator and drive generators power generation, then the electricity energy is sent to the grid on the ground by using the cable. Certainly, the frame must be fixed on the sea bed and prevent them from being washed away by tidal flow, and the frame can uses steel frame and can also use steel and reinforced concrete, which can be single-pile structure, may also be a tripod structure, may also be a floating structure, steel chains and suspended power generation equipment fixed in the sea bed. In addition, the frame must be able to withstand its own weight and the impact of tidal flow in both directions.

The current tidal stream power generation technology is still in early stages of development, but there are a number of companies have researched in relevant technological areas, especially, the United Kingdom occupy a leading position in the tidal stream power generation technology. For example, the United States, GCK Technology Ltd. developed Gorlov helical turbine, as shown in Figure 10, it belongs to the vertical axis tidal power generation has now produced a prototype, and the diameter is 1 meter, and the prototype began testing in July 2002 [25-26] [30]. The Gorlov Helical Turbine (GHT) was specifically designed for hydroelectric applications in flowing low head water courses.

AQUAMARINE POWER Ltd. based in United Kingdom, which developed the Neptune Tidal Stream, as shown in Figure 11, it belongs to the horizontal axis tidal power generation, and the capacity is 2.4MW. Neptune Tidal Stream is one of the most powerful tidal stream devices under development, and is designed to be competitive with other sources of electricity generation in the UK market [25-26][31].

Figure 10. The vertical axis turbine [30].

Figure 11. The horizontal axis turbine [31].

Tidal Energy Ltd based in United Kingdom, which developed the DeltaStream Turbine, as shown in Figure 12. The DeltaStream device is a nominal 1.2MW unit which sits on the seabed without the need for a positive anchoring system, generating electricity from three separate horizontal axis power generators mounted on a common frame. The distance of three separate generators is about 30m wide, and the triangular frame produces a low centre of gravity enabling the device to satisfy its structural stability requirements including the avoidance of overturning and sliding [25-26] [32].

VERDANT POWER Ltd. based in the USA and Canada, which developed the Free Flow Turbines, as shown in Figure 13. Verdant Power's Free Flow Kinetic Hydropower System uses three-bladed, horizontal-axis turbines deployed underwater in order to generate clean renewable energy from tidal and river streams, which is installed fully underwater and operates automatically and invisible from shore. The system does not require dams or other major civil works, and does not redirect the natural flow of the water. Simple and modular in design, the Free Flow System can be scaled to generate clean energy in a wide range of sites worldwide [25-26] [33].

Figure 12. DeltaStream Turbine [32].

(a)

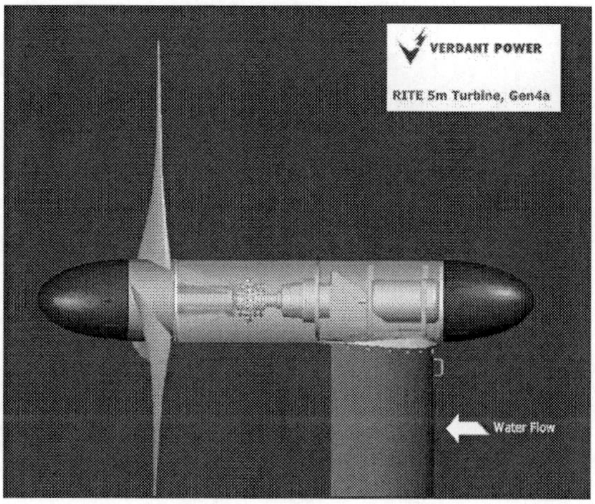

(b)

Figure 13. Free Flow Turbines [33].

SMD Hydrovision Ltd. based in United Kingdom, which developed the Tidal Stream Turbine (TidEl), as shown in Figure 14. TidEl differs from most tidal turbines design in that it floats and is restrained by mooring chains to the seabed rather than being fixed to piles driven into the sea bed. Every one TidEl turbine will incorporate two buoyant 500kW generators joined together by a cross beam giving a total power capacity of 1MW, and the rate flow is 2.3m/s, and the cut in flow is 0.7m/s. The pitch turbine blades are fixed, and the diameter is 15 meters, and three 8m blades on a 2.5m hub. The design of TidEl and its anchoring mechanism leave it free to align itself downstream of the tidal flow as the tide changes direction with no need for re-orientation electronic controls. TidEl can be installed in deep waters (below the light depth) and so barnacle and other sealife damage can be avoided reducing maintenance intervals and downtime [25-26] [34].

OpenHydro Ltd. based in Irish, which developed the Open centre turbine, as shown in Figure 15. The Open-Centre Turbine is designed to be deployed directly on the seabed. Installations will be silent and invisible from the sea surface. They will be located at depth and present no navigational hazard. Certainly, the size can be produced very big or very small. Figure 15 (a) shows the big size, which fixed to piles driven into the sea bed. Figure 15 (b) shows the small size, which can sits on the seabed by using the triangular frame [25-26] [35].

Figure 14. Tidal Stream Generator [34].

(a)

(b)

Figure 15. Open centre turbine [35].

Lunar Energy Ltd. based in United Kingdom, which developed Lunar energy tidal turbine, as shown in Figure 16. The RTT is a bi-directional horizontal axis turbine housed in a symmetrical venturi duct. The venturi draws the existing ocean currents into the RTT in order to capture and convert energy into electricity. Use of a gravity foundation will allow the RTT to be deployed quickly and with little or no seabed preparation at depths in excess of 40 meters [25-26] [36].

Pulse Generation Ltd. based in United Kingdom, which developed the Pulse tidal hydrofoil, as shown in Figure 17. Pulse Tidal technology can operate in shallower water. Traditional designs use a rotating blade to extract energy from the tidal or river streams. The diameter of generator blade is limited by the water depth and this reduces the amount of power that can be generated in shallow water. Pulse Tidal's technology overcomes this limitation by using a wide, flat machine where the blade length is not limited by water depth. According to the data from Pulse Generation Ltd., at a typical site where depth is limiting a Pulse generator will extract more than four times energy compare with a single traditional turbine [25-26] [37].

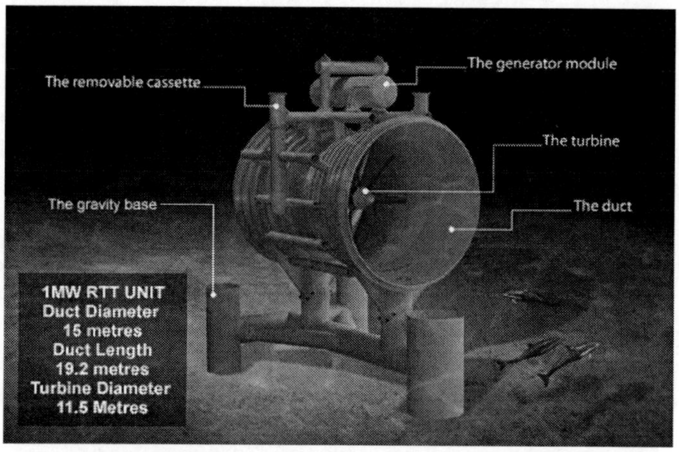

Figure 16. Lunar energy tidal turbine [36].

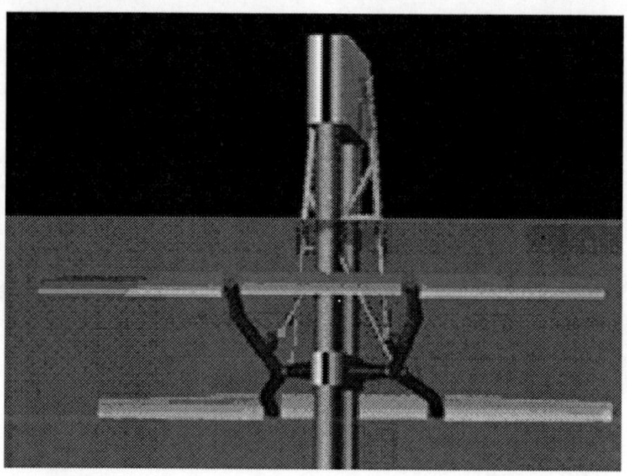

Figure 17. Pulse tidal hydrofoil [37].

Figure 18. Stingray tidal energy converter [38].

IHC Engineering Business Ltd. based in United Kingdom, which developed the Stingray tidal energy converter, as shown in Figure 18. Stingray tidal energy converter can capture tidal movement energy by using an oscillating hydrofoil. The wing like hydroplane can move in the oncoming current by the lift or drag forces, and the changing angle of attack relative to the stream can controls the oscillating motion [25-26] [38].

Ocean Flow Energy Ltd. based in United Kingdom, which developed the Evopod Tidal Turbine, horizontal axis of rotation Moored structure and Three-bladed design. Evopod uses a simple but effective mooring system that allows the free floating device to maintain optimum heading into the tidal stream. It can be accessed by boat for first line maintenance and has been developed specifically to address the need for a tidal stream device that can operate in exposed deep water sites where severe wind and waves also make up the environment, here, Figure 19 (b) shows the planform of Evopod Tidal Turbine [25-26] [39].

Blue Energy Ltd. based in the Canada, which developed the Tidal Fence Davis Hydro Turbine such as Figure 20, Turbines fixed in an array known as a tidal fence Vertical axis of rotation Three-bladed design. Cutaway graphic depicting an array of vertical-axis tidal turbines stacked and joined in series across a marine passage. Tidal currents push on vertical mounted hydrofoils that apply a torque force to rotating shafts, which are coupled to generators housed just above the water level [25-26] [40].

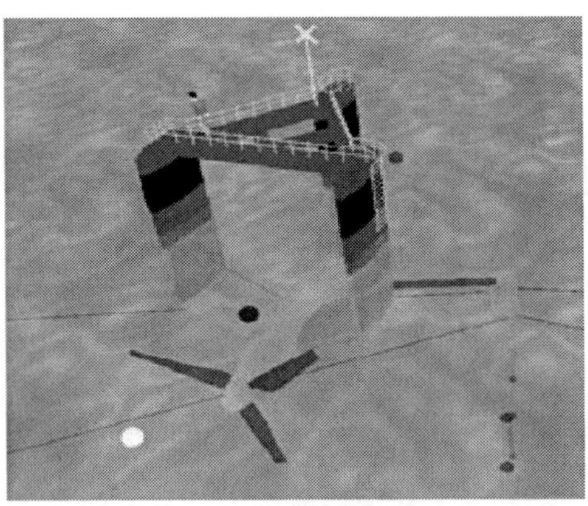

(a)

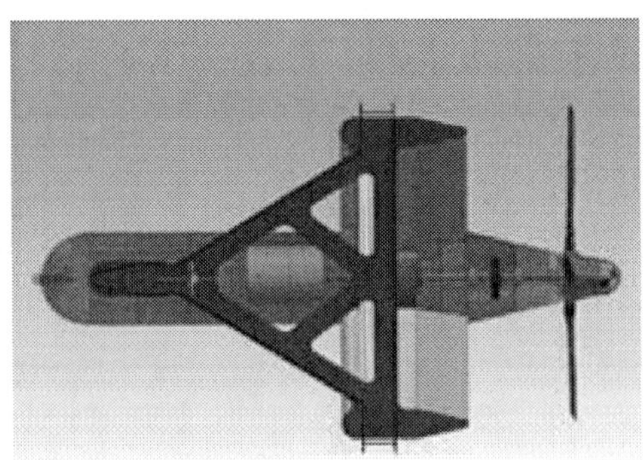

(b)

Figure 19. Evopod TidalTurbine [39].

Figure 20. Tidal Fence Davis Hydro Turbine [40].

Figure 21. Tidal Stream Turbine [41].

(a) (b)

Figure 22. SeaGen Turbine [42].

Hammerfest Strom AS based in the Norway, which developed Tidal Stream Turbine such as Figure 21, Horizontal axis of rotation Three-bladed design. In 2003 electricity was delivered to the grid from the world's first tidal turbine converting kinetic energy from the tidal water to electricity. The turbine, being a 300kw prototype called HS300, has generated both knowledge and electricity for approximately four years before it was retrieved for verification and reinstalled for further research [25-26] [41].

Marine Current Turbines Ltd. based in the United Kingdom, which developed the SeaGen such as Figure 22, Twin horizontal axis rotors and two variable-pitch blades. SeaGen has already delivered over 350 MWh into the Northern Irish electricity grid. The twin generators typically produce an average of 5MWh of electricity during the 6¼ hours of each ebb and flood tide. This is enough energy to meet the average electricity needs for 1500 UK homes. The SeaGen turbine, with its twin 16m diameter rotors, is officially accredited to OFGEM as a "UK power station" [25-26] [42].

Atlantis Resource Corporation PTE Ltd. based in the Singapore, which developed the Nereus and Solon and AK Tidal Turbines, Horizontal axis of rotation. Nereus is extremely robust, and Solon is a ducted deep and AK has two mono-directional water turbines such as shown in Figure 23 [25-26] [43].

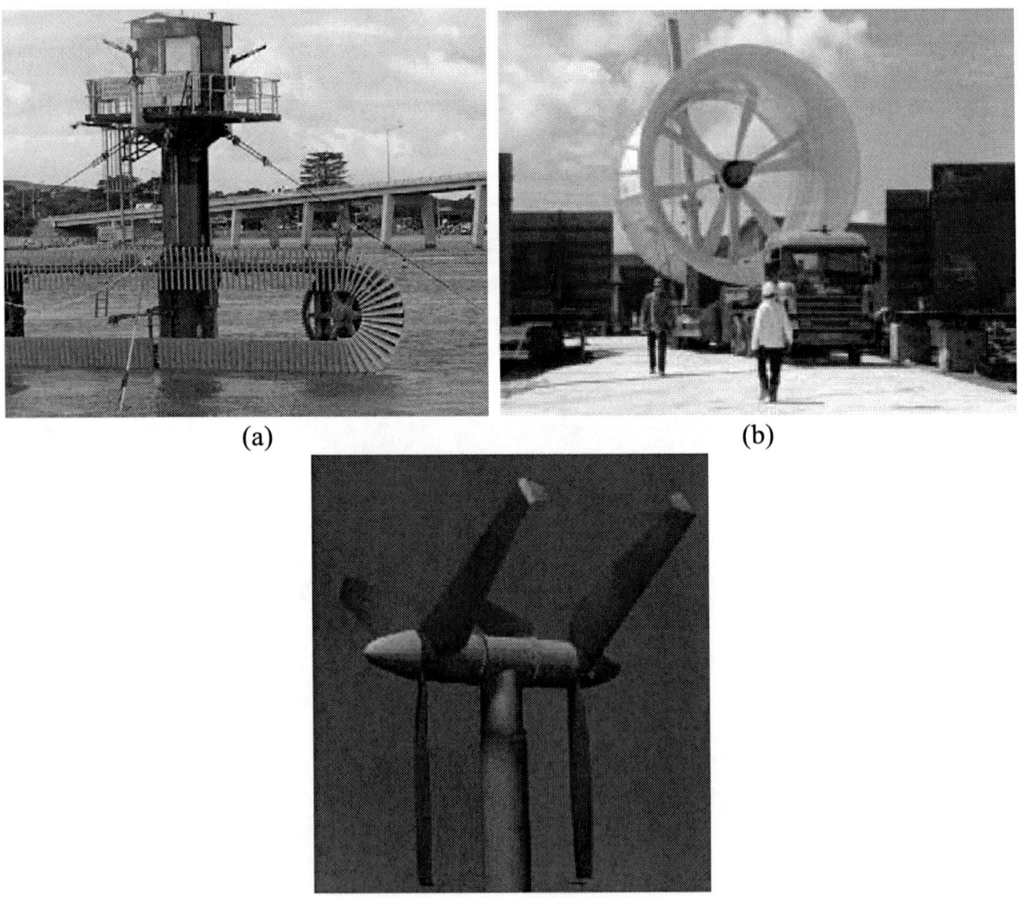

(a) (b)

Figure 23. Nereus Turbine is shown, and below is the Solon Turbine, and AK-1000 Tidal Current Turbine [43].

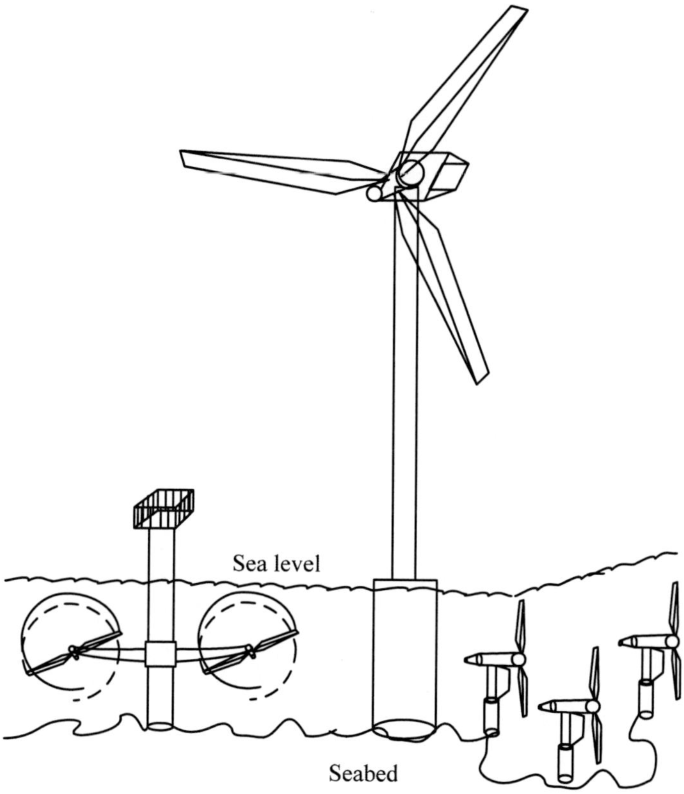

Figure 24. Combined wind energy and tidal energy power system.

In addition, the sea has abundant wind energy resources, and therefore the development of tidal stream and wind power development can be synchronized, as shown in Figure 24, namely, the erection of tidal stream power generation devices at the same time frame establishment of offshore wind turbines, lower installation costs. However, development of tidal stream is also facing some problems, such as installation, maintenance, power transmission, cost and impact on the environment and other issues. As the tidal stream power generation in general installed in the sea level several meters to tens of meters under the sea under the bed, so installing the difficulties and failures are more difficult to repair, if far from the coast long distance power transmission, transmission costs. In addition, the trend of the energy density is 832 times more wind, so the trend of power devices, blades must be strong selection of material, while the trend of velocity fluctuations will lead to more fatigue damage leaves. There are tens of meters of sea water will turn the leaves of the surrounding environment and marine life to a certain extent the impact of small-scale development that may affect the trend can not, and if large-scale development must take into account the marine environment and marine life impact.

4. Conclusions and Recommendations

Tidal power generation include the tidal power potential and kinetic energy forms electricity, tidal power advantages, including its energy is renewable, inexhaustible, but also has the laws, and can accurately predict, there is no dry season can be perennial power generation, clean, no greenhouse gas emissions, lower operating costs. Drawbacks include large investment and high cost, low-water head and power generation is not continuous and so, if the construction of dam sediment deposition may also face problems. And tidal power will have the local currents, biological and shipping an impact on marine life is also restored at the same time attached to the power generation installations, dams and water channels to reduce its service life, but also sea water corrosion of the metal must also be considered.

With the emphasis on the human right to environmental protection, reducing greenhouse gas emissions has become a consensus of all mankind, so no emissions of greenhouse gases of new energy power generation will attract the world's attention in the future, and the statistical data shows that the total tidal energy resources in the whole world is more than 3 billion kWh. Therefore, application and development of tidal energy in the future will become an important direction of the new energy development, of course, the rapid development it can also have some external force to promote, such as policy, funding, technical and human resources and so on.

i) Policy, for tidal energy development, the country's incentive policies are necessary, such as Internet access and reasonable price, tax relief, tariff reductions for imported equipment. The Government should ensure that the interests of investors and ensure a certain profit will it be possible to attract private capital to state-owned enterprises and a large number of inputs.
ii) Funding, tidal energy development requires a lot of capital investment, individuals generally can not be completed, so the government's investment is necessary, and whether the private capital attracted to the new field of energy depends on government policies, as well as profits can be guaranteed.
iii) Technology, tidal energy technology development for now is relatively mature, but it does not mean that there is no room for improvement, so the continuing importance of research related technologies, universities and technical cooperation is still essential for enterprises, governments, and business related fields in colleges and universities should be increased investment in research funding, technological progress, it could accelerate.
iv) Human resources, at present in the field of new energy research and development of talent less, the key issue is not related to universities and institutes of professional settings, only a small part of the researchers involved in related areas, so increasing the field of new energy professionals culture is very important.

At present, because a larger capital investment, development and application of tidal energy is still in its infancy, but as the world demand for electricity supply to the rapid growth of concern about environmental protection, tidal energy development will be rapid development in the future. Moreover, some developed countries have in tidal energy applications has made great achievements, such as Britain, France, Canada and so on, so it should be convinced that the development of tidal energy will have a beautiful future.

References

[1] Gao, H. & Zhang, A. (2007). *New Energy Technology and Applications*, National Defense Industry Press.

[2] Hui, J., Fang, G., Yan, W. & Shi, C. (2008). *New Energy Conversion and Control Technology*, Machine Industry Press.

[3] Chen, W. (2006). *An Introduction to Energy*, GaoLi Book Ltd.

[4] Wang, G. & Ai, D. (2006). *An Introduction to New Energy*, Chemistry Industry Press.

[5] Zhai, X., Liu, K. & Han, Q. (2003). *New Energy Technology*, Chemistry Industry Press.

[6] Wang, C., Cui, R. & Zhou, H. (2003). *New Energy Power Technology*, Chinese Electric Power Press.

[7] Huang, S. & Gao, W. (2004). *An Introduction to Energy*, Higher Education Press.

[8] Li, C. (2005). *New and Renewable Energy Technology*, Southeast University Press.

[9] World Wind Energy Association, *World Wind Energy Report 2008,* http://www.cwea.org.cn/upload/WorldWindEnergyReport2008_s.pdf, 2009.

[10] Li, J. & Wang, S. (2007). *China Solar PV Report*, China Environmental Science Press.

[11] Liu, W., Zhang, Z., Li, Y. & Zhao, P. (2007). *Wind Energy and Wind Power Technology*, Chemistry Industry Press.

[12] China News Net, China's largest offshore wind farm is proposed in Yancheng, Jiangsu Province, http://www.chinanews.com.cn/gn/news/2009/09-04/1850177.shtml.

[13] China Construction Tender Net, China's largest wind farm started in Jiuquan, Gansu province, http://www.zhaobiao.gov.cn/projectinfodetail/23839240.html.

[14] Su, F., Liu, Y. & Bai, S. (2006). The physical nature of the phenomenon of tidal, *Yan'an University (Natural Science Edition)*, Vol. 20, No.2, 47-49.

[15] Wang, C. (1999). Ancient Chinese knowledge of tides, Journal of Xinyang Normal University (Natural Science Edition), Vol. 23, No.5, 43-47.

[16] Tian, X. (1996). *Causes of tides, college physics*, Vol. 15, No.10, 24-27.

[17] Zhang, S. (1997). Tidal phenomena of the primary note, *Physics Bulletin*, No.8, 5-7.

[18] Zhao, A. & Jiao, Z. *Mechanical analysis of tidal phenomena and similar phenomena to explore*, http://down3.yestime.net/460000-469999/462039/zaj.doc.

[19] Roger, H. & Charlier, A. (2003). *"sleeper" awakes tidal current power, renewable and sustainable energy reviews*, Vol. 7, No.6, 515-529.

[20] Atlantis Resources Corporation Ltd., *High potential areas for tidal resources*, http://www. atlantisresourcescorporation.com/marine-power/global-resources.html.

[21] Roger, H. (2007). Charlier, Forty candles for the Rance River TPP tides provide renewable and sustainable power generation, renewable and sustainable energy reviews, Vol. 11, No. 9, 2032-2057.

[22] Blunden, L. S. & Bahaj, A. S. (2006). Initial evaluation of tidal stream energy resources at Portland Bill, UK, *Renewable Energy*, Vol. 31, No.2, 121-132.

[23] Rafael, M. (2009). Ferreira*, Segen F. Estefen, Alternative concept for tidal power plant with reservoir restrictions, *Renewable Energy*, Vol.34, No4., 1151-1157.

[24] Roger, H. (2003). Charlier, Sustainable co-generation from the tides: A review, Renewable and Sustainable Energy Reviews, Vol. 7, No. 3, 187-213.

[25] Fergal, O. (2010). Rourke, Fergal Boyle, Anthony Reynolds, Tidal energy update 2009, *Applied Energy*, Vol. 87, No.2, 398-409.

[26] Fergal, O. (2009). Rourke, Fergal Boyle, Anthony Reynolds, Marine current energy devices Current status and possible future applications in Ireland, *Renewable and Sustainable Energy Reviews*, Available online 30 November.

[27] Fujian (1999). *Province editorial board of local history*, Fujian province's Water log, China Social Sciences Press.

[28] Twidell, J. & Weir, T. (2006). *Renewable energy resources*. 2nd ed, Taylor & Francis.

[29] Yang, M. (2009). The economic benefits of tidal power and its mode of operation in China, *Chinese Energy*, Vol. 31, No.5, 5-8.

[30] GCK Technology Ltd. The Gorlov Helical Turbine; 2008. http://www.gck technology.com/GCK/pg2.html.

[31] Aquamarine Power Ltd. Aquamarine Power Announces Contract with ABB for Neptune; 2009. http://www.aquamarinepower.com/news-and-events/news/latest-news/view/51/aquamarine -power-announces-contract-with-abb-forneptune/.

[32] Tidal Energy Ltd. *DeltaStream Concept*; 2008. http://www.tidalenergyltd.com/ technology. htm.

[33] Verdant Power Ltd. *Verdant Power's Free Flow Turbines*; 2008. http://www. verdant tpower.com/.

[34] SMD Hydrovision Ltd. *Tidal Stream Generator*; 2008. http://www.reuk.co.uk/TidEl-Tidal Turbines.htm.

[35] OpenHydro Ltd. *OpenHydro Becomes First Tidal Energy Company to Generate Electricity onto the UK National Grid*; 2008. www.openHydro.com.

[36] Lunar Energy Ltd. *Renewables boost as Lunar Energy seals £500m deal*; 2008. http://www. lunarenergy.co.uk/News.php.

[37] Pulse Generation Ltd. *Hydrofoils or Turbines*; 2008. http://www.pulsetidal. com/?q=node/25.

[38] IHC Engineering Business Ltd. Stingray Tidal Stream Generator, http://www.engb. com/services_09a.php.

[39] Ocean Flow Energy Ltd.. *Development status*; 2008, http://www.oceanflowenergy.com/ technology.html.

[40] Blue Energy Ltd.. *Tidal power*; 2008, www.bluenergy.com.

[41] Hammerfest Strom AS. *World leading technology developed by Hammerfest Strom; 2007*, http://www.hammerfeststrom.com/content/view/58/86/lang,en/.

[42] Marine Current Turbines Ltd.. *SeaGen*, www.marineturbines.com.

[43] Atlantis Resources Corporation Ltd., *Nereus and Solon Tidal Turbines*; 2008, http://www. atlantisresourcescorporation.com/technology/.

In: Sea Level Rise, Coastal Engineering, Shorelines and Tides ISBN: 978-1-61728-655-1
Editor: Linda L. Wright, pp. 83-103 © 2011 Nova Science Publishers, Inc.

Chapter 3

THE STATISTICAL DISTRIBUTIONS OF NEARBED WAVE PRESSURE, ORBITAL VELOCITY AND WAVE PERIOD IN FINITE COASTAL WATER DEPTH

Zai-Jin You *

Coastal Unit, Department of Environment, Climate Change and Water, Dangar, NSW, Australia

Abstract

The statistical distributions of wave pressure, orbital velocity and wave period in the vicinity of the seabed are systematically determined from the comprehensive field wave data collected at 0.5m above the seabed in finite coastal water depth. The zero-crossing method is used to analyse the field data to obtain wave pressure and orbital velocity amplitudes of individual waves. It is found that the distributions of instantaneous wave pressure and orbital velocity are identical and perfectly follow the Gaussian distribution as commonly assumed. The distributions of wave pressure and orbital velocity amplitudes are also found to be almost identical and more closely follow the Weibull and modified Rayleigh distributions than the Rayleigh distribution. The distribution of wave periods is shown to follow the normal distribution. The joint distribution of wave pressure and period is also discussed. Several useful formulas are derived to compute characteristic wave pressure, orbital velocity and wave period, and the expected values of maximum wave pressure and orbital velocity. These derived formulas may be also applied to calculate characteristic wave heights based on linear wave theory.

Keywords: Irregular Wave, Orbital velocity, Wave Pressure, Wave Period, Gaussian, Rayleigh, and Weibull.

* E-mail address: bob.you@environment.nsw.gov.au

1. Introduction

When ocean water waves propagate from deep water to intermediate and shallow waters, they are able to penetrate to the seabed to induce wave orbital velocity and dynamic wave pressure in the vicinity of the seabed and transmit enormous amount of wave energy into the seabed through wave-induced bottom shear stress and nearbed dynamic wave pressure. The deep, intermediate and shallow waters are explicitly defined as $k_0 h \geq \pi$, $0.1 < k_0 h < \pi$ and $k_0 h \leq 0.1$ (You, 2008), where k_o is the deepwater wave number and h is the mean water depth. The nearbed orbital velocity or wave-induced bottom shear stress interacts with the seabed, resulting in frictional dissipation of wave energy, mobilization of bed sediment into motion, formation of different bed forms, and modification of wave bottom shear stress (You, 1994, 2005, 2007). The nearbed dynamic wave pressure is another important wave parameter for calculation of dynamic wave forces on coastal structures, determination of the stability of undersea oil and gas pipelines, and design of coastal and ocean structures. Both the wave orbital velocity and dynamic pressure are a function of wave period. A simple relationship among the wave orbital velocity, pressure and period was derived from linear wave theory by You (2008). The wave period also is another important design parameter for floating marine systems. The resonant motion of floating marine structures will occur when wave periods are close to the natural motion period of the floating marine structures. Therefore, this study is of practical importance in modelling of coastal wave hydrodynamics, wave forces, sediment seabed liquefaction, scour around coastal structures, and coastal sediment transport.

The statistical distribution of wave height has been extensively studied by numerous researchers. The Gaussian distribution of surface wave displacement from the mean water level was first found by Rudnick (1951) based on his analysis of field wave data collected in the Pacific Ocean. Longuet-Higgins (1952) first applied the Rayleigh distribution to study the statistical distribution of deepwater wave height for a narrow-banded Gaussian sea. Theoretical relationships derived from the Rayleigh distribution generally agree well with the values determined from field wave data (Thompson, 1977; Chakrabarti, 1987), even though the frequency spectra of ocean waves may not always be narrow-banded as assumed in the Rayleigh distribution. The Rayleigh distribution is found to sometimes deviate from field measurements and this deviation appears to increase with increasing wave heights and decrease as the wave spectrum becomes sharply peaked (EM, 2008). Cartwright and Longuet-Higgins (1956) and Longuet-Higgins (1980) introduced a scale parameter in the Rayleigh distribution to account for a finite-banded random sea. A more general distribution, the Weibull distribution, was also applied by Forristall (1978) to study the distribution of wave height in the Gulf of Mexico. A summary of some of the statistical distributions of wave heights was given by Vinje (1989). For a non-Gaussian ocean waves in finite water depths, Longuet-Higgins (1963) derived the pdf of surface wave displacements by directly applying the Gram-Charlier distribution. Bitner (1980) and Ochi and Wang (1984) demonstrated that the Gram-Charlier distribution generally represents the pdf of surface wave displacements in shallow waters, but this density distribution function becomes invalid for large negative displacements of surface waves. Another drawback of this approach is that the probability distribution of wave height can not be directly derived from the Gram-Charlier distribution, while the Rayleigh distribution of wave height can be deduced from the Gaussian density function of wave displacements. Stokes wave theory has also been applied to study the distribution of non-Gaussian waves by several researchers, e.g. Tayfun (1980), Guang and Sun (1997), and Hou *et al* (2006). The probability density function derived from Stokes wave theory does not have any negative values like those from the Gram-Charlier distribution. Ochi and Ahn (1994) also adopted another approach, a nonlinear system method, to study the distribution of wave heights in shallow water by applying the solution of Kac and Siegert (1947).

The distribution of wave period has not been studied as extensively as that of wave height mainly due to the assumption of a narrow-banded Gaussian sea. The frequency bandwidth of ocean waves in a narrow-banded sea should be narrow and thus the wave periods are expected to be more or less the same. However, information on statistical properties of wave periods is of coastal engineering importance for estimating a possible critical resonant condition of floating coastal structures (e.g. floating breakwater). When the natural motion period of a floating costal structure is close to wave periods, for example, the resonant motion of the floating coastal or marine structure may occur and subsequently results in damaging the structure. One attempt to determine the distribution of wave period was made by Bretschneider (1959) based on the assumption that the wavelength follows the Rayleigh distribution. A different distribution of wave period was then derived by Longuet-Higgins (1962). This distribution is found to fit well with field measurements and often used for offshore design (EM, 2008). In general, the distribution of wave period is shown to be narrower than that of wave height, and the spread lies mainly in the range of 0.5 to 2 times the mean wave period. The joint probability distribution of wave height and period was first developed by Longuet-Higgins (1975). He considered wave heights and periods in an arrow-banded random sea, and derived the joint distribution under a more suitable assumption for real sea states, i.e. wave heights and periods are correlated. The joint distribution of Longuet-Higgins (1975) was later modified by Longuet-Higgins (1983) for a finite-banded random sea.

Only a few studies have been undertaken to investigate the distribution of wave orbital velocity. Sultan (1992) directly measured wave orbital velocities close to the bed in a laboratory wave flume with a laser Doppler velocity current meter. It was found that the distribution of instantaneous orbital velocities generally obeys the Gaussian distribution and the distribution of orbital velocity amplitudes follows the Rayleigh distribution. It was also shown that the Gram-Charlier distribution is less effective than the Gaussian distribution in fitting to the measured instantaneous orbital velocities, while a modified form of the Rayleigh distribution, the Beta-Rayleigh distribution, agrees better with the measured orbital velocity amplitudes than the Rayleigh distribution. Song and Wu (2000) also studied the distributions of horizontal and vertical instantaneous orbital velocities based on a second-order random wave theory. The parameters of the distributions were estimated from the wave spectrum of Donelan and Pierson (1987). They found that the distribution of vertical orbital velocities obeys the Gaussian distribution, while the distribution of horizontal orbital velocities deviates from the Gaussian distribution when the water depth becomes less than 10m. No field data were used to confirm their findings. More recently, Wiberg *et al* (2008) also investigated the distribution of orbital velocity amplitude based on their 2-month field data collected in a 40m water depth in Hudson Shelf Valley, and they found that the histogram of wave orbital velocity amplitudes well obeys the Rayleigh distribution. They finally concluded that the distribution of the amplitude of a narrow-banded and Gaussian random process, such as surface waves and orbital velocities, should follow a Rayleigh distribution. Most recently,

You (2009) quantitatively determined the distribution of nearbed wave orbital velocity from his comprehensive field data on water particle velocities. It was found that the distribution of instantaneous orbital velocity well follows the Gaussian distribution, and that the distribution of orbital velocity amplitudes is less accurately described by the Rayleigh distribution than the Weibull distribution.

However, none of the previous studies has systematically determined the distributions of wave height or pressure, orbital velocity and wave periods from the field wave data collected simultaneously. In this study, the statistical distributions of wave pressure and wave orbital velocity will be quantitatively determined from the comprehensive field velocity data collected simultaneously and then compared with the distribution functions of the Rayleigh, the modified Rayleigh and the Weibull. Several useful formulas are also derived from the three distribution functions to calculate characteristic wave pressures and orbital velocities, and the expected values of both maximum wave pressure and orbital velocity. The distribution of wave period and the joint distribution of wave period and wave pressure are also studied based on the same field data collected.

2. Field Data

The comprehensive field data on wave pressure and water particle velocity were collected simultaneously at two study sites off MacMaster Beach on the NSW Central coast (see Figure 1). The sandy beach is about 1.5km long and located between two high sandstone headlands. The orientation of the beach is south easterly, the dominant wave direction of the NSW coast. The offshore morphology of the beach is relatively simple and there is no offshore reef (Evans et al, 2000). The inner study site in Figure 1 was located in a mean water depth of 11m and the outer site in 23m. The seabed sediment at the two sites was fine, with median grain size of $d_{50}\approx 0.2$ mm. The echo sounder and side scan sonar with DGPS were used to map the offshore bathymetry and sediment types. The offshore wave data have been collected continuously by a directional wave rider buoy deployed permanently in water depth of 85m off Sydney, about 120km south of this study site. The yearly mean significant wave period $T_{1/3}$ measured by the wave-rider buoy is about 8s. The relative water depth $k_o h$, which varies from 0.66 at the inner site to 1.38 at the outer site and falls into the range of $0.1 < k_0 h < \pi$, is intermediate based on the explicit criterion of You (2008).

Two ocean Acoustic Doppler Velocimeters (ADV), which were mounted on two separate stainless steel tripods sitting on the seabed, were used to directly measure instantaneous water particle velocities and wave pressures at 0.5m above the seabed. One instrumented tripod was deployed at the inner site in a mean water depth of 11m, and the other tripod at the outer site in a mean water depth of 23m as shown in Figure 1. The horizontal water particle velocity components (u, v) and wave pressure p, were collected at a sampling rate of 2Hz for 17.07 min every hour at the inner and outer sites, simultaneously. Nine field deployments were undertaken at the two sites over a period of about two years, and the duration of each deployment ranges from 31 to 66 days. The sampling length of 17.07 min was used to give 2048 or 2^{11} data points in each velocity or pressure burst in order to apply the Fast Fourier Transform to compute wave spectrum. The sampling rate of 2Hz was chosen to record about 10~20 velocity data points within one wave period of T=5~10s.

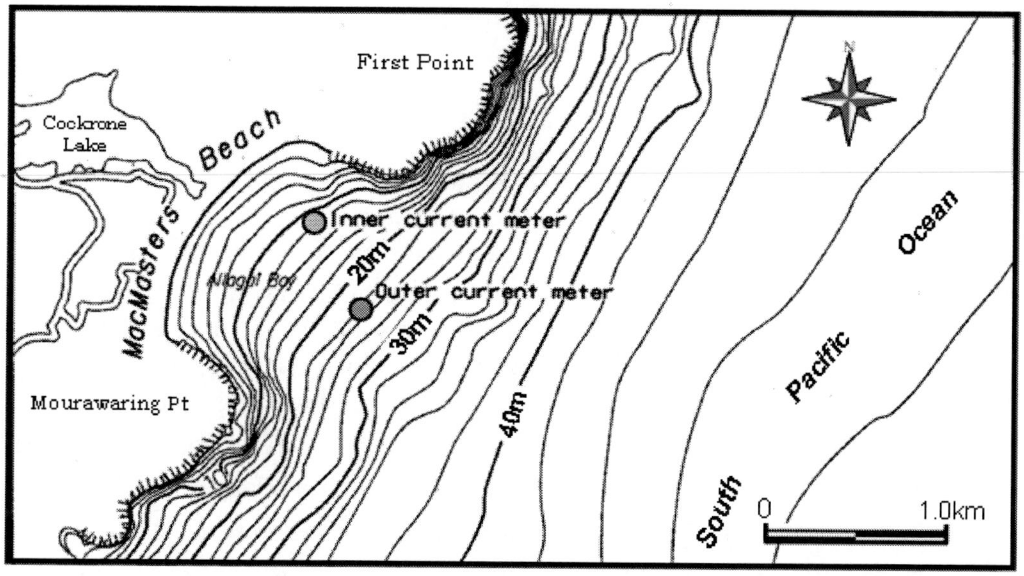

Figure 1. Locations of the study sties in the water depths of 11m and 23m (You, 2009).

3. Statistical Distributions

3.1. Instantaneous Wave Pressure and Orbital Velocity

The instantaneous horizontal water particle velocity components (u, v) and water depth h(t) were directly measured at a sampling rate of 2Hz for 17.04min every hour at the inner and outer sites, simultaneously. The instantaneous wave orbital velocity components, $\tilde{u}(t)$ and $\tilde{v}(t)$, and wave pressure, $\tilde{p}(t)$, can be then determined as

$$(\tilde{u},\tilde{v}) = (u,v) - (\overline{u},\overline{v}) \quad \textbf{and} \quad \tilde{p}(t) = h(t) - \overline{h} \,, \tag{1}$$

where $(\overline{u},\overline{v})$ are the mean horizontal velocity components and $\overline{h}$ is the mean water depth and they are all averaged over one sampling length of 17.04min. The root-mean-square values of $\tilde{u}_{rms}$, $\tilde{v}_{rms}$ and $\tilde{p}_{rms}$ are then calculated from individual segment of $\tilde{u}(t)$, $\tilde{v}(t)$ and $\tilde{p}(t)$, respectively. The velocity or pressure segments, with abnormally large values of $\tilde{u}_{rms}$, $\tilde{v}_{rms}$ or $\tilde{p}_{rms}$, are excluded in this study. The abnormally large velocities were found during the coastal storms. This may be because high suspended sediment concentrations and possibly entrained air bubbles in the vicinity of the seabed affect the readings of the ADVs.

The distributions of $\tilde{u}(t)$, $\tilde{v}(t)$ and $\tilde{p}(t)$ are calculated from the field data with the same approach. The histogram of $\tilde{p}(t)$, for example, is computed as follows: (1) computing wave pressure $\tilde{p}(t)$ from Eq.(1) and the root-mean-square pressure $\tilde{p}_{rms}$ for each pressure segment of 17.04min, (2) normalising individual wave pressure $\tilde{p}(t)$ by $\tilde{p}_{rms}$ for each pressure segment, (3) ordering the normalised values of $\tilde{p}/\tilde{p}_{rms}$ from all pressure segments to obtain a large sample of $\tilde{p}/\tilde{p}_{rms}$ in ascending order, (4) grouping the ordered values of $\tilde{p}/\tilde{p}_{rms}$ into different wave pressure classes, and (5) computing the histogram of $\tilde{p}/\tilde{p}_{rms}$ as

$$f(x) = \frac{1}{\Delta}\left(\frac{n}{m}\right), \tag{2}$$

where $x = \tilde{p}/\tilde{p}_{rms}$, m is the total number of wave pressure data collected over a period of hours to months, n is the number of wave pressure data falling into a wave pressure bin or class, and Δ is the pressure bin width. The bin width of Δ=0.2 is used in this study to divide the normalised wave pressure range (-3, 3) into 31 wave pressure classes.

The use of the normalised wave pressure $\tilde{p}/\tilde{p}_{rms}$ is to ensure that the process $\tilde{p}/\tilde{p}_{rms}$ is stationary at any time provided instantaneous wave pressure $\tilde{p}(t)$ is stationary over a short sampling length of 17.04min. Note $\tilde{p}$ may not be stationary over a long period (e.g. hours or days). Since the rms value of $\tilde{p}/\tilde{p}_{rms}$ calculated from any pressure segment of 17.07min is always equal to 1, it may not be difficult to prove that the rms value of $\tilde{p}/\tilde{p}_{rms}$ computed from a large number of wave pressure segments collected in one deployment is also equal to 1. Thus, the process of $\tilde{p}/\tilde{p}_{rms}$ is stationary at any time provided $\tilde{p}$ is stationary over one sampling period of 17.04min.

The pdfs of $\tilde{u}/\tilde{u}_{rms}$, $\tilde{v}/\tilde{v}_{rms}$ and $\tilde{p}/\tilde{p}_{rms}$ are calculated from individual burst and 24 bursts of Mac4001 and Mac5001 and also compared with the Gaussian distribution in Figure 2. It can be seen that the pdfs of $\tilde{u}/\tilde{u}_{rms}$, $\tilde{v}/\tilde{v}_{rms}$ and $\tilde{p}/\tilde{p}_{rms}$ calculated from individual burst generally follow the trend of the Gaussian distribution, while the pdfs computed from 24 bursts are almost identical and perfectly obey the Gaussian distribution. This finding is also supported by the field data collected in the other field deployments.

For engineering applications, the resultant wave orbital velocity $\tilde{U}(t)$ in the direction of wave propagation is often used to determine wave hydrodynamics, seabed roughness, sediment transport and coastal scour (You, 1994; 2005; 2006). The wave orbital velocity $\tilde{U}(t)$ may be expressed in terms of $\tilde{u}$ or $\tilde{v}$ as

$$\tilde{U}(t) = \frac{\tilde{u}}{\sin\theta} = \frac{\tilde{v}}{\cos\theta}, \tag{3}$$

where θ is the mean wave direction over a sampling length of 17.07min and may vary slowly from one burst to another one. You (2009) applied the formula of You and Yin (2001) to estimate θ in Eq.(3). Since $\tilde{U}$ is linearly proportional to $\tilde{u}$ or $\tilde{v}$ as shown in Eq.(3), the value of $\tilde{U}/\tilde{U}_{rms}$ should be independent of θ and therefore the pdf of $\tilde{U}/\tilde{U}_{rms}$ should be identical to that of $\tilde{u}/\tilde{u}_{rms}$ or $\tilde{v}/\tilde{v}_{rms}$ as shown by You (2009).

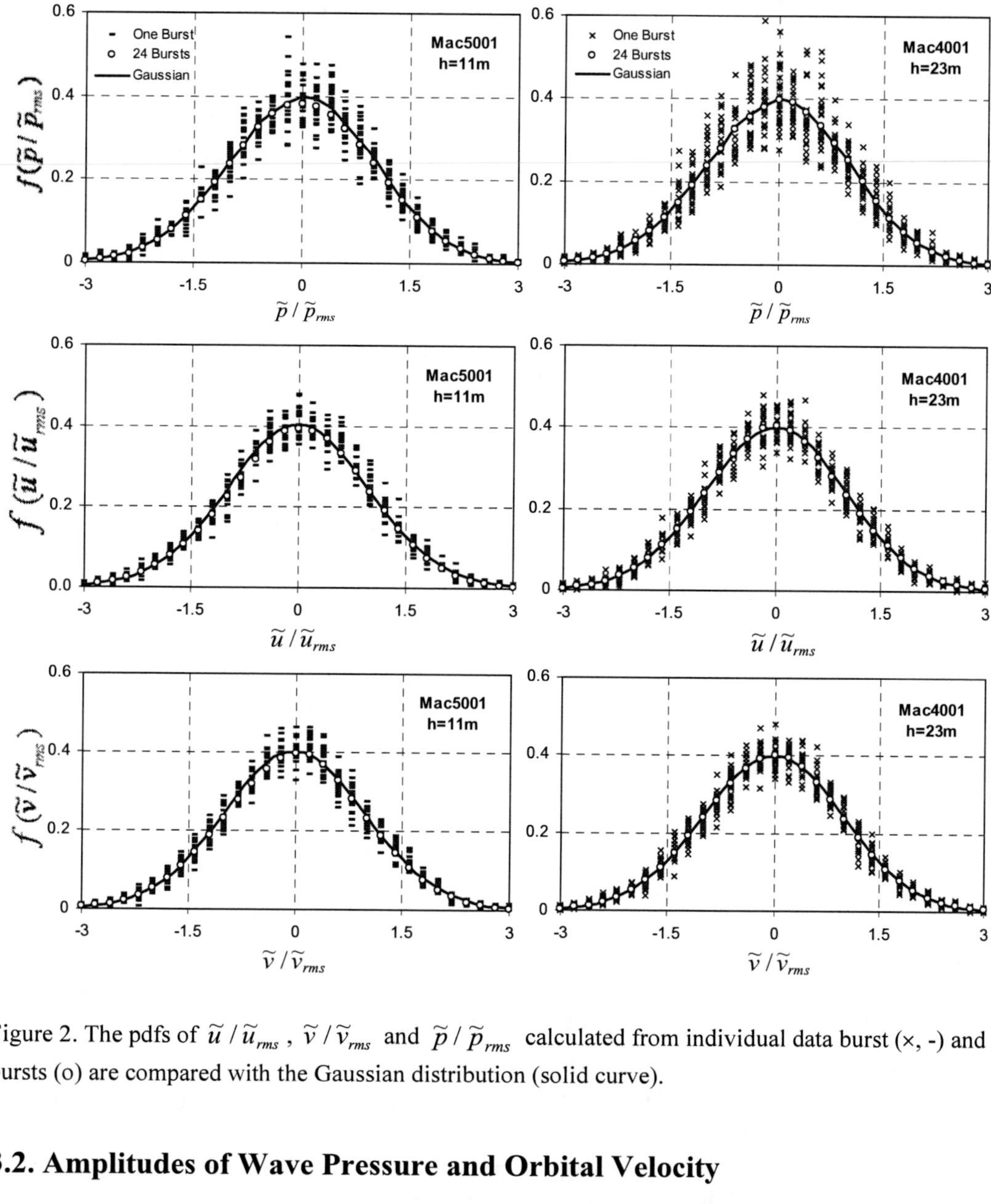

Figure 2. The pdfs of $\tilde{u}/\tilde{u}_{rms}$, $\tilde{v}/\tilde{v}_{rms}$ and $\tilde{p}/\tilde{p}_{rms}$ calculated from individual data burst (×, -) and 24 bursts (o) are compared with the Gaussian distribution (solid curve).

3.2. Amplitudes of Wave Pressure and Orbital Velocity

In engineering practice, coastal engineers are normally concerned with amplitudes of wave orbital velocity and wave pressure rather than their instantaneous values (You, 1994). In this study, the wave pressure and orbital velocity amplitudes of individual waves are analysed from time-series wave orbital velocity $\tilde{U}(t)$ and wave pressure $\tilde{p}(t)$ collected for 17.04min every hour with the zero-up crossing method. The orbital velocity amplitude U is defined as $U=0.5\times(u_{\max}-u_{\min})$, where $u_{\max}$ and $u_{\min}$ are the maximum and minimum orbital velocities of a wave. There are about 100 orbital velocity amplitudes analysed from a velocity segment, about 2400 from one-day measurement, and about 72,000 from one-month measurement. The same approach is also applied to define the wave pressure amplitude P. It should be noted here that the values of U and P were also analysed with the zero-down crossing method and are found to be nearly identical to those with the zero-up crossing method because of the symmetrical coastal waves found at the study sites.

The histograms of U/U_{rms} and P/P_{rms} are computed with the same method that is used to calculate the histograms of $\tilde{u}/\tilde{u}_{rms}$ and $\tilde{p}/\tilde{p}_{rms}$ in Figure 2. The histograms of U/U_{rms} and P/P_{rms} are computed from the field data of Mac4001 and Mac5001 and shown in Figure 3, together with the distribution functions of the Rayleigh and the Weibull. It can be seen that the distributions of U/U_{rms} and P/P_{rms} are generally identical and can be approximately described by the Rayleigh distribution

$$Q = \exp\left[-\left(\frac{X}{X_{rms}}\right)^2\right], \tag{4}$$

where Q is the probability of exceeding and X can be replaced with U or P. Eq.(4) can be also derived theoretically from the Gaussian distribution with the same approach that was used by Longuet-Higgins (1952) to derive the Rayleigh distribution of wave height under the assumption of a narrow-banded Gaussian sea.

Under a broad-banded sea, the Rayleigh distribution may be modified with an introduced scale parameter in Eq.(4)

$$Q = \exp\left[-\left(\frac{X}{C\,X_{rms}}\right)^2\right] = \exp\left[-\left(\frac{X}{X_r}\right)^2\right], \tag{5}$$

where C=0.925 was proposed by Longuet-Higgins (1980). The scale parameter C was initially introduced to account for finite wave spectral bandwidth (Massel and Sobey 2000). Mathematically, the term $C \times X_{rms}$ in Eq.(5) may be considered as a reference pressure P_r, for example, which is smaller than the rms pressure P_{rms}, but larger than the mean pressure P_1. The modified Rayleigh distribution Eq.(5) with one degree of freedom should give a better fitting to the data than the Rayleigh distribution Eq.(4) with no degree of freedom. The modified Rayleigh distribution Eq.(5) with a newly proposed value of C=0.96 in this study is found to agree better with the data than the original Rayleigh distribution Eq.(4). The value of C=0.96 is used for all field data collected in this study.

Alternatively, with introducing a shape parameter α rather than the scale parameter C in Eq.(4), the Rayleigh distribution can be also modified as

$$Q = \exp\left[-\left(\frac{X}{X_{rms}}\right)^{\alpha}\right], \tag{6}$$

which is also called an one-parameter Weibull distribution with one degree of freedom. The value of α=2.15 is determined with a linear fitting method, i.e., transforming Eq.(6) into a linear equation $y = \alpha\, x$ and then fitting a linear regression line through the data points (x, y) of Mac4001 to obtain α, where $x = \ln(X / X_{rms})$ and $y = \ln[-\ln(Q)]$. The same value of α=2.15 is used for all field data collected in this study.

It can be seen from Figure 3 that the one-parameter Weibull distribution of Eq.(6) agrees better with the data than the Rayleigh distribution of Eq.(4). Forristall (1978) also proposed a two-parameter Weibull distribution function to study the distribution of wave heights in the Gulf of Mexico and obtained α=2.13. A more general three-parameter Weibull distribution with the scale, location and shape parameters is also fitted to the data in Figure 3, but found no better than the one-parameter Weibull distribution of Eq.(6).

3.3. Wave Period

Individual wave pressure segments, which were measured for 17.04min every hour, are also analyzed to obtain the wave periods of individual waves. There are about 100 wave period data analyzed from one pressure burst, about 2400 wave period data from one-day pressure measurement, and about 72000 wave period data from one-month pressure measurement. The wave orbital velocity segments were also analyzed to obtain the wave periods of individual waves in this study. The wave periods analyzed from both the wave pressure and orbital velocity segments, which were measured simultaneously with the same sampling rate and interval, are found generally close. The wave pressure data may be still preferred to the orbital velocity data for the analysis of the wave periods because the accuracy of the measured wave pressure data are about 10 times higher than that of the measured wave orbital velocity data according to the specifications of the ADVs used in this study.

The distribution of wave period T is also computed with the same approach as described in Eq.(2). The histograms of T calculated from the data of Mac4001 and Mac5001 are shown in Figure 5, together with the distribution functions of Breschneider (1959), Longuet-Higgins (1962) and You (2009). It can be seen that the wave periods at the outer site of 23m are more narrowly distributed than those at the inner site of 11m, and the distribution functions of Bretschneider (1959) and You (2009) agree better with the histograms of T than that of Longuet-Higgins (1962).

Bretschneider (1959) derived the distribution function for the wave period under the assumption that the wave length L also follows the Rayleigh distribution. Since L is proportional to T^2 according to linear wave theory, the distribution of T^2 should also follow the Rayleigh distribution. The distribution function of Bretschneider (1959) is written as

$$f(T) = \frac{\pi}{1.2}\frac{x^3}{\overline{T}}\exp\left[-\frac{\pi}{4.8}x^4\right] \text{ or } Q = \int_T^{\infty} f(T)dT = \exp\left[-\frac{\pi}{4.8}x^4\right], \tag{7}$$

where $\overline{T}$ is the mean wave period and $x = T/\overline{T}$. A different distribution function for the wave period was also derived by Longuet-Higgins (1962) as

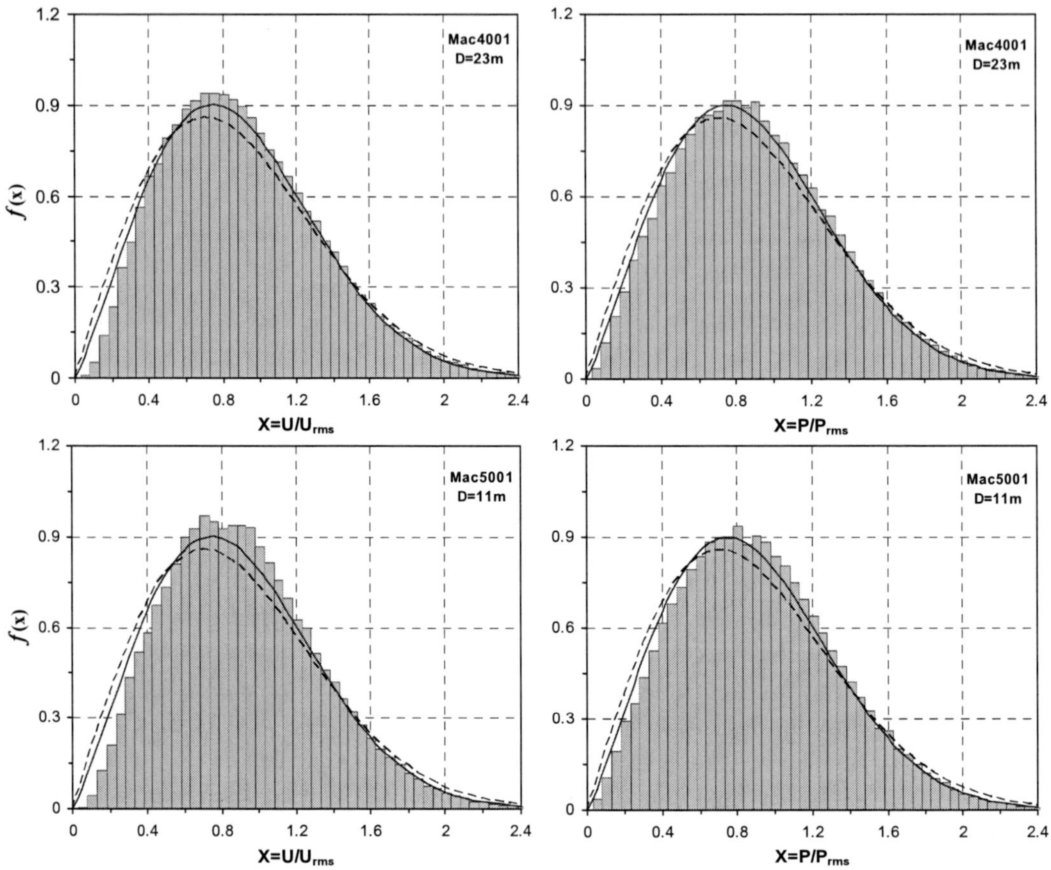

Figure 3. The histograms of U/U_{rms} and $P/P_{\rm rms}$ computed from the field data of Mac4001 and Mac5001 are compared with the Rayleigh distribution (dashed curve) and the Weibull distribution (solid curve).

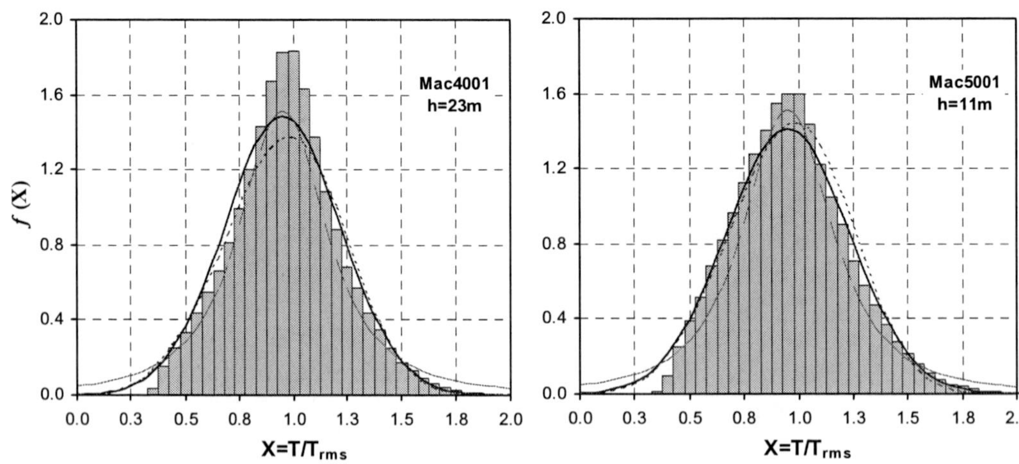

Figure 4. Histograms of wave periods measured in Mac4001 and Mac 5001 are compared with the pdfs of Bretschneider (1959) [short dashed], Longuet-Higgins (1962) [long dashed, ν=0.33] and You (2009) [solid curve].

$$f(x) = 0.5\nu^2\left[\nu^2 + (x-1)^2\right]^{-1.5} \text{ or } Q = \int_T^{\infty} f(T)dT = 0.5\left[1 - \frac{x-1}{\sqrt{\nu^2 + (x-1)^2}}\right], \quad (8)$$

where ν is the spectral width parameter. The value of ν=0.33 is used in this study, which gives the best fitting to the wave period data in Figure 4. Note that Eq.(8) does not satisfy the boundary condition of $f = 0$ at T=0 when ν≠0. When $T = \overline{T}$, f(x) is maximum and equal to $0.5/\nu$ and Q=0.5 in Eq.(8). You (2009) also proposed a normal distribution function for the wave period based on his comprehensive field data

$$f(T) = \frac{1}{C_0 T_{rms}\sqrt{\pi}} \exp\left[-\left(\frac{x - C_1}{C_0}\right)^2\right] \text{ or } Q = \int_T^{\infty} f(T)dT = 1 - \Phi\left(\frac{\sqrt{2}(x - C_1)}{C_0}\right), \quad (9)$$

where C_0=0.95, C_1 is constant for a given water depth and, and Φ is the norm distribution. The value of C_1=0.40 is determined for the inner site of 11m and C_1=0.38 for the outer site of 23m. The distribution functions of Longuet-Higgins (1962) and You (2008) have one degree of freedom, but that of Breschneider (1959) has no degree of freedom.

3.4. Joint Distribution

The distributions of wave pressure, orbital velocity and wave period have been quantitatively determined in the preceding sections. The distribution of wave pressure is shown to be almost identical to that of wave orbital velocity as shown in Figures 3 and 4. Only the joint distribution of wave pressure and period is then investigated here. Based on the wave pressure data from Mac1001, a simple plot of P versus T is made in Figure 5. It can be seen that for a given P, the range or spread of wave periods generally increases with P/P_{rms} when $P/P_{rms} < 1.0$, and then gradually decreases with P/P_{rms}, and finally approaches to $T \approx T_{rms}$. This implies that the wave periods of large waves are more narrowly distributed than those of small waves. For a given P, there exists the minimal wave period T_{min} and it may be empirically expressed as

$$\left(\frac{T}{T_{rms}}\right)_{min} = 1 - \exp\left[-\frac{P}{P_{rms}}\right], \quad (10)$$

which satisfies the boundary condition of T=0 at P=0.

Figure 6[A] also shows how the distribution of wave period is affected by different values of P/P_{rms}. It can be seen that the distribution of wave period may follow the Weibull distribution and it is moving to the right and becomes more peaked with increasing P/P_{rms} from 0.25 to 2.2. It is also shown in Figure 6[B] that the mean wave period of large waves with $P > P_{rms}$ becomes nearly constant. Based on this finding, it may be concluded here that the assumption of a narrow-banded sea is more justified for large sea waves than small ones.

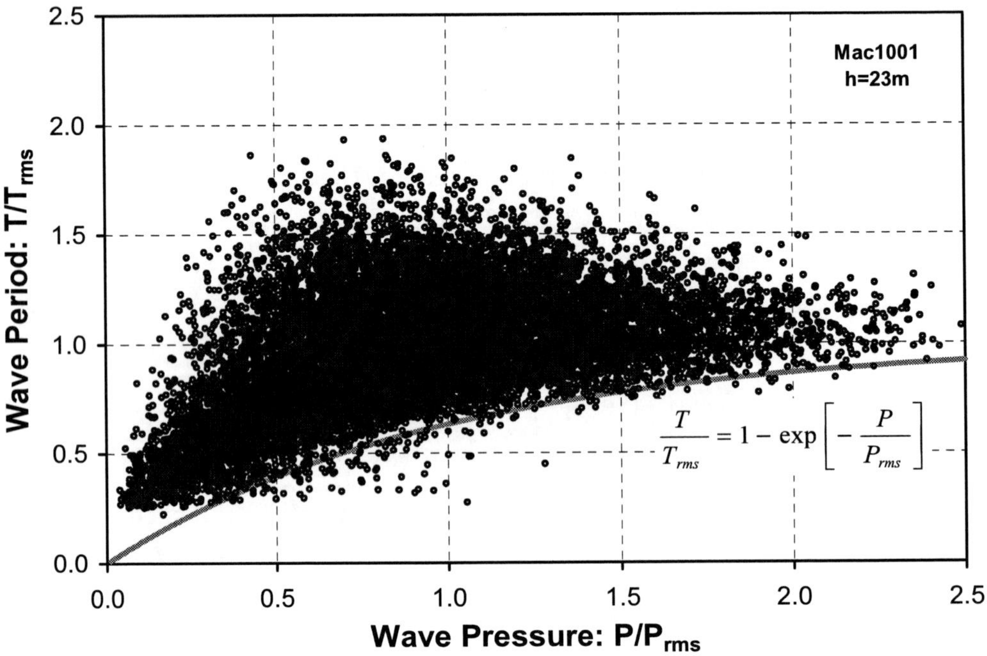

Figure 5. Correlation between P and T in Mac1001 and the empirical curve on (P, T_{min}).

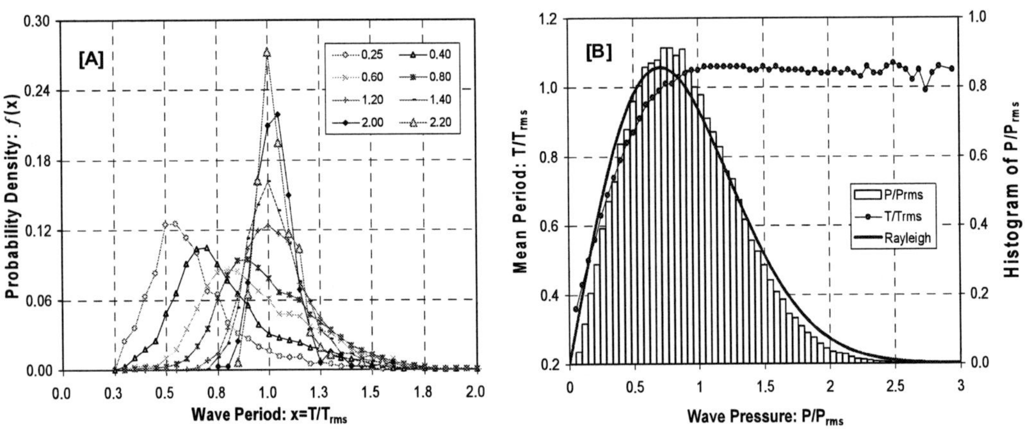

Figure 6. [A] Distribution of wave period for a given P/P_{rms} at the outer site of Mac4001, and [B] The mean wave period varies with wave pressure amplitude P.

4. Statistical Relationships

4.1. Mean of Largest Amplitudes

One of important engineering applications for this study is to estimate characteristic wave orbital velocity U_Q, wave pressure amplitude P_Q or wave height H_Q from a short wave record. The orbital velocity amplitude U_Q, for example, is defined as the average orbital velocity of n

largest orbital velocities from the total number of m orbital velocity amplitudes and $Q = n/m$ is the probability of exceeding. P_Q is also defined with the same approach as U_Q. When the probability distribution of X follows the Rayleigh distribution, X_Q can be then estimated as

$$\frac{X_Q}{X_{rms}} = \frac{\int_{\hat{X}}^{\infty} X f(X) dX}{X_{rms} Q} = \sqrt{-\ln Q} + \frac{\sqrt{\pi}}{2Q} \operatorname{erfc}\left(\sqrt{-\ln Q}\right), \tag{11}$$

where X can be replaced by U or P, $\hat{X}$ is an arbitrary value, and *erfc*(x) is the complementary error function. The relationship, $\hat{X}/X_{rms} = \sqrt{-\ln Q}$, from the Rayleigh distribution of Eq.(4) is also applied in the deviation of Eq.(11).

When $Q \rightarrow 0$, *erfc*(x) has an asymptotic solution and thus Eq.(11) may be simplified as

$$\frac{X_Q}{X_{rms}} = \sqrt{-\ln Q} + \frac{1}{2\sqrt{-\ln Q}} - \frac{(-\ln Q)^{-\frac{3}{2}}}{4} + \frac{3(-\ln Q)^{-\frac{5}{2}}}{8} \quad \text{for } Q \ll 1, \tag{12}$$

which is approximated up to second order. The last two terms in Eq.(12) are quite small and about 1% of the first two terms.

When $0 < Q \leq 1$, Eq.(11) was well approximated by You (2009) as

$$\frac{X_Q}{X_{rms}} \approx \sqrt{-\ln Q} + \frac{1}{\sqrt{-\ln Q} + \sqrt{-\ln Q + \pi/4}}, \tag{13}$$

where the inequality of *erfc*(x) was applied, see Eq.(7.1.13) of Abramowitz and Stegun (1970, pp.298). The use of Eq.(13) will lead to the maximum error of about 3% at Q=0.9 and nil error at Q=1. Eq.(13) may be preferred over Eq.(11) for the calculation of U_Q and P_Q in engineering applications because of its simplicity and required engineering accuracy.

Similarly, X_Q can also be derived from Eq.(5) of the modified Rayleigh distribution as

$$\frac{X_Q}{X_{rms}} = C \times \left[\sqrt{-\ln Q} + \frac{\sqrt{\pi}}{2Q} \operatorname{erfc}\left(\sqrt{-\ln Q}\right)\right], \tag{14}$$

and from Eq.(6) of the Weibull distribution as

$$\frac{X_Q}{X_{rms}} = (-\ln Q)^{1/\alpha} + \frac{1}{\alpha Q} \Gamma(1/\alpha, -\ln Q), \tag{15}$$

where $\Gamma(a, x)$ is the upper incomplete gamma function. Eqs.(14) and (15) all reduce to Eq.(11) when α=2 and C=1.

When $Q \rightarrow 0$, $\Gamma(a, x)$ has an asymptotic solution and Eq.(15) may be simplified as

$$\frac{X_Q}{X_{rms}} = (-\ln Q)^{\frac{1}{\alpha}} + \frac{(-\ln Q)^{\frac{1}{\alpha}-1}}{\alpha} + \frac{(-\ln Q)^{\frac{1}{\alpha}-2}}{\alpha}\left(\frac{1}{\alpha}-1\right) + \frac{(-\ln Q)^{\frac{1}{\alpha}-3}}{\alpha}\left(\frac{1}{\alpha}-1\right)\left(\frac{1}{\alpha}-2\right) \text{ for } Q<<1, \quad (16)$$

which is also approximated up to second order. The last two terms in Eq.(16) is about 1% of the first two terms and may be neglected. Eq.(16) will reduce to Eq.(11) when α=2.

In Figure 7, the measured values of U_Q/U_{rms} and P_Q/P_{rms} from Mac4001 and Mac5001 are compared with those computed from the Rayleigh and Weibull distribution functions. It can be seen that the Rayleigh distribution overestimates the amplitudes of wave pressure and orbital velocity, while the Weibull distribution agrees well with the data. The modified Rayleigh distribution is found to be nearly as accurate as the Weibull distribution for the calculation of U_Q or P_Q. For engineering applications, however, the modified Rayleigh distribution may be preferred to the Weibull distribution for the calculation of U_Q or P_Q in terms of its simplicity and required engineering accuracy.

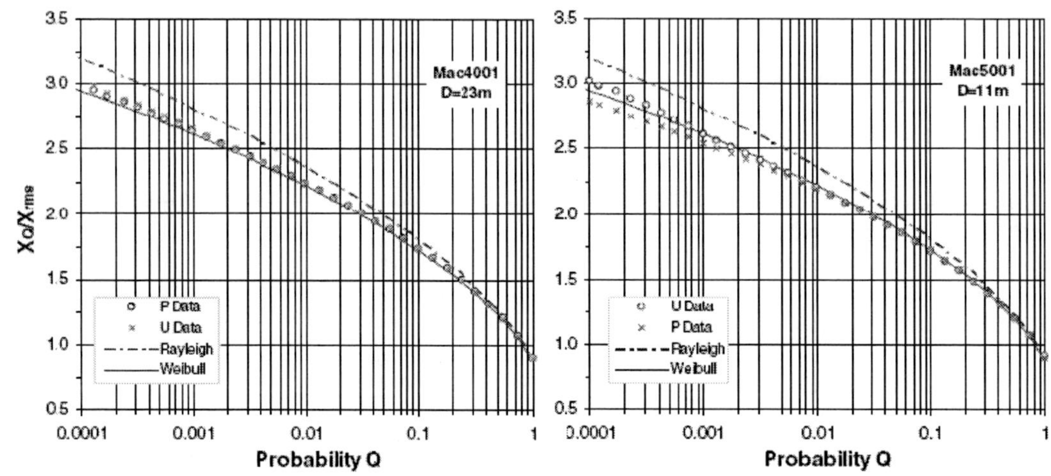

Figure 7. The measured values of U_Q/U_{rms} and P_Q/P_{rms} from Mac4001 and Mac5001 are compared with those calculated from the Rayleigh (dashed curve) and the Weibull distribution (solid curve).

4.2. Mean of Smallest Amplitudes

For some engineering applications, we may be also interested in characteristic orbital velocity $U_{(1-Q)}$ from a short wave record. For example, in studying the stability of an offshore spoil dumping ground, we often want to know the percentage of time during which the spoil dumping ground is stable or the nearbed wave orbital velocity $U_{(1-Q)}$ is always less than or equal to the threshold velocity of sediment motion on the spoil ground. $U_{(1-Q)}$ is averaged from k smallest orbital velocity amplitudes from the total number of m wave orbital velocity amplitudes and $(1-Q)=k/m$ is the cumulative probability. The orbital velocity or the pressure amplitude, $X_{(1-Q)}$, can be derived from Eq.(5) of the modified Rayleigh distribution as

$$\frac{X_{(1-Q)}}{X_{rms}} = \frac{\int_0^{\hat{X}} X f(X) dX}{X_{rms}(1-Q)} = C\left[-\sqrt{-\ln Q}\left(\frac{Q}{1-Q}\right) + \frac{\sqrt{\pi}}{2(1-Q)}\left[1-\mathrm{erfc}\left(\sqrt{-\ln Q}\right)\right]\right], \quad (17)$$

and from Eq.(6) of the Weibull distribution as

$$\frac{X_{(1-Q)}}{X_{rms}} = -(-\ln Q)^{1/\alpha}\left(\frac{Q}{1-Q}\right) + \frac{1}{\alpha(1-Q)}\left[\Gamma(1/\alpha) - \Gamma(1/\alpha, -\ln Q)\right], \quad (18)$$

where X can be replaced by U or P. Both Eqs.(17) and (18) reduce to one derived from the Rayleigh distribution when C=1 and α=2.

The values of $U_{(1\text{-}Q)}$ and $P_{(1\text{-}Q)}$ measured in Mac1001 and Mac2001 are compared with those calculated from Eqs.(17) and (18) of the modified Rayleigh and Weibull distributions. The Weibull distribution of Eq.(18) is shown to agree slightly better with the data than the modified Rayleigh distribution for the calculation of $U_{(1\text{-}Q)}$ and $P_{(1\text{-}Q)}$.

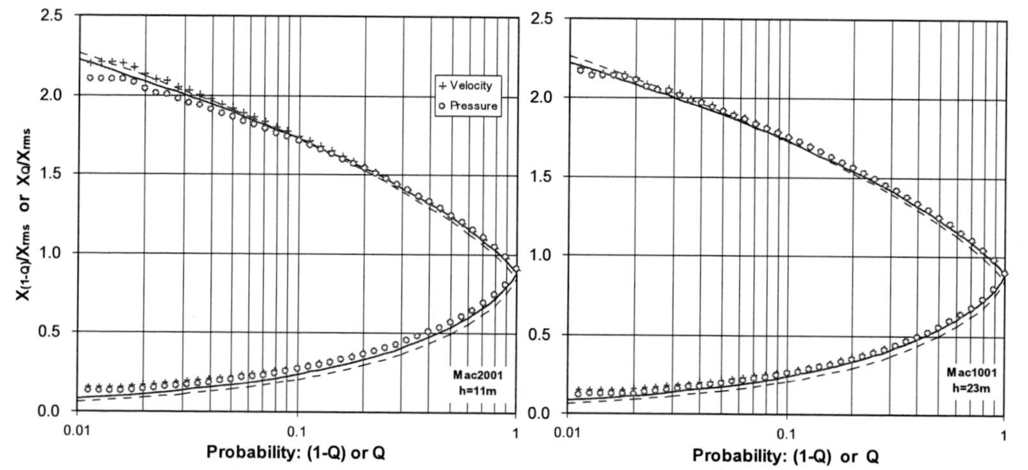

Figure 8. The measured values of $U_{(1\text{-}Q)}$, U_Q, $P_{(1\text{-}Q)}$ and P_Q from Mac1001 and Mac2001 are compared with those calculated from the modified Rayleigh (dashed line) and the Weibull distribution (solid line).

4.3. Expected Value of Maximum Amplitudes

Another important engineering application for this study is to estimate the expected value of maximum orbital velocity U_{max}, for example, in a velocity segment of finite size N. The maximum orbital velocities in different velocity segments with the same size N are expected to be different, but may follow a certain statistical distribution. The cumulative probability that all N orbital velocity amplitudes in a segment of finite size N shall be less than U_{max} may be estimated from the Nth order statistics as

$$P = (1-\Psi)^N = \left\{1-\exp\left[-\left(\frac{X_{\max}}{X_{rms}}\right)^2\right]\right\}^N, \tag{19}$$

where $X_{\max}$ can be replaced by $U_{\max}$ or $P_{\max}$, and ψ is the probability for any one of the N velocity or pressure amplitudes being less than $X_{\max}$ and may be computed from the Rayleigh distribution. The pdf of $X_{\max}$ derived from Eq.(19) is asymmetric and becomes narrowly distributed with increasing N (You, 2009). When N is quite large (e.g. N>20), Eq.(19) may be approximated as

$$P(x) = \exp\left[-\exp\left(x_0^2 - x^2\right)\right] \quad \text{or} \quad f(x) = 2x\exp\left[\left(x_0^2 - x^2\right) - \exp\left(x_0^2 - x^2\right)\right], \tag{20}$$

where $x = X_{\max}/X_{\text{rms}}$ and $x_0 = \sqrt{\ln N}$ from the Rayleigh distribution. For $N \to \infty$, the identity, $(1 + x/N)^N = \exp(x)$ is also used in the derivation of Eq.(20). Eq.(19) can be well approximated by Eq.(20) when $N \geq 20$ (You, 2009). Now, the expected value of $X_{\max}$, denoted by $E(X_{\max})$, can be estimated from Eq.(20) as

$$\frac{E(X_{\max})}{X_{rms}} = \int_0^\infty x f(x)\,dx = -\int_\infty^0 \left(x_0^2 - \ln\alpha\right)^{0.5} \exp(-\alpha)\,\mathrm{d}\alpha$$

$$\approx \int_0^\infty \left(x_0 - \frac{\ln\alpha}{2x_0} - \frac{(\ln\alpha)^2}{8x_0^3}\right) \exp(-\alpha)\,\mathrm{d}\alpha = x_0 + \frac{\gamma}{2x_0} - \frac{\left(\gamma^2 + \pi^2/6\right)}{8x_0^3} \tag{21}$$

where $x_0 = \sqrt{\ln N}$ and γ is Euler's constant (=0.5772). In the deviation of Eq.(21), the variable x is replaced by a new variable, $\alpha = \exp(x_0^2 - x^2)$, and the upper and lower limits (∞, 0) of x are then changed to (0, N) of α. Since the integrand, $x\exp(-\alpha)$, decays rapidly to zero when α>6~10, the lower and upper limits (0, N) of α may then be written as (0, ∞) in Eq.(21) provided $N \geq 20$. The term $(x_0^2 - \ln\alpha)^{0.5}$ is expanded as a binomial series of up to second order and the following identities are also used

$$\gamma = -\int_0^\infty \ln\alpha \exp(-\alpha)\,\mathrm{d}\alpha = 0.5772 \quad \text{and} \quad \int_0^\infty (\ln\alpha)^2 \exp(-\alpha)\,\mathrm{d}\alpha = \frac{\pi^2}{6} + \gamma^2. \tag{22}$$

It should be noted here that Longuet-Higgins (1952) also derived a formula similar to Eq.(21) for calculation of the expected value of maximum wave height in a sample of finite size N, but Eq.(21) has been derived differently from Eq.(59) of Longuet-Higgins (1952).

Similarly, $E(X_{\max})$ may be derived from the modified Rayleigh distribution of Eq.(5) as

$$\frac{E(X_{\max})}{X_{rms}} = C\left[x_0 + \frac{\gamma}{2x_0} - \frac{\left(\gamma^2 + \pi^2/6\right)}{8x_0^3}\right], \tag{23}$$

where $x_0 = \sqrt{\ln N}$, and from the Weibull distribution of Eq.(6) as

$$\frac{E(X_{\max})}{X_{rms}} = x_0 + \frac{\gamma}{\theta\, x_0^{\theta-1}} - \frac{(\gamma^2 + \pi^2/6)(\theta-1)}{2\theta^2\, x_0^{2\theta-1}}, \tag{24}$$

where $x_0 = (\ln N)^{1/\alpha}$ from the Weibull distribution of Eq.(6). Both Eqs.(23) and (24) reduce to Eq.(21) when C=1 and θ=2. The values of E($X_{\max}$) computed from the modified Rayleigh distribution of Eq.(23) are found to be quite close to those from the Weibull distribution of Eq.(24), but smaller than those from the Rayleigh distribution of Eq.(21). Again, the Rayleigh distribution also overestimates the expected values of E($U_{\max}$) and E($P_{\max}$).

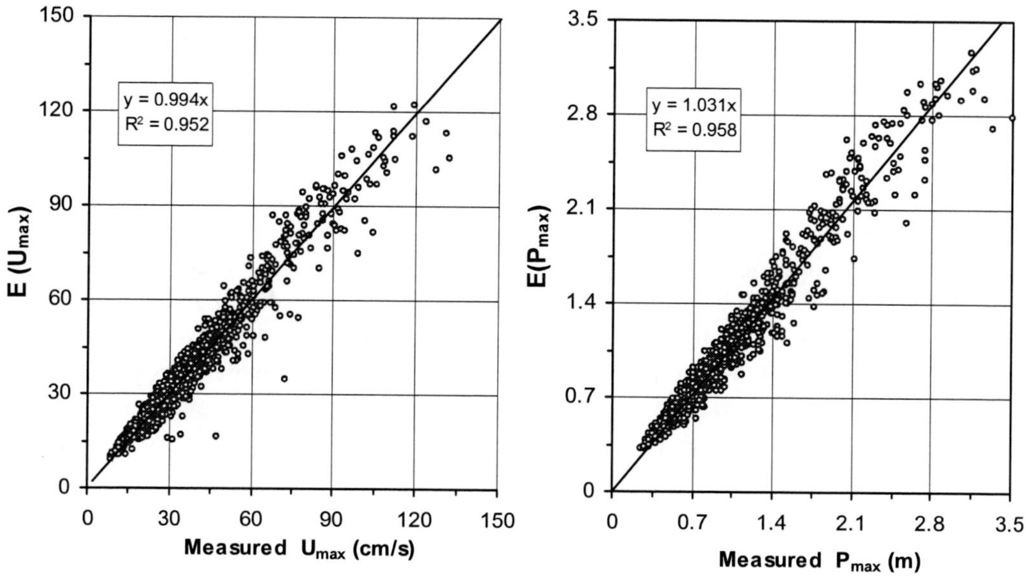

Figure 9. The maximum values of $U_{\max}$ and $P_{\max}$ analysed from individual velocity and pressure segments of Mac7001 are compared with those calculated from Eq.(21).

The values of $U_{\max}$ and $P_{\max}$ measured in Mac7001 are compared with those calculated from the Weibull distribution of Eq.(24) in Figure 9. There are about 1400 data points ($U_{\max}$, N) used for this comparison. Each velocity record of 17.07min only generates one data point. It can be seen that the expected values of $U_{\max}$ and $P_{\max}$ calculated from the Weibull distribution of Eq.(24) agree well with the data.

4.4. Mean of Longest Wave Periods

Another important objective of this study is to determine the characteristic wave period T_Q from the comprehensive wave period data collected. T_Q is defined with the same definition as U_Q. When the wave period T follows the distribution of Bretschneider (1959), T_Q can be then computed as

$$\frac{T_Q}{T_{rms}} = \frac{\int_{\hat{T}}^{\infty} T\, f(T)\,dT}{T_{rms}\, Q} = c\left[(-\ln Q)^{1/4} + \frac{1}{4Q}\Gamma\left(\tfrac{1}{4}, \ln\tfrac{1}{Q}\right)\right], \tag{25}$$

where $c = C_1 \times C_2 = 1.056$, C_1 is the ratio of the mean wave period $\overline{T}$ to the rms wave period T_{rms} and $C_2 = (4.8/\pi)^{1/4}$. The value of C_1=0.95 is determined in this study. Similarly, T_Q can be also derived from the distribution of You (2008) as

$$\frac{T_Q}{T_{rms}} = C_1 + \frac{C_0 \exp[-y]}{\sqrt{\pi}\, Q}, \tag{26}$$

where $y = \Phi^{-1}\left[(1-Q);0,\frac{1}{\sqrt{2}})\right]$ is the inverse normal distribution with the variance of $\sigma^2 = 1/2$ and the mean of $\mu = 0$, $C_1 = 0.95$ and C_0 is constant for a given water depth. Based on the field data, C_0=0.40 is determined for the inner site of h=11m and C_0=0.38 for the outer site of h=23m. The constant C is shown to slowly decrease with increasing h. Eq.(26) with one degree of freedom is expected to agree better with the data than Eq.(20) with no degree of freedom.

In Figure 10, the measured values of T_Q from Mac1001 and Mac2001 are compared with those calculated from the distributions of Bretschneider (1959) and You (2008). It can be seen that the distribution of You (2008) agrees better with the data than that of Bretschneider (1959). Based on Figures 7 and 10, the value of T_Q is shown to be much smaller than that of U_Q for a given value of Q. This indicates that the wave periods are more narrowly distributed than those of wave pressure and orbital velocity and the spread of wave period may lies mainly in the range of T=(0.5 to 1.5)$\overline{T}$. This is termed as the narrow-banded condition under which all the wave periods are in a narrow period band about $\overline{T}$.

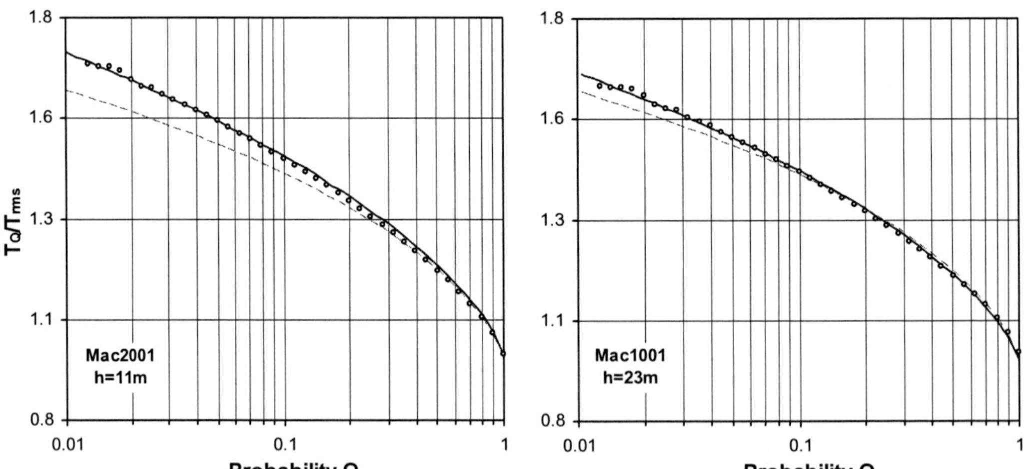

Figure 10. The measured values of T_Q from Mac1001 and Mac2001 are compared with those calculated from the distribution functions of Bretschneider (1959) and You (2008).

5. Conclusion

The statistical distribution of wave pressure, orbital velocity and wave period in the vicinity of the seabed has been quantitatively determined from the comprehensive wave pressure and velocity data collected simultaneously. The extensive and comprehensive field data on (U, P, T) were analysed from the field data with the zero-crossing method. The distributions of instantaneous wave pressure $\widetilde{p}(t)$ and orbital velocity $\widetilde{U}(t)$ are found to be identical and perfectly obey the Gaussian distribution as commonly assumed. The distributions of wave pressure and orbital velocity amplitudes are also found to be almost identical and less accurately described by the Rayleigh distribution than the modified Rayleigh and the one-parameter Weibull distribution. The distribution of wave period follows the normal distribution. The mean wave period of small waves with $P<P_{rms}$ is shown to increases with P, but the mean wave period of large waves with $P>P_{rms}$ is almost independent of P. Several useful formulas are also derived to compute the values of U_Q, $U_{(1-Q)}$, P_Q, $P_{(1-Q)}$ and T_Q, and the expected value of $E(U_{max})$ and $E(P_{max})$. These formulas can be also used to compute characteristic wave heights such as $H_{1/3}$ and $E(H_{max})$ based on linear wave theory.

References

Abramowitzm, M. & Stegun, I. A. (1970). *Handbook of Mathematical Functions*, Dover Publications, Inc., New York.

Bitner, E. M. (1980). Nonlinear effects of the statistical model of shallow water wind waves. *Applied Ocean Research*, 2, 63-73.

Bretschneider, C. L. (1959). *Wave variability and wave spectra for wind-generated gravity waves*. Tech. Memo. No.118. Beach Erosion Board, US Army Corps of Engineer, Washington, DC.

Bretschneider, C. L (1969). Wave Forecasting, Charper 11 in *Handbook of Ocean and Underwater Engineering*, MacGraw-Hill Book Co., New York.

Cartwright, D. E. & Longuet-Higgins, M. S. (1956). The statistical distribution of the maxima of a random function. *Proc R Soc London, Series A,* 247, 22-48.

Donelan, M. A. & Pierson, W. J. P. (1987). Radar scattering and equilibrium granges in wind-generated waves with application of scatterometry. *Journal Geophysical Research*, 92, 4971-5029.

Evans, P., Hanslow, D., Coutts-Smith, A. & You, Z. J. (2000). Nearshore-Inner shelf sediment exchange on the NSW central coast. *Proc. 27th Int. Conf. on Coastal Eng.*, ASCE, 3151-3164.

EM (2008). *Coastal Engineering Manual-Part II: Water Waves Mechanics* (Chapter-I). US Army Corps of Engineers.

Guan, C. L. & Sun, F. (1997). Statistical distribution of surface elevation for the fourth order nonlinear random sea waves. *Science in China* (Series D). 40, 605-612.

Forristall, G. Z. (1978). On the statistical distribution of wave heights in a storm. *Journal of Geophysical Research*, 83, 2353-2358.

Hou, Y. J. & Guo, P. F. (2006). Statistical distribution of nonlinear random wave height. *Science in China* (Series D), 49, 443-448.

Kac, M. & Siegert, A. J. F. (1947). On the theory of noise in radio receivers with square ;aw detectors. *J. Appl. Phys.*, 8, 383-397.

Longuet-Higgins, M. S. (1952). On the statistical distribution of the heights of sea waves. *Journal of Marine Research*, 11, 245-266.

Longuet-Higgins, M. S. (1962). The distribution of intervals between zeros of random function. *Phil. Trans. Roy. Soc. London*, Series A, 254, 557-599.

Longuet-Higgins, M. S. (1963). The effect of non-linearities on statistical distributions in the theory of sea waves. *J. Fluid Mech.*, 17, 459-490.

Longuet-Higgins, M. S. (1975). On the joint distribution of the periods and amplitudes of sea waves. *J. Geophys. Res.*, 80, 2688-2694.

Longuet-Higgins, M. S. (1980). On the distribution of the heights of sea waves: Some effects of nonlinearity and finite band width. *J Geophys. Res.*, 85, 1519-1523.

Longuet-Higgins, M. S. (1983). On the joint distribution of wave periods and amplitudes in a random wave field. *Proc. Roy. Soc. London*, Ser. A, 389, 241-258.

Massel, S. R. & Sobey, R. J. (2000). Distribution of the highest wave in a record. *Coastal Engineering Journal*, 42, 153-173.

Ochi, M. K. (1998). *Ocean waves: The Stochastic Approach. Ocean Technology Series* 6, Cambridge University Press, 319.

Ochi, M. K. & Ahn, K. (1994). Non-Gaussian probability distribution of coastal waves. In *Proc. 24th Coastal Engng. Conf.*, Venice, 512-525.

Ochi, M. K. & Wang, W. C. (1984). Non-Gaussian characteristics of coastal waves. In *Proc. 19th Conf. Coastal Engineering Conference*, Houston, 516-531.

Peregrine, D. H. (1972). Equations for water waves and the approximation behind them. In *Proc. 17th Coastal Engr. Conf.*, 784-800.

Rudnick, P. (1951). Correlograms for Pacific ocean waves. In *Proc. 2nd Berkeley Symp. On Math. Stat. and Probability*, 627-38.

Sultan, N. J. (1992). Irregular wave-induced velocity in shallow water. US Army Corps of Engineering, Washington DC, *Technical Report CERC-92-9*.

Sultan, N. J. & Hughes, S. A. (1993). Irregular wave-induced velocities in shallow water. *Journal Waterway, Port, Coast and Ocean Engineering, ASCE* 119, 429-447.

Song, J. B. & Wu, Y. H. (2000). Statistical distribution of water particle velocity below the surface layer for finite water depth. *Coastal Engineering*, 40, 1-19.

Thompson, E. F. (1977). *Wave climate at selected locations along US Coasts*. Technical Report TR77-1, US Army Engineer Waterways Experiment Station, Vicksburg, MS.

Tayfun, M. A. (1980). Narrow-band nonlinear sea waves. *J. Geophys. Res.*, 85, 1548-1552.

Vinje, T. (1989). The statistical distribution of wave heights in a random seaway. *Applied Ocean Research*, 11, 143-152.

Wiberg, P. L. & Sherwood, C. R. (2008). Calculating wave-generated bottom orbital velocities from surface-wave parameters. *Computers and Geosciences*, 34, 1167-1416.

You, Z. J. (1994). A simple model for current velocity profiles in combined wave-current flows. *Coastal Eng*, 23, 289-304.

You, Z. J. (1998). Initial motion of sediment in oscillatory flow. *Journal of Waterway, Port, Coastal and Ocean Engineering*, ASCE, 124, 68-72.

You, Z. J. & Yin, B. S. (2001). Determination of coastal wave direction in shallow water. 15th *Australasian Coastal and Ocean Engineering Conference*, Gold Coast, 523-527.

You, Z. J. (2005). Estimation of bed roughness from mean velocities measured at two levels near the seabed. *Continental Shelf Research*, 25, 1043-1051.

You, Z. J. & Yin, B. S. (2006). A unified criterion for initiation of sediment motion and inception of sheet flow under waves. *Sedimentology*, 53, 1181-1190.

You, Z. J. & Yin, B. S. (2007). Direct measurement of bed shear stress under water waves. *Journal of Coastal Research*, Special Issue, 50, 1132-1136.

You, Z. J. (2008). A close approximation of wave dispersion relation for direct calculation of wavelength in any coastal water depth. *Applied Ocean Research*, 30, 113-119.

You, Z. J. (2009). Statistical distribution of nearbed wave orbital velocity in intermediate coastal water depth. *Coastal Engineering*, 56, 844-852.

In: Sea Level Rise, Coastal Engineering, Shorelines and Tides ISBN: 978-1-61728-655-1
Editor: Linda L. Wright, pp. 105-122 © 2011 Nova Science Publishers, Inc.

Chapter 4

SEA LEVEL VARIATIONS ALONG THE ESTONIAN COAST OF THE BALTIC SEA

Ülo Suursaar*
Estonian Marine Institute, University of Tartu, Estonia

Abstract

The aim of this chapter is to present an overview and statistical analysis of the sea level data obtained from the main Estonian coastal tide gauges (Pärnu, Tallinn, Ristna, Narva-Jõesuu) over the period 1842–2009. Long-term variations of both mean and extreme sea level values are studied in the eastern section of the semi-enclosed and nearly tideless Baltic Sea. After correcting the sea level series to spatially varying postglacial land uplift rates (0–2.5 mm/yr), the series display increasing (1–3 mm/yr) trends, which are roughly equal to or insignificantly higher than the recent global sea level rise estimates. The increase is larger in winter, which is in accordance with similar seasonal structures of the NAO index trends. However, the remarkably steep rise in annual maximum sea levels (3–13 mm/yr) could be explained by the local response to the changing regional wind climate. Due to its windward location, the sea level variations in the semi-enclosed study area are sensitive to the ongoing intensification of cyclones and prevailing west winds. Maximum value analysis revealed that in the south-westerly exposed Pärnu Bay, two storm surge events (253 in 1967 and 275 cm in 2005) can not be predicted by means of return statistics. The parameters of maximum expected storm surges could be estimated on the basis of hydrodynamic modelling.

1. Introduction

Estonia lies in the northern hemisphere zone of moderate latitudes (58º–60º N) in the eastern section of the semi-enclosed Baltic Sea (Figure 1). The coastal sea of Estonia is heavily indented and shallow. It includes two semi-enclosed sub-basins, the Gulf of Riga and the Väinameri Sea, which are connected to each other and to the Baltic Proper through four major straits (Figure 1). With amplitudes of M_2 and K_1 waves measuring less than 5 cm, the study area is nearly tideless [e.g., Magaard & Krauss, 1966; BACC, 2008]. Given these

* E-mail address: ulo.suursaar@ut.ee

characteristics, both currents and sea level variations are predominantly meteorologically forced in the study area. The frequent occurrence of cyclones passing from west to east, and their corresponding changes in wind direction and speed, create considerable sea level fluctuations there. As an exposed and windward coast, the area is also sensitive to changes in wind conditions and storminess. While more than 95% of the sea level data fall within the range of –50 and +60 cm in relation to the Kronstadt zero benchmark (which is nearly equal to the long-term mean sea level for the Estonian coast), there are a few exceptional (up to 275 cm) storm surges on record.

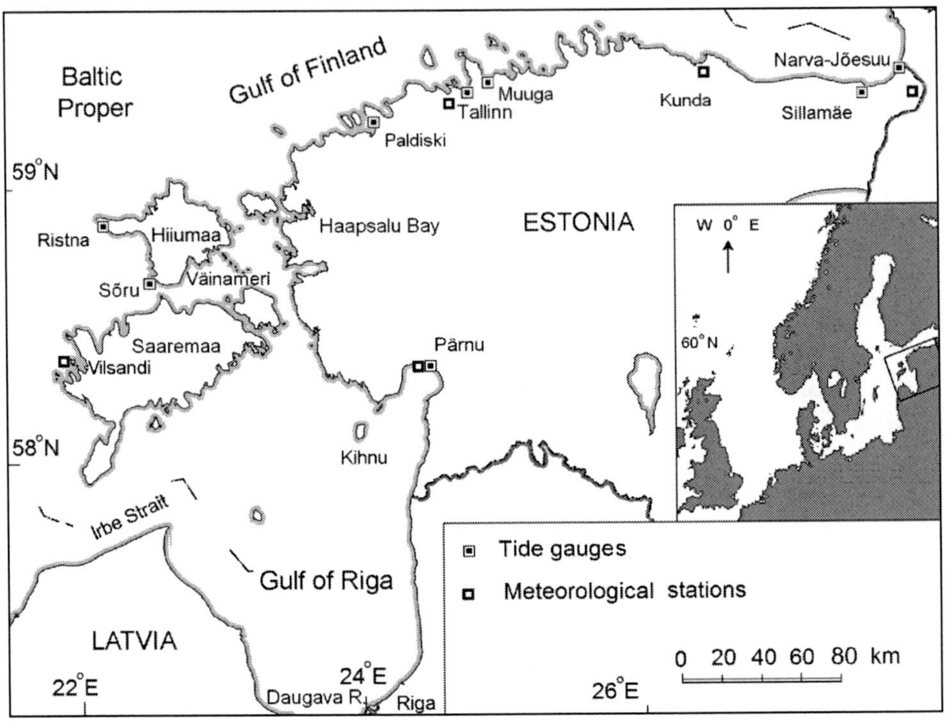

Figure 1. Map of the study area together with locations of selected meteorological and tide gauge stations.

Table 1. Monthly and annual time series of mean and extreme sea level available for the investigated tide gauges

Time series	Years covered (but may have missing months)
Tallinn (mean)	1842-1882, 1886-1917, 1923-40, 1945, 1947-96
Tallinn (extremes)	1899-1917, 1923-40, 1945, 1947-96
Narva-Jõesuu	1899-1915, 1923-43, 1945-2009
Pärnu	1923-2009
Ristna	1950-2009

Measurements of relative sea level variations by means of coastal tide gauges have more than two centuries of history in the Baltic Sea. The longest continuous sea level record in Stockholm starts in 1774 [Ekman, 1999]. In St. Petersburg, Russia, the storm surge record starts in 1703, and a more or less continuous sea level time series in 1806 [Lazarenko, 1986; Bogdanov et al., 2000]. In Tallinn, the capital city of Estonia, regular sea level measurements started in 1809. The near-continuous data sets are available from 1842 (Table 1). Estonian historical tide gauge data is certainly a valuable source for investigating various aspects of decadal sea level tendencies, as the regional conditions affecting the sea level (land uplift rate, meteorological and oceanographic conditions, configuration of coastline) are rather different along the west and east coasts of the Baltic Sea, not to speak of the whole World Ocean. On the other hand, as virtually every tide gauge record carries some kind of local imprint, an analysis of possibly a large amount of tide gauge data is needed for studying the patterns and trends of global sea level [Emery & Aubrey, 1991].

The concept of a "global sea level trend" and attempts to identify a "climate signal" in sea level rise play a central role in the current climate change research [Smith et al., 2000]. "Sea level" is understood as relative sea level throughout this chapter, if not otherwise specified. With regard to mean relative sea level of a location, there are three main factors influencing its variations: global sea level change, regional land uplift or subsidence and changes in the water balance of the particular sea. According to numerous studies [Gornitz et al., 1982; Vermeer et al., 1988; Ekman, 1999; IPCC, 2001; Church & White, 2006] the global sea level rose by 1–2 mm per year during the 20th century.

The second main factor, the radial crustal movements, which are mainly influenced by postglacial rebound (glacial isosatic adjustment) in Estonia, is of regional origin [e.g., Fjeldskaar et al., 2000] and varies locally. In the coastal areas of Estonia it varies between 0.5 and 2.8 mm/yr [Vallner et al., 1988]. Another regional sea level component appears as a result of variations in the water balance of the Baltic Sea as a whole. These variations are mainly caused by water exchange processes through the Danish Straits, and by river runoff. Both are largely controlled by atmospheric pressure patterns over the North Atlantic, the North Atlantic Oscillation (NAO) [Ekman & Stigebrandt, 1990; Lehmann et al., 2002; Johansson et al., 2001; Hagen & Feistel, 2005]. Minor factors for decadal Baltic Sea level changes are variations in salinity, temperature, precipitation and evaporation regime [e.g., Ekman & Stigebrandt, 1990; Meier et al., 2004; BACC, 2008].

In addition to the mean sea level rise and its possible acceleration [e.g., Church & White, 2006], systematic changes in extreme high waters have been reported [Langenberg et al., 1999; Lowe et al., 2001; Suursaar & Kullas, 2009]. Factors related to cyclone frequency, intensity, and their tracks are the most crucial ones for the issue. Due to site-specific features, the West Estonian bays are subject to storm surge hazard. Besides, it is supposed that due to the low-lying land, the gently sloping coasts and indented coastline, the flooding generated both by local storm surges and climatologically induced global sea level rise will increasingly affect the nearly tideless Estonian coastal zone [Kont et al., 2003; Suursaar et al., 2006].

The objectives of this study are to give an overview of Estonian historical tide gauge data, to investigate long-term changes both in mean water level and in extreme sea level events in the Baltic Sea along the Estonian coast and to discuss the possibility of estimating maximum expected future surge heights.

2. The Data

Sea level observations on the territory of Estonia have been carried out at different times at 29 locations [Raudsepp et al., 1999; Jevrejeva et al., 2001]. However, their quality and the extension vary in a great deal. Most of the time series are not long term and have a lot of gaps, especially in 1944-45. The number of tide gauges currently operated by the Estonian Meteorological and Hydrological Institute (EMHI) is 12. In addition several automatic tide gauges have been recently put into operation by ports (e.g. Muuga, Tallinn, Sõru, Sillamäe, Paldiski). Most of the tide gauges by EMHI are equipped with tide poles and have a sampling frequency of 2, 3 or 4 times a day. Such tide gauges have also a lot of gaps in their time series. Automatic tide gauges, currently operated by the EMHI, which provide hourly data, are located only at Pärnu, at Narva-Jõesuu and at Ristna (Figure 1). Measurements at the historically valuable location of the Tallinn Port were unfortunately discontinued and transferred to the Muuga Bay in 1996 due to problems related to local land subsidence and construction work at the port. The merging of the historical EMHI data with the new data by the Port of Tallinn has not yet succeeded.

In the present study, data from four tide gauge stations (including Tallinn), offering the most extensive and reliable data, were considered. We have used the digitized database of the EMHI, which include both monthly mean and extreme sea levels, as well as hourly data from selected tide gauges. While monthly and annual data are good for long-term analyses, the hourly data offers detailed information on total and short-term variability.

The height system used in Estonia and in other Baltic states of the former Soviet Union differs from the Finnish and Scandinavian tide gauging standards [Lisitzin, 1966; Johansson et al., 2001]. It is called the Baltic Height System 1977 with its reference zero-benchmark at Kronstadt near St. Petersburg [Lazarenko, 1986; Bogdanov et al., 2000]. The work on merging the different reference systems around the Baltic Sea is still ongoing. The Kronstadt zero was defined as the average sea level of that tide gauge in 1825-1840. As the result of global sea level rise, the Kronstadt zero is not the present-time mean sea level of this tide gauge. According to the data published by Bogdanov et al. [2000], the mean was 5.9 cm in 1971-1993. Instead, the mean sea level of Estonian tide gauges is roughly around the Kronstadt zero now. There is a quasipermanent sea level slope towards the Danish Straits [e.g. Lazarenko, 1986; BACC, 2008]. The zeroes of Estonian tide gauges are at least annually levelled in relation to local benchmarks. A less frequent levelling between local benchmarks and base frame is performed by national geodetic organizations [Jevrejeva et al., 2001].

Due to non-uniform land uplift in the study area, the use of fixed land level benchmarks poses specific problems for time series analysis as well as comparison of the results from different tide gauges. Data on land uplift (Table 2) for the studied stations were determined by using the map composed of the precise levelling data [Vallner et al., 1988]. In 1977-1985, Vallner together with his colleagues carried out an extensive study of repeated precise levellings, taking into account data from previous levelling campaigns (1933-1943 and 1956-1970). The accuracy of uplift rates was considered ±0.2 mm/yr then. Later, using precise gravity measurements [Sildvee, 1998] and GPS technology, regional geoid models were worked out in cooperation with the Danish Institute of Geodesy [Forsberg et al., 1996] and the vertical motion rates verified. The actual accuracy and constancy of uplift rate estimates is yet unknown.

Tendencies in sea level time series were analysed using linear regression analysis. Atmospheric circulation features were described by the NAO index which is usually described as the difference between the standardized sea level pressure anomalies between Icelandic low and Azores high and can be calculated using either data from Ponta Delgada, Lisbon [Hurrell, 1995] or Gibraltar [Jones et al., 1997]. The data of Gibraltar were used, as the NAO Gibraltar index offers slightly better results for the changes in Estonian climate [Jaagus, 2006].

3. Basic Features of Sea Level Variability

While in the World Ocean the most common source of sea level variability is astronomic tides, their role in the semi-enclosed Baltic Sea are negligible. The Danish Straits are too shallow and narrow for letting considerable flows through before the astronomically forced flow direction reverses. On the other hand, the sea is not large enough for developing its own tides. Thus, the amplitudes of the main tidal wave constituents are probably around 1-2 cm in the Baltic Proper and somewhat more in the region of straits to the North Sea [Magaard & Krauss 1966; Lazarenko, 1986]. Morevover, these are vastly overrun by the meteorologically driven variations, which include influence of wind in reaction with the coastal topography, but also so-called inverse barometer effect, and thermohaline effects being the least important.

Typically for the practically tideless water bodies, the sea level tends to very close to the mean state. According to the empirical histograms composed on the basis of hourly data from 1961-2005, some 90 % of the data falls within the relatively narrow range from -40 to 50 cm even at Pärnu (Figure 2). The distribution function is close to log-normal, with larger skewness at Pärnu and Narva-Jõesuu, and smaller at Ristna and Tallinn. Still, the sea level variability range at Pärnu is rather large (400 cm, Figures 2,3). The variability range in other Estonian tide gauges is somewhat smaller (Figure 3). The average sea level at Narva-Jõesuu was 3.7 cm over the period 1899-2009, -1.6 cm at Tallinn in 1899-1996, -2.2 cm at Ristna in 1950-2009, and 1.5 cm at Pärnu in 1923-2009.

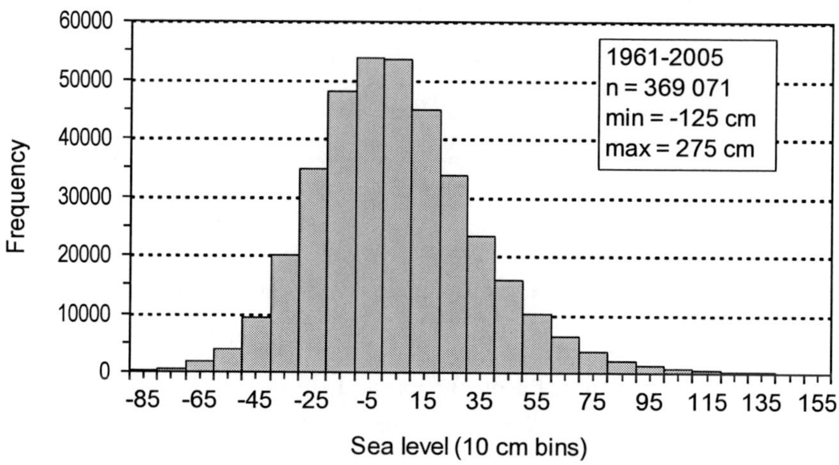

Figure 2. Empirical frequency distribution of Pärnu hourly sea level data.

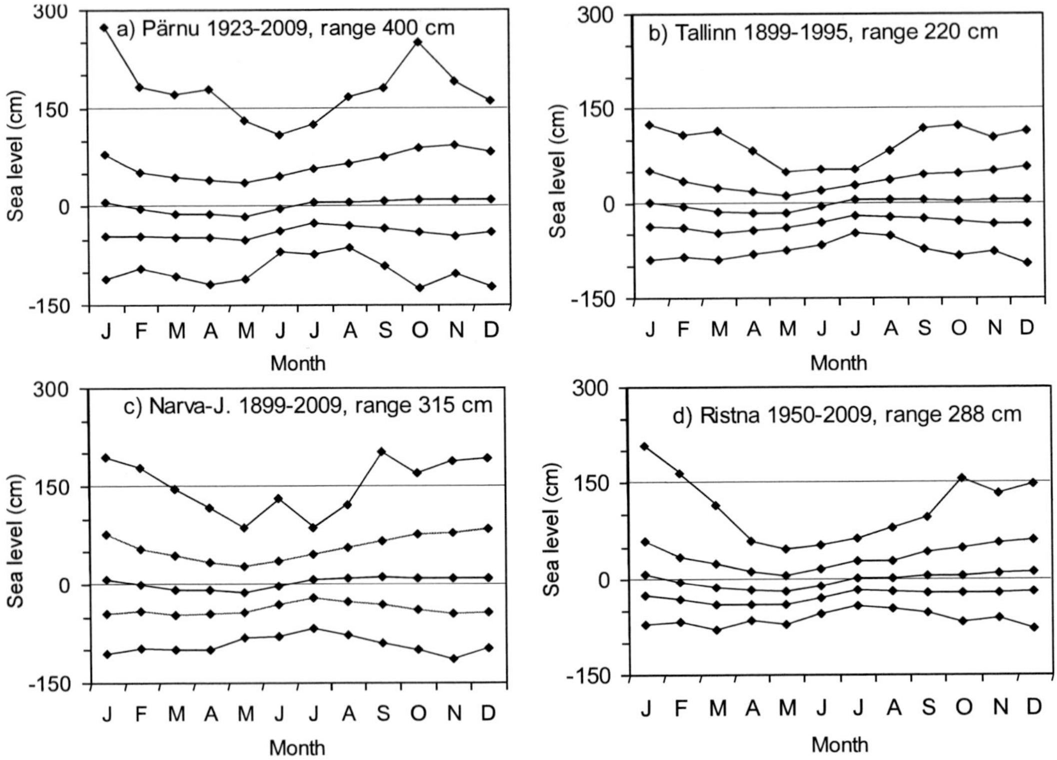

Figure 3. Seasonal variations in monthly sea level statistics at four tide gauges. Lines from top: absolute maximum, average maximum, mean, average minimum, absolute minimum.

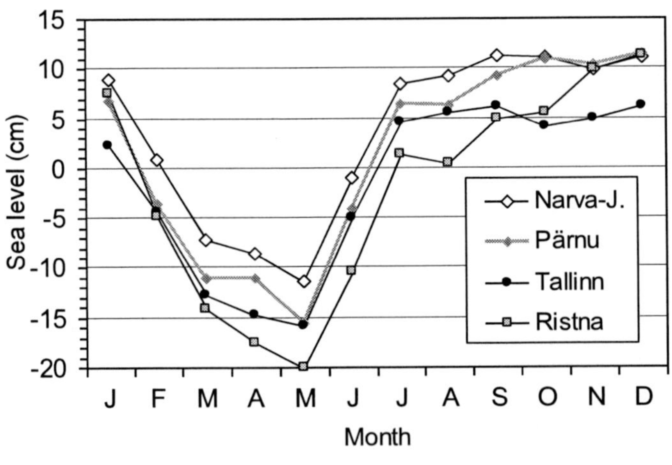

Figure 4. Seasonal variations in mean sea level at four main tide gauges of Estonia.

The main periodic component is seasonality (Figure 4). The annual amplitude (i.e., half of the annual range) was 14 cm at Pärnu, 11 cm at Tallinn, 12 cm at Narva-Jõesuu, and 16 cm at Ristna. The main reason for that is seasonality in meteorological conditions, mainly in wind conditions above the Baltic Sea. Such a pattern with high mean sea level in autumn-winter and low level in late winter or spring is common for the other parts of the Baltic Sea as well. However, in practice, this regular pattern is hardly visible, as the mean sea level may considerably deviate in different years. Also it is masked by irregular wind driven sea level variations, which make the sea level record rather "noisy" by appearance. In addition, the influence of the 27 h Baltic unimodal seiche [Wübber & Krauss, 1979] can be occasionally seen in the data of the Baltic Proper, and the 5 h seiche in the form of decaying oscillations is quite common after storm surges in the Gulf of Riga [Suursaar et al., 2006].

Both extreme high and low sea level events tend to occur in meteorologically more variable winter months (Figure 3). Summer months are typically calm in Estonia. Strong east winds lower the sea level in the Estonian coastal waters, and west winds raise it [Suursaar et al., 2002]. This applies to both short-time variations, as well as for low-frequency variations through corresponding changes in the Baltic seawater volume. The high sea level events are mainly associated with north European cyclonic storms. The centre of a powerful and fast moving cyclone should bypass Estonia to the north over the Scandinavian Peninsula and Bothnian Sea to make the local wind direction to turn from SW to NW. As a rule, the strongest winds occur a few hundred kilometres to the right from the cyclone track, reduced friction above the sea surface and the elongated shape of the Baltic Sea together with the Pärnu Bay (in case of Pärnu tide gauge) or Gulf of Finland (for Narva-Jõesuu and St. Petersburg) provides a span for surge wave to increase towards the east, as also depth diminishes and the gulf width converges [Suursaar et al., 2006]. The span for storm surges is considerably smaller at Ristna and Tallinn.

4. Long-Term Tendencies

As the result of Fennoscandian postglacial uplift, which may reach up to 8-9 mm/yr in some Finnish tide gauges, the sea level records from Finland and Sweden mainly exhibit decreasing trends [Ekman, 1999; Lisitzin, 1964; Johansson et al., 2001; Emery & Aubrey, 1991]. The land uplift is considerably smaller in Estonia. Thus, the time series representing annual mean sea levels show quite different, but still mostly increasing tendencies at Pärnu, Tallinn and Narva-Jõesuu, and yet decreasing trends at Ristna (Figures 5,6; Table 2). Moving averages of annual mean sea level series show 30-40 year cycles (Figure 5) which coincide with the similar cycles e.g. in Lithuanian [Dailidiene et al., 2006] or Finnish [Johansson et al., 2001] sea level data. These cycles, with the amplitude of about 5 cm, can influence linear trend estimates (Suursaar & Sooäär, 2007). For example, the series of Narva mean sea level for 1899-2009 starts "high" and ends "low" (Figure 5) and the linear trend estimate (Table 2) probably underestimates the real tendency. Due to the irregular nature of these cycles and shortness of the series, it is difficult to properly eliminate the cycles' influence, but some alternative (shorter) periods are considered in Table 2, bearing also in mind the longer gaps in the series. Table 2 illustrates how the trend slope estimates depend on chosen terminal points and lengths of time series.

Corrected with local uplift rates, the mean sea level rise rates for Tallinn, Narva and Ristna are roughly equal to or insignificantly higher than the most recent global estimates, about 1.7 mm/yr according to e.g. Church & White [2006] or about 1.5 mm/yr as the central value for 1-2 mm/yr estimate of 20th century sea level rise by IPCC [2001; 2007]. However, even considering possible uncertainties in data quality and land uplift rates, the trend estimates for Pärnu tide gauge seem to be significantly higher (2.4-2.7 mm/yr) than the global one (Table 2, Figure 5).

The excessive Pärnu sea level rise rate (Table 1; Figures 5-7) over the global estimates can be explained by a mechanism, which was studied in a hydrodynamic modelling experiment [Suursaar & Kullas, 2006]. These results indicate that in case of an obvious decadal trend in wind conditions the sea level change rates of a semi-enclosed basin may deviate from the global estimates. A positive trend in wind speed and storminess should result in a steeper than average sea level trend on the windward side and one that is less steep on the leeward side. Indeed, a clear increase in the westerly wind component occurred between 1950 and 1990 [Alexandersson et al., 1998; Keevallik & Rajasalu, 2001; Jaagus et al., 2008]. Although further increase in westerly winds is anticipated, it is also possible that the increase in westerlies and winter NAO in 1950–90 could be followed by a relative increase in easterlies [Lowe et al., 2001; BACC, 2008].

The Pärnu tide gauge is both regionally and locally windward, and therefore sensitive to wind climate changes. The effect is localised at the bays exposed to SW or W [Suursaar et al., 2006]. The crucial role of changes in wind climate could be confirmed by astonishingly large change rates in maximum sea levels, ranging between 3.7 and 13 mm/yr (Table 2; Figure 7), while the rates in minima are only 0.5-3.4 mm/yr. No other meteorological or hydrological forcing could yield such sea level developments [see also Langenberg et al., 1999].

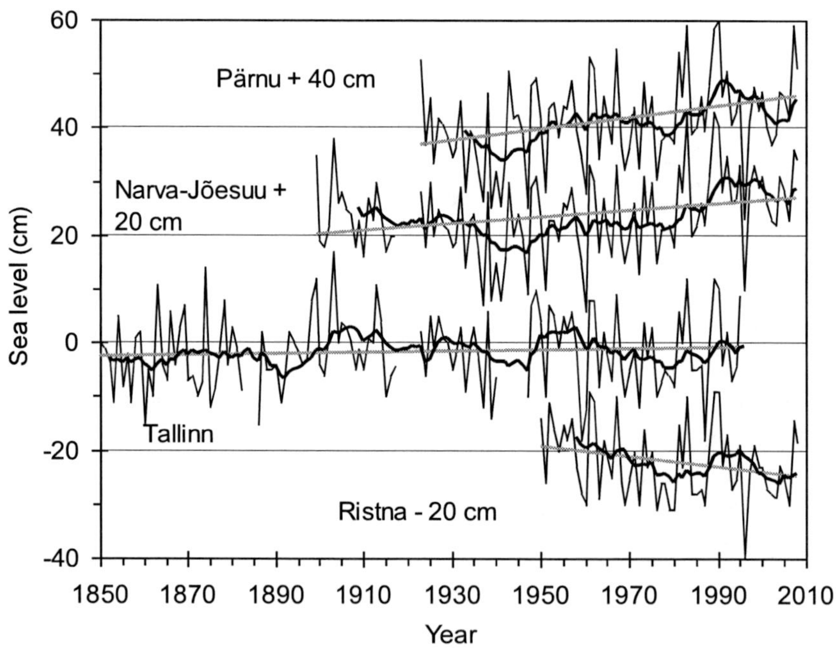

Figure 5. Decadal variations in annual mean relative sea levels together with 11-year moving averages and linear trendlines (data updated until the end of 2009). The series are not corrected with land uplift rates, and thus exhibit the isostatic land rise together with the climatological sea level rise component. The data of different tide gauges are shifted by 20 cm regarding each other for clarity.

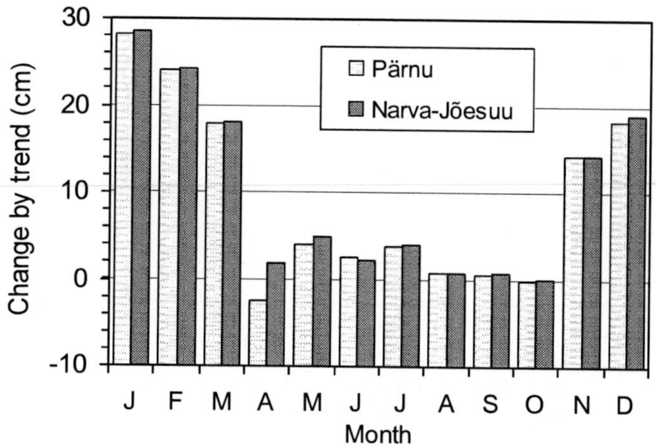

Figure 6. Seasonal structure of mean sea level changes by linear trend at Pärnu (1924-2009) and Narva-Jõesuu (1899-2009). The trend slopes are multiplied with the number of years (86 in Pärnu and 111 in Narva-Jõesuu).

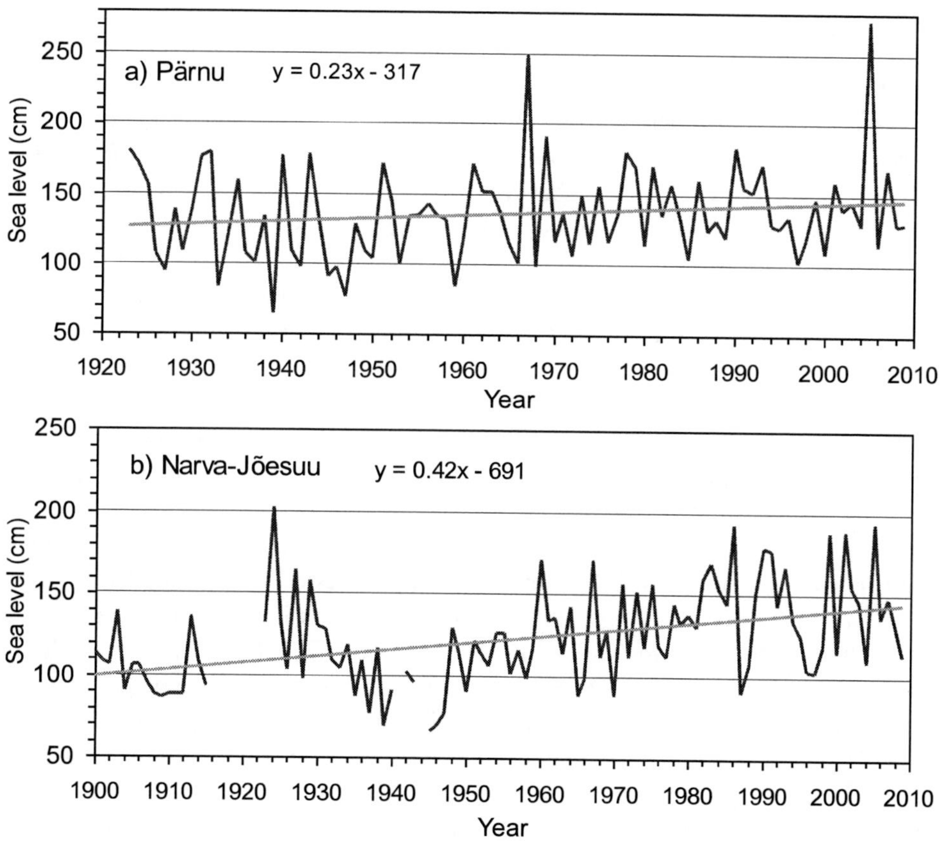

Figure 7. Decadal variations in annual maximum sea levels at Pärnu (a) and Narva-Jõesuu (b) together with linear trendlines.

Table 2. Trends for annual minimum, maximum and mean sea levels in different tide gauges and study periods. Actual sea level (ASL) rise rate is a sum of rates of relative sea level (RSL) and local land uplift (mm/yr)

Tide gauge	Period	Uplift	Min		Max		Mean	
			RSL	ASL	RSL	ASL	RSL	ASL
Pärnu	1924-2009	1.5	1.5	3.0	2.7	4.2	1.2	2.7
	1945-2009	1.5	1.9	3.4	4.4	5.9	0.9	2.4
	1939-1992	1.5	-0.1	1.4	6.1	7.6	2.1	3.6
Ristna	1950-2009	2.6	0.0	2.6	8.4	11.0	-1.0	1.6
	1950-1992	2.6	-0.7	1.9	4.5	7.1	-0.5	2.1
Narva-J.	1899-2009	0.5	0.3	0.8	4.2	4.7	0.5	1.0
	1945-2009	0.5	0.9	1.4	7.2	7.7	1.4	1.9
	1939-1992	0.5	0.0	0.5	12.5	13.0	2.5	3.0
Tallinn	1899-1995	1.8	0.1	1.9	1.9	3.7	-0.3	1.5
	1947-1995	1.8	-1.0	0.8	3.9	5.7	-0.3	1.5
	1945-1992	1.8	-0.7	1.1	4.5	6.3	-0.5	1.3

Positive sea level trends in annual time series appear due to the more positive trends in winter (December to March) sea level, since during the summer such trends are less steep or even negative (Figure 6). The significantly higher mean sea level rise in winter correlates with increased local storminess during the same months and with the greater intensity of westerlies in winter, as described by the NAO-index (Figure 8) [Suursaar & Sooäär 2007; Jaagus et al., 2008]. The existence of this type of time variable relationship between sea level and atmospheric circulation is common for other Baltic Sea level data sets as well [Andersson, 2002; Hagen & Feistel, 2005; Dailidiene et al., 2006]. The NAO index is not a single forcing for sea level, it rather manifests through a complex of forcing factors such as stronger westerly airflow, higher winter temperature, higher precipitation rate, lower ice extent, etc. The main hydrodynamic mechanism for the Baltic Sea level applies through the more intensive water inflow through the Danish Straits as a result of sea level differences in the transition area between the North and Baltic Seas, which in turn appears due to stronger westerlies and more frequent storms [Ekman & Stigebrandt, 1999; Meier et al., 2004; BACC, 2008].

Water exchange processes both in the semi-enclosed Gulf of Riga and in the Danish Straits are sensitive to similar wind conditions. Intense westerly and windstorms cause mean volume surplus within the Baltic Sea and sea level inclination in the fjord-like Baltic Sea. The effect is duplicated on a smaller scale within the Gulf of Riga. Model runs also showed that every change in long-term wind regime (change in average wind speed, variability, or directional distribution) has an effect on the established sea level equilibrium [Suursaar & Kullas, 2006]. The reaction depends on coastline configuration and bottom topography, and it is strongest in the far ends of windward bays, like Pärnu Bay. Increase in the western wind component by a magnitude of 2-3 m/s reported for the last 50 years [Alexandersson et al., 1998; Keevallik & Rajasalu, 2001] may account for up to a 5 cm rise in annual mean sea level at Pärnu [Suursaar & Kullas, 2006].

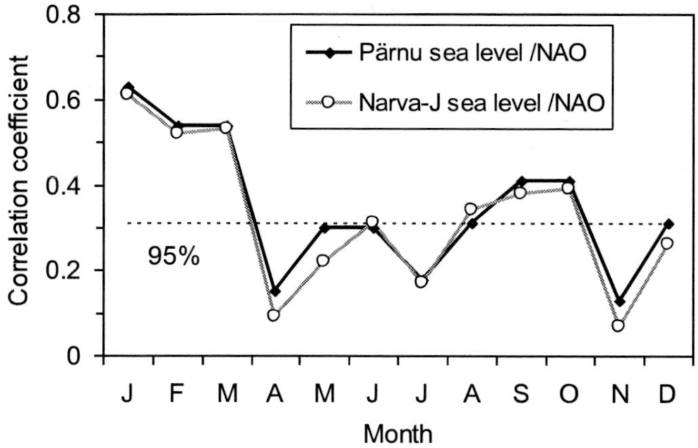

Figure 8. Seasonal variations in correlations of sea level at Pärnu and Narva with NAO index, shown with the 95% significance level.

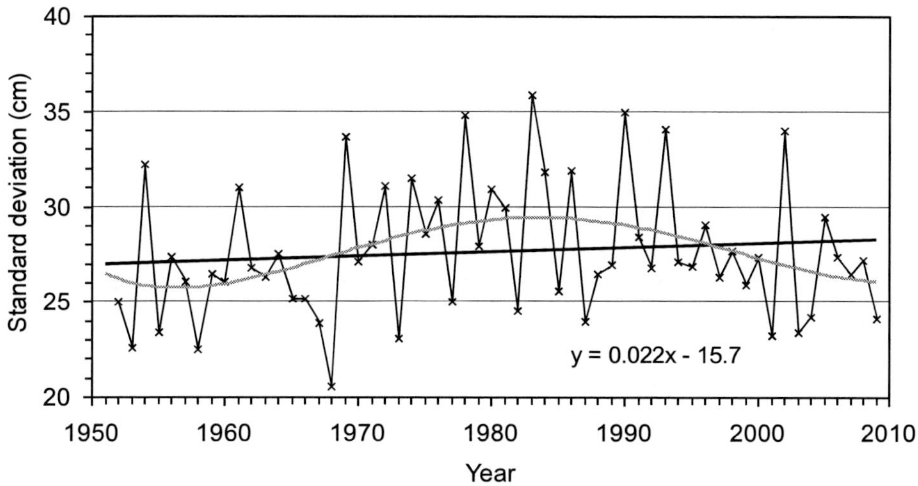

Figure 9. Decadal variations in standard deviations of annual samples of hourly sea level data at Pärnu.

Significant seasonal structure of the trends should lead to enhancement of seasonal signal in sea level records. The tendency has already been reported for various parts of the Baltic Sea [Ekman & Stigebrandt, 1990; Andersson, 2002; Suursaar & Sooäär, 2007] and it has increased over the last decades. As a result of enhanced seasonal signal, as well as increased annual maxima, overall variability has also increased. The sea level variability expressed as annual standard deviations, calculated on the basis of hourly data, show a statistically significant rise by 1.5 cm in 1952-2009 (Figure 9). The average standard deviation was 27.6 with variability range 20.6-35.9 cm at Pärnu.

5. Extreme Events

Because of the semi-enclosed configuration of the study area and the presence of shallow bays exposed to the direction of the strongest possible winds, some of the highest Baltic storm surges are registered in the coastal waters of West Estonia. However, considering the shallow sea, low-lying hypsometric curve, and virtual absence of tides, the local residents are not always sufficiently prepared for such, nevertheless rare, events. Trend analyses have shown that storm surges are becoming higher both in Estonia [Suursaar et al., 2006], as well as in Western Europe [Lowe et al., 2001].

Sea level rise of 1.5 m is usually critical in the coastal waters of Estonia, and 2 m is catastrophic. Low levels, on the other hand, disturb the operation of ports and can cancel ferry connections between the Estonian mainland and the West Estonian Archipelago. The lowest ever sea level, -125 cm, was recorded at Pärnu on 9 December 1959 after a month with strong (20 m/s) east winds during an anticyclonic blockage pattern [Suursaar et al., 2002]. The general course of the event was smooth; it took a long time to lower the background sea level of the whole Baltic Sea as a result of outflow through the Danish Straits.

The highest storm surges in the Baltic Sea, on the other hand, are relatively rapid and connected to the cyclonic storms. Most notably, since the founding of St.Petersburg as the capital of Russia in 1703 there have been about 300 floodings due to the sea level rise above the critical value of 160 cm [Bogdanov et al., 2000]. The worst one occurred in 1824 (+423 cm), killing 569 people, the second highest was in 1924 (+380 cm). In 1970 it was decided to build a 25-km dam with huge closable gates for protecting the city. Construction started in 1978, but it is not finished yet.

Over the period 1923-2009, the Pärnu sea level record identifies 30 individual events higher than the critical value of 150 cm, 24 of which occurred between the months of October–March and 6 between the months of April–September. The two highest sea level events off the Estonian coast (since 1923) were both registered at Pärnu: 253 cm on 19 October 1967 and 275 cm on 9 January 2005 (Figure 10). The event on 9 January 2005 was produced by the cyclone Gudrun, which reached the power of a hurricane according to mean wind speed measurements (up to 34 m/s) in Denmark. As Gudrun's eye passed 300 km north of Estonia, heading from SW to NE, it created very strong SW winds (and later W–NW winds once it passed) reaching an average of 28 m/s over one hour period. SW is the wind direction that is most effective in producing a storm surge in Pärnu Bay [Suursaar et al., 2006]. Another factor contributing to Gudrun's storm surge was the relatively high background sea level in the Baltic (70 cm) at the time. The higher background sea level was the result of strong cyclonic activity during the preceding month, which had the effect of pushing additional water through the Danish Straits into the Baltic Sea.

It is important to have an idea of the possible extreme sea levels we can ever expect. There are basically two possibilities for such estimation, either through dynamical or statistical modelling. To find empirical return periods, the observation period was divided by the (ranking) number of events. The plots against the corresponding theoretical Gumbel distributions [Suursaar & Sooäär, 2007] showed a more or less a satisfactory fit in case of three tide gauges, but failure in case of Pärnu maxima (Figure 11). No extreme value distributions could probably describe or predict the two extreme sea level events of 253 cm or 275 cm. There is a common misbelief that as soon as we discover the "right" theoretical distribution function, we can extrapolate the return period outside the length of the empirical series. Evidently, an estimation of certain extreme events in dependence on long return periods can be valid for certain established climatological equilibrium and changes in meteorology will lead to future changes in the return periods of extreme events. However, the changes in empirical response data cannot follow the ongoing climatological changes fast enough, as extreme events are rare anyhow. Thus, the existing bulk of the data does not carry sufficient statistical information about possible extreme sea level values at Pärnu, as the most extreme values seem to appear "outside" the system.

Even before 2005, 253 cm was considered as a value significantly outside of such limits with a theoretical recurrence period of some 300–1000 years, but it was repeated and surpassed just 38 years later. The graph (Figure 11) is similar to a graph for gust wind speed records that is contaminated by occasional tornados [Cheng & Yeung, 2002], or sea level data that includes abnormal tsunami events. These two anomalous and outlying values at Pärnu were nevertheless caused by normal storms. Due to the specific configuration of the Gulf of Riga and Pärnu Bay, the sea level is proportional to the wind speed in the power of 2.4 [Suursaar et al., 2002; 2006]. At the upper range of wind speeds, a slight incremental increase in wind speed yields an exponentially higher incremental increase in storm surge level. The third crucial factor for the two highest Pärnu surges was the high (60-90 cm) Baltic background sea level. Probability for an outstanding Pärnu storm surge therefore appears as a product of probabilities of these three events: a suitable wind speed and direction, and a high boundary sea level.

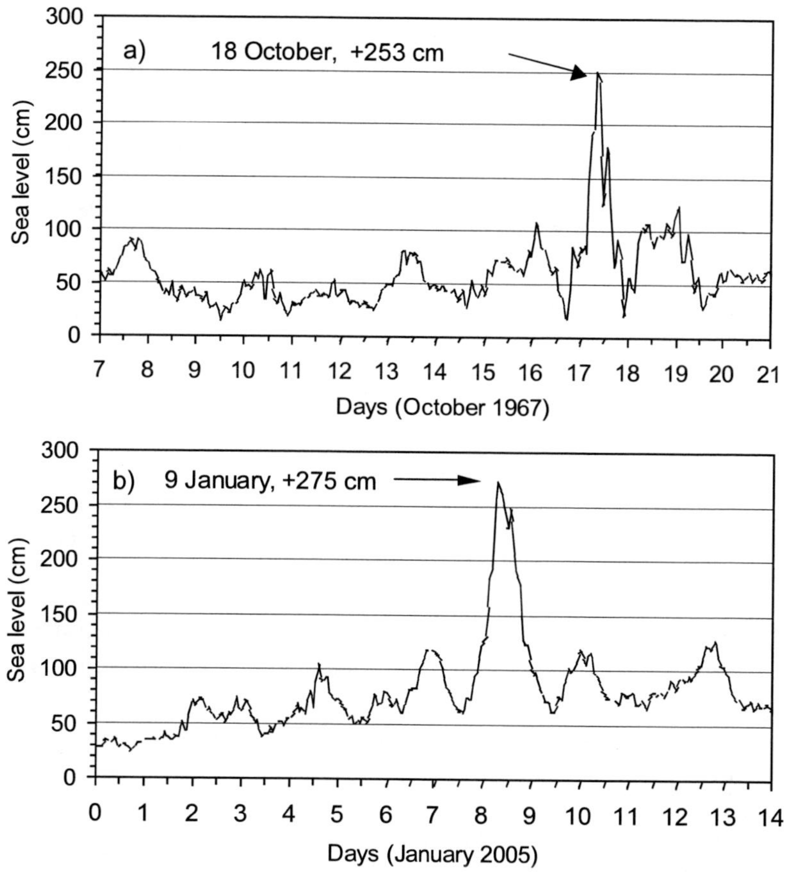

Figure 10. The two historical highest storm surges at Pärnu on 18 October 1967 (a) and on 9 January 2005 (b). The background Baltic Sea level was rather high due to series of cyclones preceding the hurricanes. The secondary sea level peaks around 100 cm represented "ordinary" cyclones with sustained wind speeds 10-20 m/s, whereas the two maximum events had 25-30 m/s of SW-W winds.

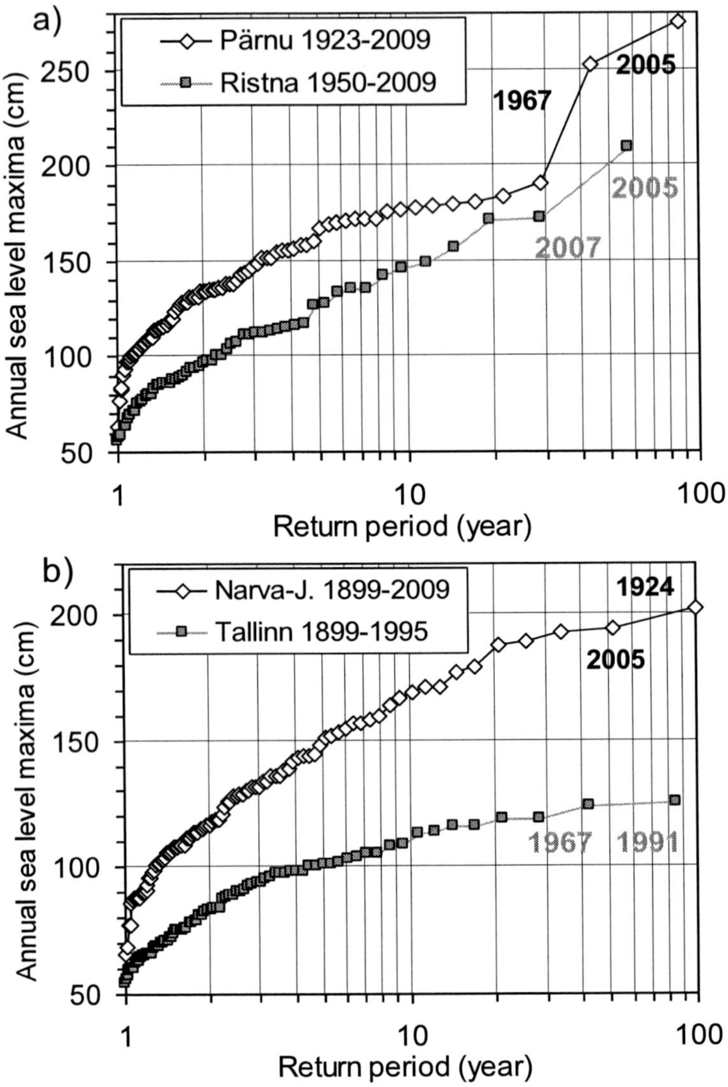

Figure 11. Empirical return periods based on annual maximum sea level data at Pärnu and Ristna (a), and Narva-Jõesuu and Tallinn (b).

According to sensitivity runs of the hydrodynamic model [Suursaar et al., 2006; Suursaar & Kullas, 2006] and considering the 30 m/s sustained wind speed (which has occurred about 1-2 times in the last 50 years), the direction of SW (which is quite usual, if the cyclone passes Estonia from the north, heading from SW to NE), and the Baltic mean sea level of 70 cm for at least a few days (which happens once or twice almost every autumn or winter), the maximum sea level may easily reach 310 cm at Pärnu. At Tallinn, Narva-Jõesuu and Ristna, the possible extremes can also be predicted statistically, provided we know the future global sea level rise rate, which is still a major issue in a global scientific debate, and not discussed in this case study.

6. Conclusion

Sea level variations in the eastern section of the semi-enclosed and practically tideless Baltic Sea are predominantly meteorologically forced, which include a mean seasonal cycle with the amplitude of 12–16 cm and occasional, but very high sea level events.

Trend analysis of the Estonian sea level data from the period 1842–2009 showed that, at Tallinn, Ristna and Narva-Jõesuu tide gauges, the mean sea level rise rates (corrected with land uplift rates) were 1–3 mm/yr. Besides, trend slope estimates considerably depended on the chosen terminal points of the series. Nevertheless, the trends at Pärnu (2.4–3.6 mm/yr) were probably steeper than the mean global sea level rise estimates for the same periods. The sea level increase mainly occurs in winter, which is in good accordance with similar seasonal structures of the NAO trends, and local storminess data. In the Estonian coastal waters the excessive rise in mean sea level and particularly in maxima (3–13 mm/yr) could be explained by the local sea level response to the changing regional wind climate. Sea level variations in the semi-enclosed study area with windward location are sensitive to the current increase in the westerly and intensification of cyclones. The results indicate that in case of an obvious decadal trend in wind conditions the sea level change rates of a semi-enclosed basin may deviate from the global estimates. A positive trend in wind speed and storminess should yield steeper than average sea level trend in windward side, and less steep in leeward side (and vice versa in case of a negative wind speed trend).

In the Pärnu Bay the statistical fit of maximum values distributions for 1923-2009 are inconsistent with the data from the two highest storm surges of 253 cm (in 1967) and 275 cm (in 2005). In statistical terms they appear as outliers, or elements of some other population. On the basis of a hydrodynamic modelling study, the height of the maximum expected storm surges could be estimated at 310 cm during the current climatological epoch, without considering the possible global sea level rise acceleration.

Acknowledgments

The study was supported by the Estonian Science Foundation through grant No. 7609 and by the Estonian target financed grant No.0104s08. The author is grateful to the Estonian Meteorological and Hydrological Institute for providing the sea level data.

References

Alexandersson, H., Schmidt, T., Iden, K. & Tuomenvirta, H. (1998). Long-term variations of the storm climate over NW Europe. *The Global Atmos. Ocean. Syst.*, 6, 97-120.

Andersson, H. C. (2002). Influence of long-term regional and large-scale atmospheric circulation on the Baltic Sea level. *Tellus,* 54A, 76-88.

BACC (2008). *Assessment of Climate Change for the Baltic Sea Basin*, Springer-Verlag, Berlin, Heidelberg, 473.

Bogdanov, V. I., Medvedev, M. Yu., Solodov, V. A., Trapeznikov, Yu. A., Troshkov, G. A. & Trubitsina, A. A. (2000). Mean monthly series of sea level observations (1777-1993) at the Kronstadt gauge. *Reports of the Finnish Geodetic Institute* 2000:1, Kirkkonummi, Finland.

Church, J. A. & White, N. J. (2006). A 20th century acceleration in global sea-level rise. *Geophysical Research Letters,* 33, L01602.

Cheng, E. & Yeung, C. (2002). Generalized extreme gust wind speeds distributions. *Journal of Wind Engineering and Industrial Aerodynamics,* 90, 1657-1669.

Dailidiene, I., Davuliene, L., Tilickis, B., Stankevičius, A. & Myrberg, K. (2006). Sea level variability at the Lithuanian coast of the Baltic Sea. *Boreal Env. Res.,* 11, 109-121.

Ekman, M. (1999). Climate changes detected through the world's longest sea level series. *Global and Planetary Change,* 21, 215-224.

Ekman, M. & Stigebrandt, A. (1990). Secular change of the seasonal-variation in sea-level and the pole tide in the Baltic Sea. *Journal of Geophysical Research-Oceans,* 95, 5379-5385.

Emery, K. O. & Aubrey, D. G. (1991). Sea levels, land levels, and tide gauges. Springer-Verlag, New York, Berlin, Heidelberg, 237.

Fjeldskaar, W., Lindholm, C., Dehls, J. F. & Fjeldskaar, I. (2000). Postglacial uplift, neotectonics and seismicity in Fennoscandia. *Quaternary Science Reviews*, 19, 1413-1422,

Forsberg, R., Solheim, D. & Kaminskis, J. (1996). Geoid of the Nordic and Baltic area from gravimetry and satellite altimetry. *Proc. Int. Symposium on Gravity, Geoid and Marine Geodesy, Tokyo*, 540-548.

Gornitz, V., Lebedeff, S. & Hansen, J. (1982). Global sea level trend in the past century. *Science,* 215, 1611-1614.

Hagen, E. & Feistel, R. (2005). Climatic turning points and regime shifts in the Baltic Sea region: the Baltic winter index (WIBIX) 1659-2002. *Boreal Env. Res.*, 10, 211-224.

Hurrell, J. W. (1995). Decadal trends in the North Atlantic Oscillation: regional temperatures and precipitation. *Science,* 269, 676-679.

IPCC (2001). *Climate change 2001: Synthesis Report.* Cambridge University Press, 408.

IPCC (2007). *Climate change 2007: The physical science basis,* Contribution of Working Group I to the Fourth Assessment report of the Intergovernmentla Panel on Climate In: S., Change, D., Solomon, M., Qin, Z., Manning, M., Chen, K. B., Marquis, M. & Averyt, H. L. Tignor, Miller (eds.), Cambridge Univ. Press, New York, 996.

Jaagus, J. (2006). Climatic changes in Estonia during the second half of the 20th century in relationship with changes in large-scale atmospheric circulation. *Theoretical and Applied Climatology,* 83, 77-88.

Jaagus, J., Post, P. & Tomingas, O. (2008). Changes in storminess on the western coast of Estonia in relation to large-scale atmospheric circulation. *Climate Research,* 36, 29-40.

Jevrejeva, S., Rüdja, A. & Mäkinen, J. (2001). Postglacial rebound in Fennoscandia: new results from Estonian tide gauges. In: *Gravity, geoid and geodynamics* (ed. M. Siders), IAG Symposia 123, Springer, 193-198.

Johansson, M., Boman, H., Kahma, K. K. & Launiainen, J. (2001). Trends in sea level variability in the Baltic Sea. *Boreal Env. Res.,* 6, 159-179.

Jones, P. D., Jónsson, T. & Wheeler, D. (1997). Extension to the North Atlantic Oscillation using early instrumental pressure observations from Gibraltar and South-West Iceland. *Int. J. Climatol.*, 17, 1433-1450.

Keevallik, S. & Rajasalu, R. (2001). Winds on the 500 hPa isobaric level over Estonia (1953-1998). *Phys. Chem Earth (B),* 26, 425-429.

Kont, A., Jaagus, J. & Aunap, R. (2003). Climate change scenarios and the effect of sea-level rise for Estonia. *Global and Planetary Change,* 36, 1-15.

Langenberg, H., Pfizenmayer, A., von Storch, H. & Sündermann, J. (1999). Storm-related sea level variations along the North Sea coast: natural variability and anthropogenic change. *Cont. Shelf Res.*, 19, 821-842.

Lazarenko, N. N. (1986). Variations of mean level and water volume of the Baltic Sea. In: Water Balance of the Baltic Sea. *Baltic Sea Environm. Proc.* No. 16, 64-80.

Lehmann, A., Krauss, W. & Hinrichesen, H. H. (2002). Effects of remote and local atmospheric forcing on circulation and upwelling in the Baltic Sea. *Tellus,* 54A, 299-316.

Lisitzin, E. (1964). Contribution to the knowledge of land uplift along the Finnish coast. *Fennia,* 89, 22.

Lowe, J. A., Gregory, J. M. & Flather, R. A. (2001). Changes in the occurrence of storm surges around the United Kingdom under a future climate scenarion using a dynamic storm surge model driven by the Hadley Centre climate models. *Climate Dynamics,* 18, 179-188.

Magaard, L. & Krauss, W. (1966). Spektren der Wasserstandsschwankungen der Ostsee im Jahre 1958. *Kieler Meeresforsch.*, 22, 155-162.

Meier, H. E. M., Broman, B. & Kjellström, E. (2004). Simulated sea level in past and future climates of the Baltic Sea. *Climate Research,* 27, 59-75.

Raudsepp, U., Toompuu, A. & Kõuts, T. (1999). A stochastic model for the sea level in the Estonian coastal area. *Journal of Marine Systems,* 22, 69-87.

Sildvee, H. (1998). Gravity measurements of Estonia. *Reports of the Finnish Geodetic Institute,* 98, 3, Masala, Finland.

Smith, D., Raper, S. B., Zerbini S. & Sanchez-Arcilla, A. (2000). *EUR* 19337 – *Sea level change and coastal processes – implications for Europe*. Office for Official Publications of the European Communities: Luxembourg, 247.

Suursaar, Ü. & Kullas, T. (2006). Influence of wind climate changes on the mean sea level and current regime in the coastal waters of west Estonia, Baltic Sea. *Oceanologia,* 48, 361-383.

Suursaar, Ü. & Kullas, T. (2009). Decadal changes in wave climate and sea level regime: the main causes of the recent intensification of coastal geomorphic processes along the coasts of Western Estonia? In: *Coastal Processes. WIT Transactions on Ecology and Environment,* Vol. 126, 105-116.

Suursaar, Ü., Kullas, T. & Otsmann, M. (2002). A model study of sea level variations in the Gulf of Riga and the Väinameri Sea. *Cont. Shelf Res.,* 22, 2001-2019.

Suursaar, Ü., Kullas, T., Otsmann, M., Saaremäe, I., Kuik, J. & Merilain, M. (2006). Cyclone Gudrun in January 2005 and modelling its hydrodynamic consequences in the Estonian coastal waters. *Boreal Env. Res.,* 11, 143-159.

Suursaar, Ü. & Sooäär J. (2007). Decadal variations in mean and extreme sea level values along the Estonian coast of the Baltic Sea. *Tellus A*, 59, 249-260.

Vallner, L., Sildvee, H. & Torim, A. (1988). Recent crustal movements in Estonia. *Journal of Geodynamics,* 9, 215-223.

Vermeer, M., Kakkuri, J., Mälkki, P., Boman, H., Kahma, K. K. & Leppäranta, M. (1988). Land uplift and sea level variability spectrum using fully measured monthly means of tide gauge readings. *Finnish Marine Research,* 256, 3-75.

Wübber, C. & Krauss, W. (1979). The two-dimensional seiches of the Baltic Sea. *Oceanol. Acta*, 2, 435-446.

In: Sea Level Rise, Coastal Engineering, Shorelines and Tides ISBN: 978-1-61728-655-1
Editor: Linda L. Wright, pp. 123-137 © 2011 Nova Science Publishers, Inc.

Chapter 5

POTENTIAL IMPACTS OF SEA-LEVEL RISE ON THE ATLANTIC AND THE GULF OF MEXICO COAST OF THE UNITED STATES

Shuang-Ye Wu[1,*] and Raymond Najjar[2]
[1]University of Dayton, Dayton, Ohio, USA
[2]Pennsylvania State University, Pennsylvania, USA

Abstract

We made projections of relative sea-level rise, horizontal inundation, and the associated impacts on people and infrastructure in the Atlantic and the Gulf of Mexico coast of the United States. We first estimated the relative sea-level rise by 2100 in the study area by using (1) rates of subsidence in different localities derived from historical tide gauge observations and (2) rates of global sea-level rise from the semi-empirical model developed by Rahmstorf (2007). The range of sea-level rise projections mainly reflects equal contributions of spatial variability (due to subsidence) and the range of projected greenhouse gas scenarios, the range of projected temperature change (from global climate models) for a given scenario, and uncertainty in the semi-empirical model itself. Using the USGS 30-m National Elevation Dataset, projected sea-level rise, and local subsidence estimates, we mapped the horizontal inundation in the region, and examine the different land use categories in the inundated zones. Finally, we overlaid the inundation maps with census block data to determine the number of people and housing units located in the area vulnerable to future sea-level rise. Analysis results were summarized by state to indicate spatial variability and facilitate information usage by stake holders. In total, 31-53 thousand km^2 of land will be inundated depending on different sea-level rise scenarios; 3.8 - 6.6 million people and 1.8 - 3.2 million housing units are at risk of inundation by future sea-level rise. The risk is not distributed evenly. In general, northern states are less vulnerable because of relatively small future sea-level rise and the steep coastal terrain, whereas southern states are more vulnerable because of the higher sea-level rise, low elevation and flat terrain profile. However, some northern states such as New York and New Jersey have highly developed coast. As a result, even moderate amount of inundation area may lead to large number of people and housing units at risk of future sea-level rise.

---

*E-mail address: shuang-ye.wu@notes.udayton.edu, Tel: 1-937-229-1720, Fax: 1-937-229-2889, University of Dayton, 300 College Park Ave, Dayton, Oh 45469 (Corresponding author)

Introduction

Sea-level rise is accelerated by global warming, which leads to thermal expansion of the ocean, melting of land-based glaciers, and increasing ice and water discharge from major ice sheets (Rignot and Kanagaratnam, 2006; IPCC, 2007). Interest in sea-level rise stems from the likely profound impact that it will have on coastal communities. Increases in sea level have already led to inundation of low-lying coastal regions, more frequent flooding, worsening beach erosion, and loss of coastal wetland; with sea-level rise expecting to continue acceleration, it is likely that these impacts will worsen substantially (Mclean et al., 2001; Scavia et al., 2002, CCSP, 2009).

The impacts of sea-level rise on the Atlantic and Gulf of Mexico Coasts of the United States are expected to be severe for two main reasons. First, the rate of relative sea-level rise in the majority of the Atlantic Coast and all of the Gulf Coast is higher than the global average, primarily a result of local subsidence driven by post-glacial rebound (Davis and Mitrovica, 1996; Engelhart et al., 2009). In addition, groundwater withdrawal and off-shore oil and gas extraction in the Gulf of Mexico further exacerbate the problem (Davis, 1987). Historical sea-level rise rates for the 20th century ranges from 2-6 mm per year for the Atlantic Coast and from 2-10 mm per year for the Gulf of Mexico Coast (Zervas 2001), whereas global-mean sea level rose at a rate of 1.7 mm per year during the same time period (Church and White 2006). The second reason why sea-level rise is particularly hazardous for this region is that the area is characterized by high population density and rapid development on the coast. According to US Census 2000, 45% of US population lives in the 18 coastal states along the Atlantic Ocean and the Gulf of Mexico. 40% of the population in these states live in the coastal counties, which only make up 15% of their total land area.

In their 4th Assessment Report (AR4), the Intergovernmental Panel on Climate Change (IPCC) projected that global-mean sea level will likely rise between 0.19 and 0.59 m by the end of the 21st Century. Their assessment excluded any rapid changes in ice flow from Greenland and Antarctica. In a comprehensive study, Rignot and Kanagaratnam (2006) showed the Greenland ice sheet loss rate in the past decade to have increased from 90 to 200 cubic km per year, corresponding to an extra 0.25 to 0.55 mm per year of sea-level rise. Therefore, IPCC projections are likely to be an underestimate of future sea-level rise. A more realistic estimate is given by Rahmstorf (2007), who developed a semi-empirical model based on past observations. Applying this model, he projected a sea-level rise of 0.5-1.4 m above the 1990 level in 2100. The range accounts for the range of projected greenhouse gas scenarios, the range of projected temperature change (from global climate models) for a given scenario, and uncertainty in the semi-empirical model itself.

In projecting the regional impact of sea-level rise, two additional factors should be considered: (1) vertical land movement, as discussed above, and (2) the regionality of absolute sea-level rise resulting from changing ocean circulation patterns (Yin et al., 2009) and gravitational changes in ice sheets (Milne, 2009). Yin et al. (2009) showed that climate models consistently project decreases in the strength of the Gulf Stream with increasing atmospheric CO_2, a result of the freshening and warming of high-latitude surface waters. This circulation change leads to a substantial rise in sea level along the coast of the Northeast U.S., roughly 0.2 m by 2100. Numerical modeling also suggests that gravitational changes associated with melting ice sheets can result in strongly regional sea-level rise distributions (summarized in Milne et al., 2009). For example, melting of the Antarctic ice sheet causes sea-level rise along the Atlantic and Gulf Coasts to be 10-20% greater than the global average. On the other hand, Greenland ice sheet melting leads to sea-level rise that is 10% (along the Gulf Coast) to about 50% (along the Northeast U.S. coast) lower than global-mean sea-level rise.

In order to understand the broad-scale ramifications of accelerated sea-level rise on human society, vertical sea-level rise projections need to first be translated into estimates of inundation (horizontal area flooded) and second into estimates of the amount of people, property and other infrastructure exposed to this inundation. Although many small-scale case studies on the impacts of sea-level rise exist (Gornitz et al., 2002; McInnes et al., 2003; Wu et al., 2002), extensive studies covering large areas are rare. The most recent large-scale study was conducted by Titus and Richman (2001), who produced maps of all land that exists below 1.5 m and between 1.5 and 3.5 m in each US state on the Atlantic and Gulf coasts. The 1.5-m (5-foot) contour was chosen because it was the lowest elevation that could be consistently resolved over large regions with the available data at the time. These maps have their limitations for assessing potential impacts of sea-level rise, however, because projections of sea-level rise for this century are in general below 1.5 m. Moreover, there are no large-scale studies that quantify impacts of sea-level rise on people, property and infrastructure. The US Climate Change Science Program (CCSP) recently published a synthesis report on coastal sensitivity to sea-level rise (CCSP, 2009). Although comprehensive in its approach, this report focuses only on the Mid-Atlantic Region, representing a stretch of coast from the State of New York to Virginia, excluding Chesapeake Bay.

Over the last several years, the United States Geological Survey (USGS) refined its national digital elevation models (DEMs), producing them at horizontal resolutions of 30 m for the whole US and 10 m for selected areas. At the same time, many coastal states have commissioned the collection of high resolution data obtained by Light Detection and Ranging (LIDAR). With this improvement of available data, we made a revised estimation of the potential impact of sea-level rise on the Middle and Upper Atlantic Region of the United States (Wu et al. 2009). Using the LIDAR data, we found that the DEMs, despite their stated average accuracy of 7 m, had much higher accuracy in coastal regions, suitable for estimating inundation and its impacts. Further, we found that the regionality in sea-level rise and slope has a significant impact on inundation. Finally, using sea-level projections from the IPCC's Third Assessment Report (which are similar to the AR4 projections), we estimated a substantial risk to people and infrastructure in the Mid- and Upper-Atlantic Region of U.S. (Maine through Virginia) as a result of inundation.

The present study builds on our earlier work in two major ways. First, we revise the global mean sea-level projections upward based on the accumulated evidence that the IPCC projections are too low. Second, we extend our analysis to coastal states in the Southeast U.S. (North Carolina to Texas). The full study area covers 18 states and the District of Columbia, and the coastal portion includes some of the biggest metropolitan areas of the nation, such as New York, Philadelphia, Baltimore, Washington DC, Norfolk, Miami, New Orleans and Houston. Even though the region is dominated by forest and agricultural land use, the coastal counties have a much higher percentage of land devoted to commercial, industrial and residential uses. This pattern of population and land use distribution means that with sea-level rise, greater amounts of people, property and infrastructure will be exposed to either direct inundation or higher risk of flooding.

Three major sections follow. Section 2 reviews the methods and data used in this study. Section 3 presents the sea-level projections, inundation estimates, and their impacts. Section 4 provides a summary of our findings.

2. Methodology and Data Sources

2.1. Sea-Level Rise Projection

Sea-level projections are made for the year 2100 and are composed of two main components: global and local. The global component is based on the semi-empirical model of Rahmstorf (2007). Low and high projections are taken to be the range given by Rahmstorf (2007), 0.5 and 1.4 m, respectively. We also include a medium projection of 1.0 m. As in Wu et al. (2009), the local component is assumed to be constant in the future, and is calculated to be the difference between local and global historical rates. Local rates are taken from the 1950-1999 estimates based on continuous measurements (Zervas, 2001) whereas the global rate is taken to be 1.8 ± 0.3 mm yr^{-1} based on the analysis of Church and White (2006) over the second half of the 20th century. Standard errors in the local trends for most records are approximately 10%.

For the inundation calculations, the local component is assumed to be fixed for each state, with the exception of Florida, and we compute this to be the arithmetic average of stations in each state. We divided Florida into three sections because of its exceedingly long shoreline and its pivotal location between the Atlantic Ocean and the Gulf of Mexico. The Florida East section includes the east side of the state with the Florida Keys; the Florida West section includes the coastal areas from Collier to Taylor County; and the Florida Panhandle section includes the rest of the state. Four states - New Hampshire, Mississippi, Alabama, and Louisiana - do not have continuous records for the second half of the 20th century. For these states, we used trends based on discontinuous records (Table 4 in Zervas, 2009): for New Hampshire, we used the Seavey Island (southern Maine) record; for Alabama and Mississippi, we used the Dauphin Island (Alabama) record, and for Louisiana, we used the Grand Isle (Louisiana) record.

2.2. DEM Inundation and Impact Estimates

Based on the mean projections of sea-level rise for each state (or section in Florida), we used DEMs to model the area that will be inundated under different future sea-level rise scenarios. Before we calculated the inundation area, we first adjusted the vertical difference between the North America Vertical Datum (NAVD) 1988 (which the DEMs are based on) and the mean sea level (which future sea-level rise is based on). We then filled in the depressions in DEMs so that a depression will only be inundated when surrounding higher grounds are. After the inundation areas were mapped and calculated, they were then divided by the length of the coastline of the state to roughly estimate the average distance of coastline retreat due to inundation only, if no shoreline protection measures were present.

In the final step, we estimated the potential impacts of sea-level rise in terms of land use, people and housing units that will be affected. We also added the state average tidal height on top of the sea-level rise values to project future high tide water levels, assuming some areas will become future tidal land even if not permanently inundated by sea-level rise. People and infrastructure in those areas will also be affected. Based on resulting values in future high water level, we then used DEMs to model the areas that will be inundated at high tide under different future sea-level rise scenarios. After that, we overlaid the inundation map with land cover data (National Land Cover Dataset 2001) to assess the amount of areas of different land cover types that will be exposed to risk of inundation owing to sea-level rise. Finally, we overlaid the inundation map with Census 2000 Block data to estimate the number of people and housing units that are at risk of inundation by future sea-level rise. In this step, census blocks whose centroids fall inside the inundation zones are summed up for each state to calculate the total number of people and housing units at risk.

Table 1. State averages of sea-level rise and local component of sea-level rise

State	No. of Stations	Mean Sea Level Rise (mm yr^{-1})[a]	Mean of local component of sea-level rise (mm yr^{-1})[b]
Maine	3	1.58 ± 0.20	-0.22 ± 0.36
New Hampshire*	1	1.76 ± 0.15	-0.04 ± 0.33
Massachusetts	2	2.21 ± 0.15	0.41 ± 0.33
Rhode Island	2	2.09 ± 0.17	0.29 ± 0.34
Connecticut	1	1.99 ± 0.22	0.19 ± 0.37
New York	2	2.66 ± 0.23	0.86 ± 0.38
New Jersey	2	3.94 ± 0.26	2.14 ± 0.40
Delaware	1	3.04 ± 0.29	1.24 ± 0.42
Washington DC	1	3.15 ± 0.35	1.35 ± 0.46
Maryland	3	3.09 ± 0.24	1.29 ± 0.38
Virginia	2	4.22 ± 0.29	2.42 ± 0.42
North Carolina	1	2.76 ± 0.34	0.96 ± 0.45
South Carolina	1	3.05 ± 0.27	1.25 ± 0.40
Georgia	1	3.43 ± 0.28	1.63 ± 0.41
Florida (East)	3	2.57 ± 0.27	0.86 ± 0.40
Florida (West)	2	2.20 ± 0.21	0.87 ± 0.37
Florida (Panhandle)	1	2.00 ± 0.27	0.20 ± 0.40
Alabama*	1	2.98 ± 0.44	1.18 ± 0.53
Mississippi**	0	2.98 ± 0.44	1.18 ± 0.53
Louisiana*	1	9.24 ± 0.30	7.44 ± 0.42
Texas	3	5.42 ± 0.38	3.62 ± 0.48

[a] Standard error is average of standard errors of the individual stations. We assume residuals of linear fit are correlated for each state.

[b] Standard error is calculated assuming residuals of global and local linear trends are uncorrelated.

Table 2. Projected sea-level rise for Atlantic and Gulf states

State	Sea-level Rise by 2100 (mm)			High Tide
	Low	medium	High	(mm)
Maine	478	928	1378	2056
New Hampshire	496	946	1396	1344
Massachusetts	541	991	1441	959
Rhode Island	529	979	1429	716
Connecticut	519	969	1419	460
New York	586	1036	1486	575
New Jersey	714	1164	1614	768
Delaware	624	1074	1524	741
Washington DC	635	1085	1535	494
Maryland	629	1079	1529	233
Virginia	742	1192	1642	428
North Carolina	596	1046	1496	683
South Carolina	625	1075	1525	866
Georgia	663	1113	1563	1122
Florida (East)	586	1036	1486	894
Florida (West)	587	1037	1487	404
Florida (Panhandle)	520	970	1420	379
Alabama	618	1068	1518	419
Mississippi	618	1068	1518	419
Louisiana	1244	1694	2144	324
Texas	862	1312	1762	133

3. Results

3.1. Sea-Level Rise Projection

Figure 1 shows the historical rates of sea-level rise at each of the stations in the study area. There is considerable spatial variability in these rates, with relative maxima appearing in the Southern Mid-Atlantic Region and the central Gulf Coast. Only four of the 34 stations have historic rates of sea-level rise that are less than the global average. Table 1 shows state averages of the historical sea-level rise and the local component of sea-level rise. We see that all states except Maine and New Hampshire have local components greater than zero.

Projections of sea-level change by 2100 for the medium scenario are shown for each site in Figure 1. Projections for most sites are 5-20% greater than the global mean, but a few sites (in Texas and Louisiana) have local contributions that are 30-70% of the global mean. Table 2 shows projections by state for all of the scenarios considered.

Table 3. Inundation area under low, medium and high sea-level rise scenarios at mean sea level (MSL) and high tide (HT) (km^2)

State	Low		Medium		High	
	MSL	HT	MSL	HT	MSL	HT
Maine	98	212	124	245	150	275
New Hampshire	6	22	12	25	16	28
Massachusetts	114	209	164	292	206	345
Rhode Island	19	28	25	36	31	44
Connecticut	40	53	52	61	61	81
New York	129	201	183	280	263	329
New Jersey	256	766	375	934	832	1,085
Delaware	113	250	155	485	437	533
Washington DC	2.8	3.6	3.5	4.4	4.3	5.4
Maryland	1,022	1,260	1,421	1,566	1,688	1,830
Virginia	322	654	666	1,330	1,345	1,610
North Carolina	3,466	5,512	5,034	6,328	5,894	7,174
South Carolina	302	1,587	889	2,251	1,601	2,502
Georgia	848	2,049	1,554	2,172	1,986	2,424
Florida (East)	3,540	6,655	4,541	8,487	6,681	9,628
Florida (West)	1,244	1,903	1,993	3,089	3,123	3,557
Florida (Panhandle)	164	425	690	991	1,080	1,315
Alabama	196	392	399	698	709	834
Mississippi	148	259	264	390	431	513
Louisiana	18,642	20,214	20,637	21,615	21,939	22,862
Texas	687	821	2,359	4,276	4,944	5,166
Total	31,357	43,476	41,540	55,555	53,423	62,141

3.2. Impact of Sea-Level Rise

Table 3 shows the estimates of area inundation at mean sea level and at high tide by state under different sea-level rise scenarios.

The inundation area presented in Table 3, however, is difficult to compare owing to the difference in the size of the state, particularly their difference in shoreline lengths. We calculated the area of inundation per length of shoreline (Table 4). This is a simple, albeit crude, representation of average amount of shoreline retreat if there were no man-made shoreline protection structures and procedures. The complexity of coastal wetland change in response to sea-level rise is also ignored. However, this provides us with some first-order assessment of each state's vulnerability to future sea-level rise. It could be seen from Table 4 that Louisiana stands out as the most vulnerable state to sea-level rise, as its shoreline may retreat up to 12 km (by 2100), on average, in response to future sea-level rise. This is owing to the fact that Louisiana has the greatest historical sea-level rise and a very flat profile, and low elevation in general. States from Maine to New York have less shoreline retreat because of a relatively small amount of sea-level rise, as their local components of sea-level rise are very small, or even negative. Also, most of these states have relatively steep coastal topography, which makes them relatively insensitive to sea level rise. This can be shown from the fact that the differences in the amount of shoreline retreat between low and high sea-level rise scenarios are in general less than 0.2 km for northern states. Southern states, however, suffer the double impacts of higher relative SLR and flat terrain. Therefore, the same amount of 0.9 m difference between low and high sea-level rise scenarios results in 1-3 km difference in shoreline retreat for these states. It should be noted that northern states in general have higher tidal range, which leads to greater increase in inundation area at high tide. However, even with this greater increase, they are still in general less vulnerable to sea-level rise than states that are located in the mid- and south-Atlantic and the Gulf coast.

In most coastal regions, inundated area would extend no more than a few kilometers inland, and is too small to see on a map that covers the whole study area. Therefore, we divided the Atlantic and Gulf coast into 6 sections so that sufficient details can be mapped to show the inundation areas under medium and high sea-level rise scenarios. These maps are presented in Figures 3-8.

Table 4. Inundation area per length of shoreline under different SLR scenarios[1] (km^2 km^{-1})

	Shoreline	Sea-level Rise Scenario		
State	**Length[a] (km)**	**low**	**medium**	**High**
Maine	706	0.14	0.18	0.21
New Hampshire	52	0.11	0.23	0.31
Massachusetts	645	0.18	0.25	0.32
Rhode Island	183	0.10	0.14	0.17
Connecticut	172	0.23	0.30	0.36
New York	664	0.19	0.28	0.40
New Jersey	469	0.55	0.80	1.77
Delaware	166	0.68	0.93	2.64
Washington DC	-	-	-	-
Maryland	1089	0.94	1.30	1.55
Virginia	1268	0.25	0.53	1.06
North Carolina	1860	1.86	2.71	3.17
South Carolina	616	0.49	1.44	2.60
Georgia	421	2.01	3.69	4.71
Florida (East)	1362	2.60	3.33	4.90
Florida (West)	1163	1.07	1.71	2.68
Florida (Panhandle)	984	0.17	0.70	1.10
Alabama	231	0.85	1.73	3.07
Mississippi	156	0.95	1.69	2.76
Louisiana	1781	10.47	11.59	12.32
Texas	2342	0.29	1.01	2.11

[a] The length of general coastline is derived from map at 1:3,000,000 scale.

Table 5. Area of inundated developed land and its percentage to total inundation area at high tide under low, medium and high sea-level rise scenarios

Developed Land:	Sea-level Rise Scenarios					
	Low		Medium		High	
State	Area (km^2)	% of Total	Area (km^2)	% of Total	Area (km^2)	% of Total
Maine	27	13	31	13	35	13
New Hampshire	3	16	4	18	5	19
Massachusetts	45	21	61	21	76	22
Rhode Island	6	23	9	25	13	28
Connecticut	19	35	23	37	30	37
New York	98	49	143	51	174	53
New Jersey	112	15	151	16	193	18
Delaware	17	7	35	7	41	8
Washington DC	2	66	3	69	4	73
Maryland	53	4	72	5	91	5
Virginia	58	9	96	7	139	9
North Carolina	146	3	199	3	253	4
South Carolina	42	3	75	3	90	4
Georgia	38	2	50	2	72	3
Florida (East)	906	14	1351	16	1609	17
Florida (West)	369	19	710	23	880	25
Florida (Panhandle)	16	4	41	4	78	6
Alabama	16	4	25	4	38	5
Mississippi	20	8	39	10	68	13
Louisiana	767	4	921	4	1065	5
Texas	55	7	310	7	409	8
Total	2815	6	4347	8	5360	9

The impact of sea-level rise does not only depend on the total amount of inundation area. In order to assess the potential impacts on human society, it is necessary to examine how land vulnerable to future sea-level rise is being used today, and how many people and housing units are located in the vulnerable areas. Tables 5-8 respectively present the amount of developed land, forest, agricultural land and wetland and their percentage of total inundation area for each state. It can be seen that the majority of the inundation area is covered by coastal wetland, around 80%. However, as the sea-level rise scenario increases from low to high, the percentage of wetland decreases whereas the percentage for developed, forest and agricultural lands all increase. Different states have different land use makeup for the potential inundation area. States with more developed coasts, such as the New England states, New York, New Jersey, and Florida, have relatively higher percentages of developed land exposed to inundation risk of future sea-level rise.

Table 9 and 10 present the number of people and housing units located in potential inundation area under low, medium and high sea-level rise scenarios for each of the coastal states on the Atlantic and Gulf coast. In terms of total at-risk population and housing units, the impact of future sea-level rise is greatest in Florida and Louisiana, which is to be expected. Louisiana has by far the largest area vulnerable to inundation because of its high subsidence rate and flat terrain. Florida has the longest coastline, which is highly developed

(19% of the inundation area is developed land at the moment for east coast of Florida). For states with highly developed coasts, such as New York and New Jersey, the number of people and housing units that are vulnerable to sea-level rise can be disproportionately larger even though the inundation area in these states may not seem that significant. For example, New York ranked 12th out of 19 states (including DC) in terms of potential inundation area, but ranked 3rd in terms of number of people and housing units at risk of inundation by future sea-level rise. This is because the fraction of inundated area that is developed is much higher in New York, about 50%, compared to 2-9% for the southern states (excluding Florida). Similarly, New Jersey ranked 10th in terms of potential inundation area, but ranked 4th in terms of at-risk population and housing units. Georgia, on the other hand, is ranked 5th in terms of inundation area, but only 11th in terms of impacts on people and houses, because its coastal region is relatively undeveloped. The impact of sea-level rise is relatively small for New England states even though they have relatively developed coast. This is largely due to the fact that these states have in general small inundation areas because of low (or negative) subsidence rate and steep coastal terrain.

Table 6. Area of inundated forest land and its percentage to total inundation area at high tide under low, medium and high sea-level rise scenarios

Forest Land:	Sea-level Rise Scenarios					
	Low		Medium		High	
State	Area (km^2)	% of Total	Area (km^2)	% of Total	Area (km^2)	% of Total
Maine	36	17	46	19	56	20
New Hampshire	1	6	2	7	2	8
Massachusetts	7	4	11	4	15	4
Rhode Island	1	3	1	3	1	3
Connecticut	5	9	6	10	8	10
New York	5	3	11	4	14	4
New Jersey	35	5	55	6	82	8
Delaware	13	5	45	9	56	11
Washington DC	0	8	0	8	0	7
Maryland	154	12	258	17	347	19
Virginia	105	16	289	22	383	24
North Carolina	569	10	722	11	875	12
South Carolina	120	8	226	10	287	11
Georgia	141	7	189	9	280	12
Florida (East)	183	3	241	3	293	3
Florida (West)	52	3	167	5	241	7
Florida (Panhandle)	22	5	73	7	144	11
Alabama	10	2	23	3	43	5
Mississippi	5	2	20	5	41	8
Louisiana	37	0	51	0	64	0
Texas	2	0	11	0	20	0
Total	1506	3	2447	4	3253	5

Table 7. Area of inundated agricultural land and its percentage to total inundation area at high tide under low, medium and high sea-level rise scenarios

Agricultural Land:	Sea-level Rise Scenarios					
	Low		Medium		High	
State	Area (km^2)	% of Total	Area (km^2)	% of Total	Area (km^2)	% of Total
Maine	5	2	7	3	9	3
New Hampshire	0	2	0	2	1	2
Massachusetts	8	4	13	4	16	5
Rhode Island	1	3	1	4	2	4
Connecticut	1	2	1	2	2	2
New York	3	2	7	3	10	3
New Jersey	29	4	44	5	59	5
Delaware	20	8	69	14	86	16
Washington DC	0	2	0	2	0	2
Maryland	115	9	225	14	335	18
Virginia	75	11	210	16	285	18
North Carolina	783	14	1103	17	1502	21
South Carolina	60	4	101	5	118	5
Georgia	22	1	28	1	43	2
Florida (East)	119	2	192	2	288	3
Florida (West)	16	1	62	2	95	3
Florida (Panhandle)	2	0	6	1	12	1
Alabama	2	1	7	1	10	1
Mississippi	1	0	3	1	6	1
Louisiana	2097	10	2723	13	3290	14
Texas	90	11	588	14	888	17
Total	3451	8	5392	10	7057	11

Table 8. Area of inundated wetland and its percentage to total inundation area at high tide under low, medium and high sea-level rise scenarios

Wetland:	Sea-level Rise Scenarios					
	Low		Medium		High	
State	Area (km^2)	% of Total	Area (km^2)	% of Total	Area (km^2)	% of Total
Maine	113	53	128	53	141	51
New Hampshire	16	75	18	71	19	68
Massachusetts	107	51	161	55	186	54
Rhode Island	13	47	17	47	20	45
Connecticut	25	47	28	45	36	45
New York	72	36	88	32	95	29
New Jersey	569	74	658	70	724	67
Delaware	191	76	320	66	332	62
Washington DC	1	19	1	17	1	14
Maryland	883	70	942	60	979	53
Virginia	391	60	683	51	740	46
North Carolina	3861	70	4112	65	4313	60
South Carolina	1329	84	1798	80	1951	78

Table 8. (Continued)

Wetland:	Sea-level Rise Scenarios					
	Low		Medium		High	
State	Area (km^2)	% of Total	Area (km^2)	% of Total	Area (km^2)	% of Total
Georgia	1824	89	1880	87	2001	83
Florida (East)	5408	81	6658	78	7388	77
Florida (West)	1580	83	2371	77	2703	76
Florida (Panhandle)	368	86	837	84	1029	78
Alabama	351	90	624	89	719	86
Mississippi	229	88	320	82	386	75
Louisiana	17109	85	17704	82	18218	80
Texas	452	55	2957	69	3367	65
Total	34892	80	42304	76	45348	73

Table 9. Number of people in inundation area under low, medium and high sea-level rise scenarios

State	Low	medium	high
Maine	10,738	12,454	14,641
New Hampshire	4,079	4,751	5,425
Massachusetts	38,525	56,275	74,105
Rhode Island	4,558	5,374	6,824
Connecticut	17,222	22,017	28,325
New York	184,902	282,379	360,337
New Jersey	142,315	208,827	269,347
Delaware	6,076	13,832	17,149
Washington DC	0	0	114
Maryland	37,042	53,263	70,964
Virginia	98,538	160,514	226,024
North Carolina	74,406	103,469	134,526
South Carolina	41,379	66,465	77,176
Georgia	38,433	45,809	59,815
Florida	1,898,097	3,077,280	3,674,503
Alabama	2,990	6,532	11,258
Mississippi	7,343	15,175	29,089
Louisiana	1,146,947	1,269,856	1,383,075
Texas	18,390	123,268	162,656
Total	3,771,980	5,527,540	6,605,239

It should be noted that this assessment is made based on elevation only. Although the maps may indicate a certain level of risk to future sea-level rise, they do not depict future shorelines, since there are a variety of other factors that will affect the shape of future shorelines, such as coastal erosion, wetland accretion, and the existence of man-made shoreline protection structures/projects, to name but a few. These maps represent a useful first-order assessment, based on which more detailed analyses can be conducted at finer spatial scales.

Table 10. Number of housing units in future inundation area under low, medium and high sea-level rise scenarios

State	low	medium	high
Maine	10,247	12,079	13,559
New Hampshire	2,970	3,701	4,277
Massachusetts	24,452	33,417	42,730
Rhode Island	3,282	3,712	4,732
Connecticut	9,007	11,311	14,552
New York	80,826	120,894	152,583
New Jersey	93,180	136,764	180,438
Delaware	6,829	14,739	17,920
Washington DC	0	0	94
Maryland	23,045	34,156	45,354
Virginia	39,331	67,133	94,507
North Carolina	49,147	69,055	89,086
South Carolina	25,680	42,801	49,499
Georgia	18,667	22,305	28,768
Florida	952,004	1,501,069	1,766,494
Alabama	4,029	7,114	10,304
Mississippi	4,178	8,034	14,366
Louisiana	477,183	530,948	578,075
Texas	13,459	69,339	89,847
Total	1,837,516	2,688,571	3,197,185

Conclusions

We have used historical tide gauge records, estimates of global historic sea-level rise, and future global sea-level rise projections based on Rahmstorf (2007) to make projections of relative sea-level rise for the Atlantic and Gulf of Mexico Coast the United States. Our projections based on medium global sea-level rise scenario, ranges from 0.9 to 1.7 m, depending on location, which reflects different rates of subsidence. In general, relative sea-level rise will be less in the northern and more in the southern portion of the Coast. Maine has the least amount projected future sea-level rise. Louisiana has the greatest projected sea-level rise, largely owing to its unusually high rate of subsidence. In additional to natural causes, such as post-glacial rebound, it is likely that oil and gas exploration has contributed to this high subsidence rate.

Sea-level rise is predicted to have profound impacts on the Atlantic and Gulf Coast. It is likely to inundate 41 thousand km^2 of land at mean sea-level, ranging from 31 to 53 thousand km^2 depending on scenarios. Inundation could increase to 55 thousand km^2 (43-62 thousand km^2) at high tide. Because of the high level of development and high population density in the region, this can affect as many as 5.5 million people living in low lying areas, ranging from 3.8 to 6.6 million depending on scenarios. Southern states in general suffer more area of inundation because of higher rate of relative sea-level rise and relatively flat terrain. However, some northern states such as New York and New Jersey are also highly vulnerable to future sea-level rise because their coastal areas are highly developed and densely populated.

There are several reasons why this assessment may underestimate the societal risk from future sea-level rise. First, there is great uncertainty in how large ice sheets may contribute to global sea-level rise in the future. More and more evidence points to accelerated rates of water and ice discharge from major ice sheets (Rignot and Kanagaratnam 2006, Joughin, 2006; Kerr, 2004, Krabill et al. 2000). This may well increase the rate of future sea-level rise. Second, the assessment does not take into account coastal erosion, which may greatly accelerate with sea-level rise, contributing to a greater amount of land lost to higher sea level. Third, this study does not take into account future population growth or a change in land use patterns. Both factors may increase the number of people and amount of infrastructure vulnerable to future sea-level rise. All these factors require further detailed research in order to refine the impact assessment of future sea-level rise.

Lastly, this study only considers the impact of sea level rise in terms of inundation. Sea level rise has other important impacts that will affect coastal communities. These include elevated storm surges, loss of coastal wetland, and salt water intrusion, to name just few. These are beyond the scope of this study, but are important components to consider if a comprehensive impact study of sea-level rise is conducted.

References

CCSP (2009) *Coastal Sensitivity to Sea-Level Rise: A Focus on the Mid-Atlantic Region.* A report by the U.S. Climate Change Science Program and the Subcommittee on Global Change Research. [James G. Titus (Coordinating Lead Author), K. Eric Anderson, Donald R. Cahoon, Dean B. Gesch, Stephen K. Gill, Benjamin T. Gutierrez, E. Robert Thieler, and S. Jeffress Williams (Lead Authors)]. U.S. Environmental Protection Agency, Washington D.C., USA, 320 pp.

Church, J. A., White, N. J., Coleman, R., Lambeck, K. & Mitroviga, J. X. (2004). Estimates of the regional distribution of sea-level rise over the 1950-2000 period. *J. Climate*, 17, 2609-2625.

Church, J. A. & White, N. J . (2006). A 20th century acceleration in global sea-level rise. *Geophys. Res. Lett.*, 33, L01602, doi: 10.1029/2005GL024826.

Davis G. H. (1987). Land subsidence and sea level rise on the Atlantic Coastal Plain of the United States. *Environmental Geology*, 10(2), 67-80.

Davis, J. L. & Mitrovica, J. X. (1996) Glacial isostatic adjustment and the anomalous tide gauge record of eastern *North America. Nature*, 379, 331-333.

Engelhart, S. E., Horton, B. P., Douglas, B. C., Peltier, W. R. & Törnqvist, T. E. (2009). Spatial variability of late Holocene and 20th century sea-level rise along the Atlantic coast of the United States. *Geology*, 37, 1115-1118.

Gornitz, V., Couch, S. & Hartig, E. K. (2002). Impacts of sea-level rise in the New York City metropolitan area. *Global and Planetary Changes*, 32, 61-88.

IPCC (2007). *Climate Change* 2007*: The Physical Science Basis.* Contribution of Working Group I to the Fourth Assessment Report of the Intergovernmental Panel on Climate Change, S., Solomon, D., Qin, M., Manning, Z., Chen, M., Marquis, K. B., Averyt, M.

Tignor, & H. L. Miller, (eds.) Cambridge University Press, Cambridge, United Kingdom and New York, NY, USA.

Joughin, I. (2006). Greenland rumbles louder as glaciers accelerate. *Science* 24 March, 311, 1719-1720.

Kerr, R. A. (2004). A Bit of icy Antarctica is sliding toward the sea. *Science* 24 September, 305, 1897.

Krabill, W; et al. (21 July 2000). "Greenland Ice Sheet: High-Elevation Balance and Peripheral Thinning". *Science* 289 (5478): 428–430. doi:10.1126/science.289.5478.428. PMID 10903198.

Mclean, R. F., Tsyban, A., Burkett, V., Codignotto, J. O., Forbes, D. L., Mimura, N., Beamish, R. J. & Ittekkot, V. (2001). Chapter 6: Coastal Zones and Marine Ecosystems, In: McCarthy JJ et al (eds) *Climate Change 2001: Impacts, Adaptation and Vulnerability*. Cambridge University Press, New York, NY, 343-379.

McInnes, K. L., Walsh, K. J., Hubbert, G. D. & Beer, T. (2003). Impact of sea-level rise and storm surges on a coastal community. *Natural Hazards* , 30(2), 187-207.

Milne, G. A., Gehrels, W. R., Hughes, C. W. & Tamisiea, M. E. (2009). Identifying the causes of sea-level change. *Nature Geoscience*, 2, 471-478.

Rahmstorf, S. (2007). A semi-empirical approach to projecting future sea-level rise. *Science*, 315, 368-370.

Rignot, E. & Kanagaratnam (2006). "Changes in the Velocity Structure of the Greenland Ice Sheet". *Science* 311 (5763): 986 et seq.. doi:10.1126/science.1121381. PMID 16484490.

Scavia, D. et al (2002) Climate change impacts on U.S. coastal and marine ecosystems. *Estuaries*, 25, 149-164.

Titus, J. G. & Richman, C. (2001). Maps of lands vulnerable to sea-level rise: modeled elevation along the US Atlantic and Gulf coasts. *Climate Research*, 18, 205-228.

Wu, S. Y., Yarnal, B. & Fisher, A. (2002). Vulnerability of coastal communities to sea-level rise, a case study of Cape May County, New Jersey. *Climate Research*, 22(3), 255-270.

Wu, S. Y., Najjar, R. & Siewert, J. (2009). Potential Impacts of Sea-level Rise on the Mid- and Upper-Atlantic Region of the United States. *Climatic Change,* 95.

Yin, J., Schlesinger, M. E. & Stouffer, R. J. (2009) Model projections of rapid sea-level rise on the northeast coast of the United States. *Nature Geoscience*, 2, 262-266.

Zervas, C. (2001). *Sea Level Variations of the United States 1854-1999*. NOAA Technical Report NOS CO-OPS 36.

Zervas, C. (2009) *Sea level variations of the United States 1854–2006*. NOAA Technical Report NOS CO-OPS 053.

In: Sea Level Rise, Coastal Engineering, Shorelines and Tides ISBN: 978-1-61728-655-1
Editor: Linda L. Wright, pp. 139-153 © 2011 Nova Science Publishers, Inc.

Chapter 6

SHORELINES IN COLOMBIA (SOUTH AMERICA)

Jesus Olivero-Verbel**, *María Velez de Lopez* *and Katia Noguera-Oviedo
Environmental and Computational Chemistry Group. Faculty of Pharmaceutical Sciences. Campus of Zaragocilla. University of Cartagena, Cartagena, Colombia

Abstract

Shorelines are essential resources for almost all South American countries. However, the lack of adequate planning and management are accelerating their destruction. In Colombia, significant issues for coastal management and sustainable development include overpopulation, sewage discharges/microbial contamination, destruction of mangrove forests, industrial/port development, shrimp farming, recreation projects, sedimentation and erosion; among others. Moreover, it is quite clear that several shoreline cities, in particular at the north coast of Colombia where the most populated are found, must act quickly to start developing viable and long lasting strategies to adapt to the impacts of climate change. Similarly, it is quite urgent to promote actions to balance the conflicting interests of development with the preservation of the shoreline natural resources. This includes creating strong education programs, shoreline use/development ordinances that are environmentally-friendly and derived from technical advisory committees, campaigns to massify the use of eolic and solar energy, and conservation plans to protect and enhance the environment while encouraging public access and adequate use of the shorelines for current and future generations.

Introduction

Colombia is the only South American country that has coastlines on two oceans, 1937 km on the north (Atlantic ocean) and 1576 km of coastline on the south coast (Pacific ocean) [1]. It has a maritime jurisdictional area of 928660 km^2 (589560 km^2, on the Caribbean and 339100 km^2 on the Pacific), which is relatively equal to that of its land territory (1141748 km^2), corresponding to 44.8% of the country area (total area: 2070408 km^2) [2]. According to

*E-mail address: jesusolivero@yahoo.com, joliverov@unicartagena.edu.co. Tel: 57-5-6698179, Fax: 57-5-6698323, Environmental and Computational Chemistry Group, Faculty of Pharmaceutical Sciences, Campus of Zaragocilla, University of Cartagena, Cartagena, Colombia (Corresponding author)

the latest national census (2005), the population located directly on the coastal zone is 4584214 inhabitants, representing 10% of the total population [1]. Colombian coasts include three main areas with different biophysical, social, economic and cultural elements that make them substantially different geographical units. The Caribbean (Atlantic), the Pacific, and the insular region (the San Andrés and Providencia Department in the Caribbean sea, and the Malpelo and Gorgona Islands in the Pacific Ocean)[3].

The Colombian Caribbean region, located on the Northwestern South America, has been classified as an area of low seismicity and low tsunami risk [2]. The region's climate is influenced by the geographical position of the inter-tropical convergence zone (ITCZ) [4] and by the movement of the American monsoon system. In addition, the Northern Colombian Caribbean, between Guajira and Cartagena, can be affected by the Caribbean hurricane season [5].The Caribbean region, through its water systems, is the recipient of all the resulting environmental impacts of natural and anthropogenic processes in the Colombian Andean region, which together with the Caribbean region itself, houses about 90% of the population, and supports the highest degree of transformation of the natural basis [6]. The Caribbean basin is formed by large rivers flowing into the sea forming deltas, in particular the Magdalena, Rancheria, Sinú and Atrato Rivers [5].

Unlike the Caribbean Coast, the Pacific is characterized as a zone of high tectonic activity, based on the convergence of the South American plate and the Nazca oceanic plate (converging at a rate of 54 mm/year) [4]. Therefore, the Pacific Coast is subject to severe natural disasters including earthquakes, destructive tsunamis, flooding and soil liquefaction [7]. Characterized by a humid tropical climate, high cloudiness and pluviosity (Average rainfall in the entire Pacific basin ranges from 2000 to 12700 mm/yr [4]), this region gives rise to a wide vegetation. This coast is also influenced by the ITCZ, as well as with its proximity to the ocean masses and climate events like El Niño [3], and also with tides that vary between 3.8 and 5 m, much more than those registered in the Caribbean[8][9]. The hydrological network of this basin is composed of many short and torrential rivers, being the most important the Mira, San Juan and Baudó Rivers [3].

The Atlantic insular coastal zone, formed by the San Andrés archipelago, consisting of three Islands and some Islets, so called “cayos” (cays) (San Andrés, Providence and Santa Catalina), was declared as a Biosphere reserve in 2000 [10] . It is characterized by the presence of coral reefs, marine pastures, mangrove and coastal beaches [8] and dunes [11]. The principal aims of this review were (1) to examine the current conditions and uses of Colombian shorelines, (2) to detail those factors that impact these ecosystems, and (3), to describe possible alternatives that could be useful to improve their sustainability.

Ecosystems and Socioeconomic Status of Coastal Areas

In its coastal, maritime and insular areas, Colombia has complex ecosystems with great biological diversity. These are represented by coral reefs (2900 km^2), mangroves (3900.1 km^2), transition forests, marine pastures (only in the Caribbean ocean, about 432.23 km^2), coastal lagoons and estuaries (Ciénaga Grande de Santa Marta, la Virgen marsh and Cartagena Bay), beaches and rocky littorals (approximately 1037 km^2) as well as some cliffs [1]. 11 out of 51 protected areas in Colombia belong to coastal ecosystems, 7 of them are located on the Atlantic with a total area of 27077 Km^2, and 4 in the Pacific, corresponding to

2348 Km^2 [12]. Moreover, the country has three Ramsar sites, two of which are located on the coastal shoreline [12]. Numerous species have limited distribution in the area of Chocó creating extraordinary 'beta' diversity, or significant variations in species from one locality to another and many of those species are not found anywhere else in the world [13]. In fact, most of the plants and animal species of Chocó are yet to be discovered, between 7000 and 8000 out of the 45000 species registered in Colombia are found in the Chocó [14]. Endemic plant species are more than 2000 while endemic birds comprise more than 100 species, which represent one of the highest endemism levels in the Planet [14]. Chocó contains one of the last pristine stretches of coastline in tropical America. This coastline is a stopping point for migratory humpback whales and the shoreline is a critical feeding, wintering and stopover site for millions of migratory shorebirds. Mangrove forests protect the coast from erosion while providing nursery for young fish that feed in and depend on nutrient rich waters around mangroves. The region's marine area is home to an abundance of fish species and marine mammal populations.

In general, the states (departments) belonging to the coastal areas of Colombia, in particular the Pacific, possess low economical development [12], due in part to the lack of industries and inadequate road infrastructure. To have access to most of this territory it is necessary to be transported on small boats using the extensive network of rivers that run through the region, which has few significant urban settlements, and no major industries to promote economic activities [7][8]. The economy is based on activities aimed to provide daily subsistence, predominantly timber harvesting, fishing, shipping, gold mining, and in some parts of the coastal zone, eco-tourism and shrimp aquaculture have become important [7][8].

Colombian coasts possess many key ecosystems that are critical to control shoreline degradation, being some of the most important mangrove forests, marine pastures and coral reefs.

Mangrove forests have been highly impacted during the last years in the country (Figure 1). According to the National Geographic Institute Agustín Codazzi (ww w.igac.gov.co), during the 60´s mangrove areas reached 5013 km^2, and in 1996 according to a study of the Environment Ministry of Colombia these were 2927.24 km^2 in the Pacific and 863.1km^2 in the Atlantic, for a total of 3780.34 km^2 [15]. More recently, in 2004 [16], estimates showed that those numbers had changed to 2830 km2 and 880 km2, respectively, for a total of 3710 km^2, including island territories. Currently (2010), it has been calculated that mangrove forests cover an approximate area of 3009.07 km^2; 675.04 km^2 in the Caribbean and 2334.03 Km^2 in the Pacific, especially at the Department of Nariño [11]. Mangrove species identified in the Atlantic coast include *Rhizophora mangle, Avicennia germinans, Laguncularia racemosa, Conocarpus erecta, Pellicciera rhizophorae*, whereas for the Pacific, in addition to those previously mentioned it is possible to find *Mora oleífera, Avicennia bicolor y Rhizophora racemosa.* Mangrove ecosystem is distributed along the Colombian shorelines (Figures 3 and 4), and although it must be protected under the current legislation, in some places, near 25% is in critical condition or death [15][16], in particular at the Cienaga Grande de Santa Marta and the Dique Channel delta.

The future of mangroves forest in Colombia is similar to that in the rest of tropical regions of the world. Extensive wood extraction, aquaculture, in particular shrimp farming, civil constructions that alter hydrodynamics of the water interchange with rivers and the ocean, mainly roads and ports, are some of the challenges for mangrove survival [6][15]. It is clear that among governmental authorities and the community there is no awareness about the environmental services provided by this ecosystem, and importantly, the role they might play during sea level rise in the global warming era. In addition, poverty is a powerful factor that threatens mangrove ecosystems, as it provides food, energy and wood for housing. Therefore, sustainable development approaches must be undertaken in order to guarantee both the mangrove and people survival.

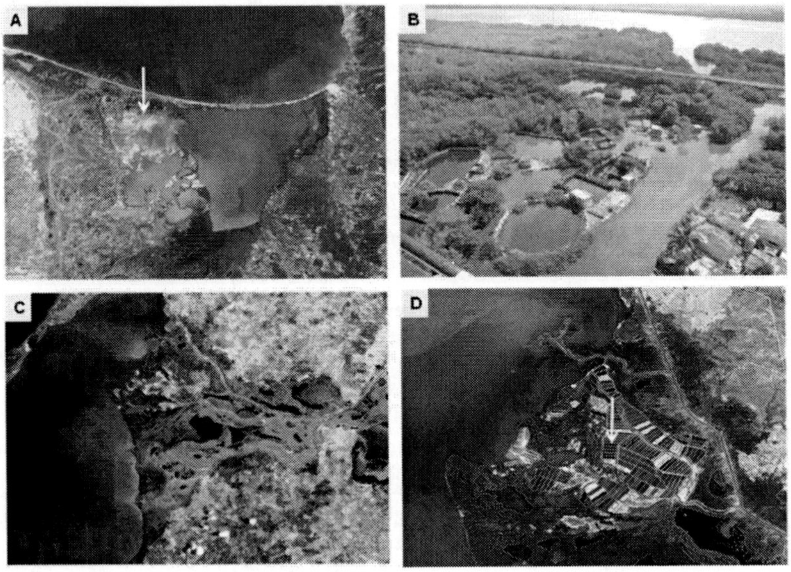

Source: U.S. Geological Survey and Rafael Vergara Navarro. 2009 (With Permission)

Figure 1. Impact of shoreline developments of mangrove ecosystems in Colombia. A. Road construction between Barranquilla and Santa Marta destroyed extensive mangrove areas. B. Fish aquaculture at La Virgen Marsh (Cartagena) is a common practice to appropriate land. Landscape at the Dique Channel (Magdalena River connection with the sea) delta before (C) and after (D) shrimp farming.

Marine pastures are only found in the Atlantic Ocean, and most of them (80.2%) are located specifically in front of the Guajira peninsula. This ecosystem serves as a refuge for many species and it helps to control the energy transferred by tides and waves, helping sediment stabilization, creating environments of low energy and protecting the shoreline [1].

Coral reefs are an important feature of some areas of the coastal shoreline in Colombia, It covers an area of 1091 km^2, representing 0.4% of the total existent in the planet. 76.5% of the Colombian coral reefs are located in San Andrés, Providencia and Santa Catalina islands, and the rest 23.5 % on the continental part of the Caribbean Coast, most of it between San Bernardo and Rosario islands, with near 220 km^2 of living corals [11]. Only a small fraction is located in the Pacific (15 km^2), near the Gorgona island [1]. Coral reef destruction in the country is an ongoing process [1], basically promoted by pollution, overfishing, sedimentation and global warming [17]. Several whitening processes in Colombian coral reefs, have been taken place, and as a result of this problem, several species such *Acropora sp.* have almost disappeared [18]. This ecosystem is critical for the control of erosion on beaches, decreasing the energy carried by the waves [15], and its preservation is fundamental to control flooding, as a source of fish for local communities [17].

Shoreline Uses and Impacts

Port Activities

Colombia has nearly 150 ports on both coasts [19], and they mobilize near 90% of the commercial trade flow of products between Colombia and the rest of the world. Main ports of the Atlantic are located in Puerto Bolívar (Guajira), Santa Marta, Ciénaga, Barranquilla, Cartagena, Coveñas, Urabá and San Andrés, whereas in the Pacific, Buenaventura and Tumaco are the most representative [12]. As in 2007, most of the port activities in the country were represented by coal (71%) and oil (20%) exportations from Puerto Bolívar/Santa Marta, and Coveñas, respectively; being the rest 9% composed of dried, liquid and refrigerated products, as well as other types of cargo [20]. On the pacific, Buenaventura is considered the most important port. During 2009 it moved a total of 10006035 tons [21]. The port of Tumaco, smaller than that of Buenaventura, is used to export oil through the Transandino Pipeline, that runs 310 km from Orito (Department of Putumayo), as well as fishing products and general cargo [20].

Tumaco and Buenaventura bays receive pollutants from transport and use of petroleum and derived compounds. Most of this contamination comes from fishing and cargo ships [8], many of which seem not to be under the control of the MARPOL convention (International Convention for the Prevention of Pollution From Ships); accordingly, they might be source of several harmful substances for the marine environment such as oil, sewage, solid hazardous waste; oily bilge water, and ballast water, among others [20]. Due to the expanding maritime traffic, port areas are of high risk of pollution from oil spills and marine garbage derived from this activity [12][20].

In general, civil constructions built on beaches and dunes to protect the port infrastructure itself and the nearby areas, have contributed to shoreline erosion, altering the transport and sediment distribution that reaches the littoral [5]. Moreover, mangrove destruction, changes in the pattern of marsh systems, dredging, and more recently, the controversial use of sacred places for indigenous people, are growing problems related to port construction and development. In recent years, an important percentage of Colombian economy has been based on coal mining and exportation through different locations along the shoreline. The legislation flexibility, and the lack of determination of environmental protection agencies, have created a mayor problem in several ports, as a result of coal dust release into the environment during transportation or transfer into ships. The process has deeply impacted the tourism in cities such as Santa Marta, and even industries located nearby the ports have been affected in Cartagena. Although the required technology to avoid these problems does exist and it is not relatively expensive, a better knowledge, about those treasures that are being lost or damaged, is quite necessary to add value to those things that shape our identity, our future.

Despite all impacts generated by the port industry, it should be pointed out that these activities have helped the development of several other sectors, critical for the economical growth of the country, such as industry, mining, fishing and tourism, among others [6].

Tourism

This is one of the most important and expanding economical activities in Colombia. According to the Ministry for Trade, Industry and Tourism, during 2009, 1353760 tourists arrived into Cartagena, Santa Marta and San Andrés Island, 10.7% more than in 2008, many of them (281490 passengers) in 176 ships from all over the world [22]. The development of tourism in these places at the Caribbean coast is due to several factors. First, this region is strategically located and it has an enormous variety of cultural and biological resources. The city of Cartagena, for instance, the most visited in the country, was declared "Historical Patrimony of Humanity" by UNESCO in 1985. Together with Santa Marta, these cities offer to the visitants magnificent colonial buildings, modern hotel infrastructure, beaches, and nearby national parks, such as Rosario Island (Cartagena), and the Tayrona Park and the Sierra Nevada of Santa Marta (Santa Marta). Second, this coast is less prone to be affected by tsunamis, as in the Pacific, as well as by hurricanes. Today, although touristic development is one of the leading factors that affect shoreline ecosystems in Colombia, the sector itself is threaten by urban sewage and residues from several industries that directly impact the littoral [12], weakening the tourism-derived offer. Major coastal cities do not have sewage treatment plants [20]. Santa Marta city has the only existing submarine outlet in Colombia, and it is located relatively far from shore. Cartagena city is building a similar one, but in both cases, the "sea" is considered the major "treatment" approach for these residues, an inappropriate concept.

Figure 2. Impact of sea level rise in Cartagena City, Colombia. A) Flooding of the Plaza de la Aduana, historical center. B) Flooding at the touristic sector of Bocagrande during the rainy season and high tides. Source: Rafael Vergara Navarro, 2009.

Each day bigger shoreline extensions are been used to built urbanistic projects, several of them without appropriate assessment of their impact on the landscape and ecosystems. This is particularly extensive in places around touristic cities such as Cartagena, Barranquilla and Santa Marta. In some cases, houses are located just few meters away from shore. Simultaneously, many natural flooding places are being filled with material, or the mangrove ecosystem is destroyed or illegally occupied, in order to create new beaches or suitable places for residential and touristic purposes. At the Atlantic coast, where most Colombian touristic cities are located, global warming is taking his toll. During the last years, in Cartagena, for instance, several touristic places suffer from flooding (Figure 2), sometimes receiving water directly from the shoreline, and quite frequently, the process occurs by water transfer through the sewage collecting system.

The Pacific coast, with a tourism business less developed, posses an immense potential for its development, basically due to the natural resources, lack of urban impact and extensive possibilities of ecological tourism [6].

Fisheries & aquaculture

Coastal fishing is one of the main economical activities linked to marine resources in Colombia. It has been estimated that these coasts have at least 2000 fish species [1], and just few of them, such as tuna, shrimp, squid and crabs, are likely to be used for commerce [12]. According to the Colombian Institute for Rural Development (INCODER), the country has a fishing potential equivalent to 224000 tons, 75% found in the Pacific, and the rest 25% in the Atlantic Ocean [23]. During the last two decades, fishing in the Pacific, in particular tuna, represented near 85% of the total marine fish production in the country, this last reaching in average 160578 tons during 1997-2007 [12]. However, currently the scenario is quite different and fish production has been focusing on farming instead of extraction from natural sources, as has been happening internationally. In fact, according to FAO, for the 2006, global fish production reached 110 million tons, and 45% of it came from aquiculture [24][25].

In Colombian coastal shorelines, aquaculture has been growing fast in the last decades, being the white shrimp (Penaeus vannamei), one of the most representative farming species. Although by 2005, 81% of the shrimp production in Colombia came from aquaculture [26], it is known that lately this activity has started to decrease its economic role as a source of income for the country; and the sector is facing enormous challenges, in particular low international prices and problems related to climate-derived factors, and illnesses such as the Taura syndrome and the white spot.

Shorelines as Pollutant Sinks

Urban sewage, industrial activities, and sedimentation are some of the most important causes of shoreline pollution in Colombia. However, scientific data is really scarce and few governmental agencies deal with most of the basic research, but without having enough resources to accomplish useful and significant characterizations of the problem. According to the National Department of Statistics (DANE), sewage systems along the coasts only covers 47.6% of the population [20], and during 2008, near 848274 m^3/day of sewage were disposed into the ocean through different rivers: 139223 m^3/day into the Pacific and 709051 m^3/day into the Atlantic [1], impacting water quality and strategic ecosystems such as coral reefs, mangroves and coastal lagoons.

The Caribbean coast of Colombia receives sewage from at least 31 different municipalities along the coast. Specifically 57358 ton/year of biological oxygen demand (DBO5), 136567 ton/year chemical oxygen demand (DQO), 2731 ton/year of nutrients and 95597 ton/year of suspended solids [20]. According to INVEMAR, the cities of Barranquilla, Cartagena, Santa Marta and Riohacha, contributed the most. On the Pacific Coast, on the other hand, the pollutant load is mainly provided by Buenaventura and Tumaco (84%), with 7975 ton/year of DBO5, 18988 ton/year DQO, 380 ton/year of nitrogen, 152 ton/year of phosphates and 13292 ton/year of total suspended solids [20].

As previously mentioned, very little is known about chemical pollution in the shorelines of Colombia. However, perfluorooctanesulfonates (PFOS) and its metabolite, the perfluorooctanoic acid (PFOA), have been detected in people living in Cartagena [27], at concentrations similar to those quantified in Japanese and Americans [28]. In Cartagena Bay, the most industrialized site along Colombian coasts, these chemicals have also been found in fish and birds [29], at concentrations greater than those found in less urbanized sites. Together with these new pollutants, polycyclic aromatic hydrocarbons (PAHs) have been measured in sediments and fish from the bay, pointing out the impact of petrochemical and refinery-derived activities on the estuary ecosystem [30]. Pollution in Colombian shorelines is not only organic. Mercury, a highly toxic heavy metal, has been found in different matrices in Cartagena Bay, most of it, geographically correlated with an extinct chlor-alkali plant located on the bay [31][32].

More recently, some evidences have shown that the health of Colombian marine ecosystems, at least those nearby to Cartagena, is not entirely satisfactory from a biological point of view. For instance, several fish species have been found highly parasitized with nematodes [33] and trematodes. In the last case, fish species such as *Mugil incilis* can be found infected with tens of thousands cysts per specimen [34].

Besides chemical and biological pollution, sedimentation and erosion are greatly contributing to shoreline degradation.

Sedimentation and erosion produce morphological changes in the coastal shoreline, that will depend on factors such as geology, hydrodynamic agents (wind, currents, tides), tectonic movements and man-induced processes, such as climate change/sea level rise [5]. In coastal areas sedimentation results from material loads carried by rivers. In the Atlantic coast, Magdalena River is the most important, and its sediment load is so high that it is considered among the first 10 in the world (144×10^6 tons/year) [35]. For comparison, the Amazon River, with a basin area 25.4 fold greater has a sediment load only of 1200×10^6 tons/year [36]. The situation of Magdalena River is the result of extensive deforestation in its basin (40% in a 20-year period)[35][37], as well as a consequence of mining activities (Cauca River basin), and lack of good and environmentally-friendly practices for land use. Besides the Magdalena River, the Atrato and the Sinu Rivers also carry important load sediments with 11×10^6 and 6.1 $\times 10^6$ tons/year, respectively [35]. Some examples of sediment plumes that impact the Colombian shorelines are shown in Figure 3.

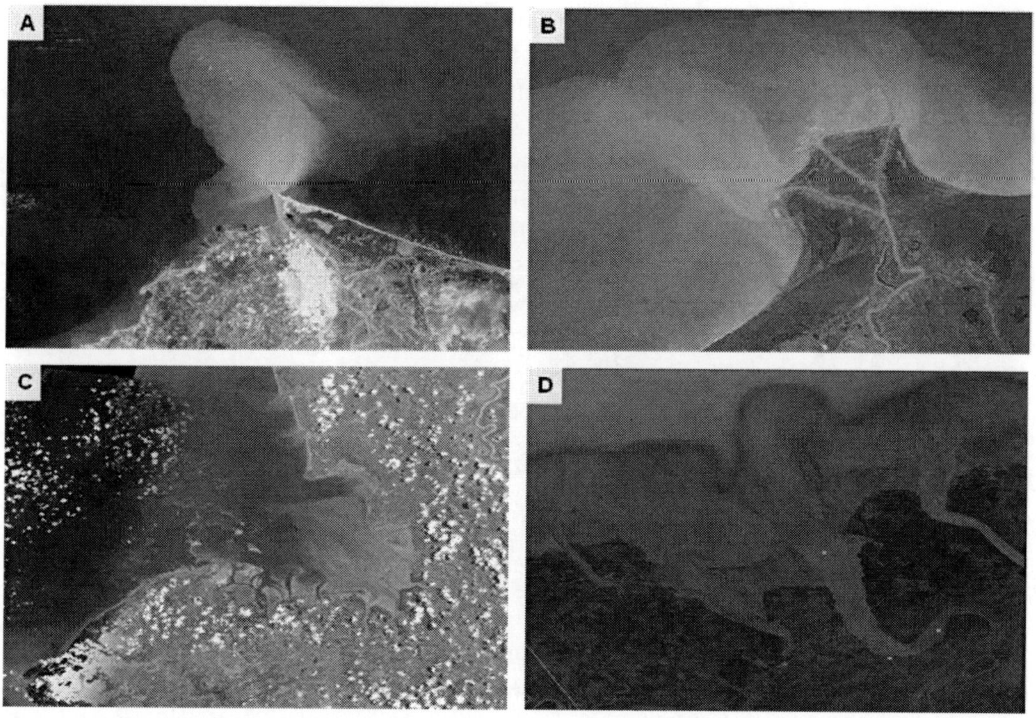

Figura 3. Sediment plume from several river deltas. A. Magdalena, River, Atlantic. B. Sinú River, Atlantic. C. Túmaco Bay, D. Guapi River. Landsat photograph courtesy of U.S. Geological Survey.

A similar scenario is also observed in the Pacific. Just as an example, the San Juan River, with $16.6x10^3$ km^2 basin, and an annual average precipitation of 7.277 mm, has the biggest water (2550 m^3/s) and sediment ($16x10^6$ ton/year) discharge in the whole West coast of South America [38], although the Patia River is getting close with $14x10^6$ ton/year for a basin area of $17x10^3$ Km^2 [35][39].

Despite the large amount of sediments carried by the rivers to the sea, during the last years, erosive processes have prevailed over sedimentation [40, 41, and 42]. In fact, on the Caribbean coast the shoreline has receded as a result of not significant sediment-dependent beach formation processes [5], in part due to small size of sediments particles that are easily moved away [41]. The shoreline has also been mined to illegally extract materials for construction. This practice is common along the Atlantic coast, in particular at places such as "Los Morros", nearby Cartagena [5]. Simultaneously with these activities, mangrove destruction and the development of urban settlements have left the littoral without the necessary sand to build beaches, and therefore, erosive processes take place.

The two coasts of Colombia have different littoral-forming substrates, and therefore, they play an important role in the susceptibility to erosion. The Atlantic coast has few hard substrates, most of them located in Santa Marta [5]. On the other hand, the Pacific coast has two distinct geomorphological characteristics derived from the closeness to the mountains: the high and the low coasts, being the second more likely to suffer erosion due to the presence of a variety of sedimentary environments [42].

The increase in the number and strength of hurricanes, tropical storms and high-swelling seas, as a result of global warming, together with the rising in the sea level, calculated in 22 cm during the last 50 years, has lead to an increase in coastal erosion, in particular at the Caribbean coast [5]. All along the shoreline, this coast shows evidence of ongoing erosion processes, with rates of erosion and sedimentation greater than 10 m/year [40]. In some areas this degradations is occurring very fast, such as along the left-hand margin where the Magdalena river joins the sea. At this place, the shoreline has moved around 600 m onto the continent, equivalent to a rate of 60 m/year [40].

On the Pacific Coast, near 22% (345 km) of the shoreline presents ongoing erosion process on the littoral, although this value could be greater when considering river-related erosion [1]. Previous measurements have shown that this process has accelerated during the last decade [43], in part due to land masses collapse and sinking after earthquakes (up to 1.6 m), and also by the sea level increase caused by El Niño (up to 35 cm during 1997-1998) [7]. As this situation will continue as a consequence of the climate change, some towns have been obligated to move deeper into the continent, as it has happened in the Choncho on the Pacific [7] and Pueblo Nuevo in the Atlantic [40].

The Future

Global warming changes along the planet will force necessary actions to protect Colombian shorelines against sea level rise, in order to decrease the impact of this phenomenon on the national economy. Solutions must be clever and not rely only on heavy civil infrastructure construction and/or population displacement. Nature should be called for help using the best of its knowledge, and sustainable development needs to account on preservation of mangroves, coral reefs and any other shoreline biodiversity spots. High risk flooding/erosion areas need to be detected, and appropriate planning with all interested actors, from the local people to the government and the scientific community, has to be undertaken to avoid further irreversible damage to the shoreline. This process must start as soon as possible and not only should be focused on the shoreline. Pollution and sediments carried by rivers from their basins require control to decrease their effects on coastal areas. In those cases where assembling of constructions would be necessary to protect the shoreline, sunlight, wave and eolic energy projects should be considered as an integral part of the facilities, and also on those new projects undertaken elsewhere around the coastal line. This will transfer part of the energy force acting on the shoreline to the local economy, or help on the maintenance of the site. There is no excuse to continue wasting these non-pollution types of energy in a temperature-rising planet.

It is a fact that few policy-makers are generally aware of the importance of preserving shoreline ecosystems, and this thinking is also shared by developers and constructors, who believe this is just a place for ports, hotels/resorts and beaches. Therefore, the survival of shoreline ecosystems will depend on the adequate selection of very well qualified and honest technical advisory committees, the active participation of the community on policy- and decision-making processes, on efforts to avoid wasting energy available on site (waves, wind, and sunlight) and on massive and well designed environmental education programs for all academic and administrative levels.

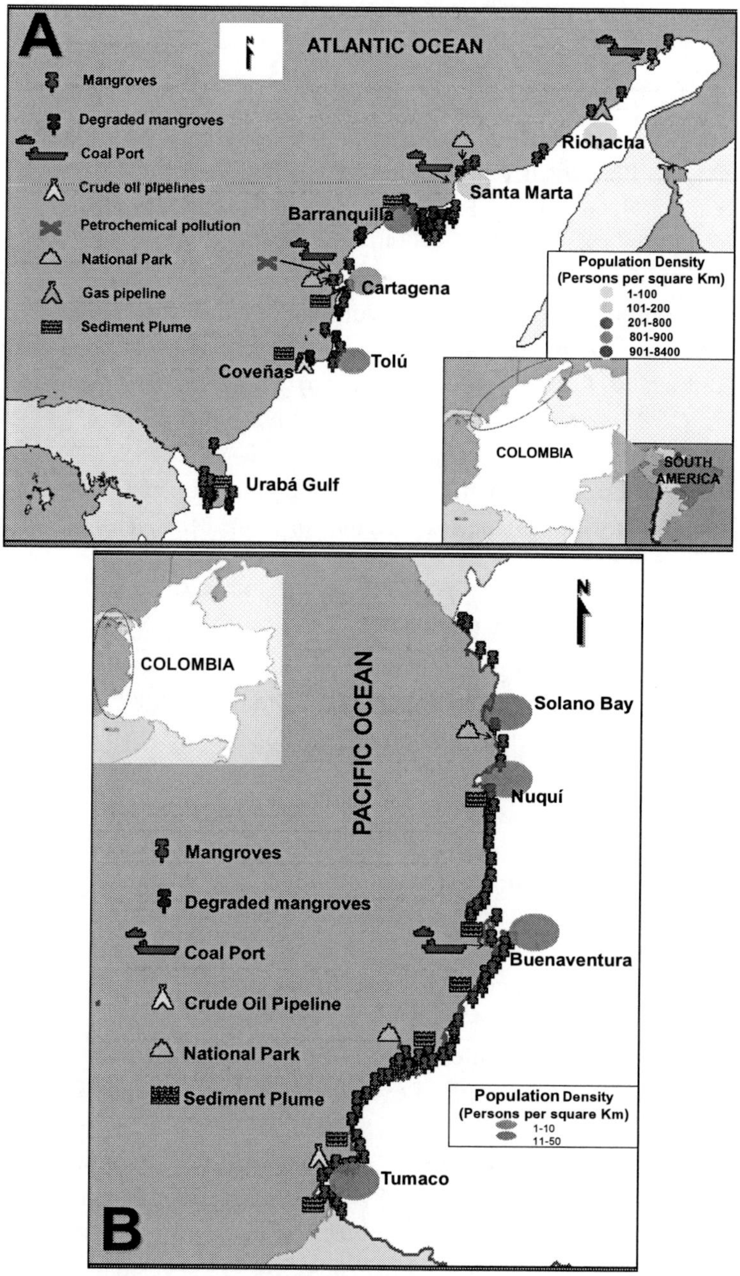

Figure 4. Map of Colombian coasts showing strategic locations and some examples of factors that impact the quality of shorelines. A. Atlantic, B. Pacific.

Colombian shorelines are still of magnificent environmental value, although they face enormous challenges due to many potential impacts from diverse sources (See Figure 4), and as well as many other tropical ecosystems, those resources should be considered as classrooms. Learning from these unique habits represents excellent opportunities to explore many frontiers of our own life. Using them under sustainable criteria is not only our responsibility, it is also the field where our survival and development battle to shape our destiny.

References

[1] INVEMAR -Instituto de Investigaciones Marinas y Costeras. *Informe del Estado de los Ambientes y Recursos Marinos y Costeros en Colombia: Año 2008*. Santa Marta: Serie de Publicaciones Periódicas No. 8. Santa Marta, 2009.

[2] DNP- Dirección Nacional de Planeación. *Elementos básicos para el Manejo Integrado de Zonas Costeras*. ISBN: 978-958-8340-36-4. Bogotá. 2008.

[3] INVEMAR. (2003). Programa holandés de asistencia para estudios de cambio climático, Colombia: Definición de la vulnerabilidad de los sistemas bio-geofísicos y socioeconómicos debido a un cambio en el nivel del mar en la zona costera colombiana (Caribe Continental, Caribe insular y Pacifico) y medidas de adaptación. VII, *Resumen Ejecutivo y CD- Atlas digital*. Programa de Investigación para la Gestión Marina y Costera – GEZ. Santa Marta, Colombia. ISBN: 958 – 97264 -2-9.

[4] Restrepo, J. D & López, S. A. (2008). Morphodynamics of the Pacific and Caribbean deltas of Colombia, South America. *Journal of South American Earth Sciences*, 25, 1-21

[5] Posada, B. & Henao, W. (2008). Diagnóstico de la erosión en la zona costera del Caribe colombiano. Santa Marta: INVEMAR, *Serie Publicaciones Especiales* No. 13.

[6] MMA.Política nacional ambiental para el desarrollo sostenible de los espacios oceánicos y las zonas costeras e insulares de Colombia. 2000. Available from: http://www1.minambiente.gov.co/viceministerios/ambiente/dir_ecosistemas/2.%20Ordenacion%20manejo%20y%20restauracion/Espacios%20Oceanicos/Politica%20Zonas%20Costeras.PDF.

[7] Correa, I. V & González, J. L. (2000). Coastal erosion and village relocation: a Colombian case study. *Ocean & Coastal Management*, 43, 51- 64.

[8] Tejada, C., Castro, L., Navarrete; Cardona, T., Otero, L., Afanador, F., Mogollón, A. & Pedroza, W. (2003). Panorama de la Contaminación Marina del Pacífico Colombiano. Centro Control Contaminación del Pacífico Colombiano. Ed. DIMAR.2003 Serie Publicaciones Especiales Vol. 3, *San Andrés de Tumaco*, 120.

[9] Pabón, J. D., Rojas, P. J., Montealegre, J. E., Robertson, K., Ceballos, J. L., Martínez, N. (2001). El océano. In: Pablo Leyva. El medio ambiente en Colombia. Second Edition Bogotá. *IDEAM*. 190-227.

[10] Friedlander, A., Sladek, J., Sanchez, J., Appeldoorn, R., Usseglio, P., McCormick, C., Bejarano, S. & Mitchell-Chui, A. (2003). Designing effective marine protected areas in Seaflower biosphere reserve, Colombia, based on biological and sociological information. *Conservation Biology*, 17, 1769-1784

[11] Invemar. Invemar: 46 años de investigación de los recursos y ambientes marinos y oceánicos de Colombia. 2009. Available from: http://www.invemar.org.co/noticias.jsp?id= 3769&pagina=2&idcat=122 Last accessed February 2/2010.

[12] DNP-Departamento Nacional de Planeación. "2019 Visión Colombia II Centenario. Aprovechar el territorio marino-costero en forma eficiente y sostenible". Propuesta para discusión. Bogota: *Excelsior Impresores*., 2007.

[13] WWF–World Wildlife Fund. Chocó-Darién moist forests (NT0115). 2001. Available from: http://www.worldwildlife.org/wildworld/profiles/terrestrial/nt/nt0115_full.html.

[14] WRM–World Rainforest Movement. The Colombian Chocó: A megadiverse region in a megadiverse country . . . for how long?. WRM's bulletin Nº 44. 2001. Available from World Rainforest movement http://www.wrm.org.uy/bulletin/44/Colombia.html.

[15] Garay, J. (1999). Modulo problemática ambiental, contaminación marina y cambio climático global en las zonas costeras. Corporación Universitaria Tecnológica de Bolívar- Universidad Industrial de Santander. *Especialización en Ingeniería Ambiental. Cartagena, Junio*, 1999, 98.

[16] Sanchez–Paez H. et al. (2000). *Hacia la recuperación de los manglares el Caribe colombiano*. Minambiente/OIMT/ACOFORE. Santa fé de Bogotá. D.C. *294*.

[17] Fernandez, L. (2007). "Efecto de los cambios globales sobre los humedales de Iberoamérica": una iniciativa para la transferencia de conocimientos en materia de protección, conservación y gestión sostenible de los humedales de Iberoamérica. In: Serie El Agua en Iberoamérica. *"Efecto cambios globales sobre los humedales de Iberoamérica"*. Volumen *12*. Buenos Aires: RED CYTED 406RT0285, 5-34.

[18] Alvarado, M. E. & Corchuelo, C. (1992). Los nutrientes, la temperatura y la salinidad provenientes del Canal del Dique como factores de deterioro en el Parque Nacional Natural Corales del Rosario–CCO. Memorias del VIII Seminario Nacional Ciencias del Mar. Santa Marta, Colombia, 277-287.

[19] Vilora, J. Ciudades portuarias del Caribe colombiano: propuestas para competir en una economía globalizada. Cartagena: Banco de la República – Centro de estudios económicos regionales (CEER). 2006.

[20] Redcam-Red de Vigilancia de la Calidad Ambiental Marina en Colombia 2008. Diagnóstico y Evaluación de la calidad ambiental Marina en el Caribe y Pacifico Colombiano. 2008. Avalilable from: http://www.invemar.org.co/redcostera1/invemar/docs/Informe_R
EDCAM_2009.pdf.

[21] SPRBUN-Sociedad Portuaria Regional de Buenaventura. Información para comercio exterior. 2009. Estadísticas de exportaciones e importaciones para el año 2009. Available from: http://www.sprbun.com/informacion-para-comercio-exterior/estadisticas/acumulado_ano.php.Last accessed February 12, 2010.

[22] PROEXPORT. Informe Consolidado de Exportaciones, Turismo e Inversión Extranjera. 2010/01/26. Available from: http://www.proexport.com.co/VbeContent/NewsDetail.asp?ID=9971&IDCompany=16.

[23] Ministerio de Agricultura y Desarrollo Rural. INCODER, CCI – Corporacion Colombia Internacional. Pesca y acuicultura Colombia, 2006. 138p. Available from: http://www.agronet.gov.co/www/docs_agronet/200833174541_InformePescaCompleto.pdf.

[24] FAO. El estado mundial de la pesca y la acuicultura 2008. 2009. Organización de las Naciones Unidas para la Agricultura y la Alimentación. Roma. ISSN 1020-5500. Available from: http://www.fao.org/docrep/011/i0250s/i0250s00.HTM. Last accessed: February 13, 2010.

[25] FAO. Visión general del sector acuícola nacional Colombia. 2010. Available from: http://www.fao.org/fishery/countrysector/naso_colombia/es. Last accessed: February 19, 2010.

[26] Minagricultura - Ministerio de Agricultura y Desarrollo Rural. Observatorio de agrocadenas de Colombia. 2005. La cadena del camarón de pesca en Colombia. Una

mirada global de su estructura y dinámica 1991 – 2005. Available from http://www.agrocadenas.gov.co, Last accessed Febreary 15, 2010.

[27] Kannan, K., Corsolini, S., Falandysz, J., Fillmann, G., Kumar, K. S., Loganathan, B. G., Mohd, M. A., Olivero, J., Van Wouwe, N., Yang, J. H. & Aldoust, K. M. (2004). Perfluorooctanesulfonate and related fluorochemicals in human blood from several countries. *Environmental Science and Technology*, 38, 4489-4495.

[28] Kärrman, A., Van Bavel, B., Järnberg, U., Hardell, L. & Lindström, G. (2006). Perfluorinated chemicals in relation to other persistent organic pollutants in human blood. *Chemosphere*, 64, 1582-1591.

[29] Olivero, J., Tao, L., Johnson, B., Guette, J., Baldiris, R., O'byrne, I. & Kannan, K. (2006). Perfluorooctane sulfonate and related fluorochemicals in biological samples from the north coast of Colombia. *Environmental Pollution.*, 142, 367-372.

[30] Johnson, B., Olivero, J., Lu, S., Guette, J., Baldiris, R., O'Byrne, I., Aldous, K., Addink, R. & Kannan, K. (2008). Polycyclic aromatic hydrocarbons and their hydroxylated metab-olites in fish bile and sediments from coastal waters of Colombia. *Environmental Pollution.*, 151, 452-459.

[31] Olivero, J., Johnson, B., Baldiris, R., Güette, J., Magallanes, E., Vanegas, L. & Kunihiko, N. (2008). Human and crab exposure to mercury in the Caribbean coastal shoreline of Colombia: Impact from an abandoned chlor-alkali plant. *Environment International.*, 34, 476-482.

[32] Alonso, D., Pineda, P., Olivero, J., González, H. & Campos, N. (2000) .Mercury levels in muscle of two fish species and sediments from the Cartagena Bay and the Ciénaga Grande de Santa Marta, Colombia. *Environmental Pollution,* 109, 157-163.

[33] Olivero, J., Baldiris, R. & Arroyo, B. (2005). Nematode infection in *Mugil incilis* (Lisa) from Cartagena Bay and Totumo Marsh, north of Colombia. *The Journal of Parasitology*, 91, 1109-1112.

[34] Galván, D., Olivero, J. & Barrios, L. (2010). Occurrence of Ascocotyle (Phagicola) longa Ransom, 1920 (Digenea: Heterophyidae) in *Mugil incilis* from Cartagena Bay, Colombia. *Veterinary Parasitology.*, 168, 31-35.

[35] Restrepo, J. D & Kjerfvea, B. (2000 a) . Magdalena river: interannual variability (1975–1995) and revised water discharge and sediment load estimates. *Journal o Hydrology*, 235, 137-149.

[36] Milliman, J. D & Syvitski, J. P. M. (1992). Geomorphic/tectonic control of sediment discharge to the ocean: The importance of small mountainous rivers. *The Journal of Geology*, 100, 525-544.

[37] Restrepo, J. D & Syvitski, J. P. M. (2006). Assessing the effect of natural controls and land use on sediment yield in a major Andean river: The Magdalena drainage basin, Colombia. *Ambio*, 35, 65-74

[38] Restrepo, J. D & Kjerfve, B. (2000 b). Water discharge and sediment load from the western slopes of the Colombian Andes with focus on Rio San Juan. *The Journal of Geology,* 108, 17-33.

[39] Restrepo, J. (2005). Los sedimentos del rio magdalena: reflejo de la crisis ambiental. Fondo editorial Universidad EAFIT, *Medellín*.

[40] Molina, A., Molina, C., Thomas, Y. & Molina, L. (2001). Comportamiento de la línea de Costa del Caribe Colombiano (Sector entre barranquilla, desde bocas de ceniza hasta la flecha de galerazamba 1935–1996). *CIOH. Boletín científico* No. 19, 68-79.

[41] Rangel, N. & Posada, B. (2005). Geomorfología y procesos erosivos en la costa norte del departamento de Córdoba, Caribe colombiano (sector paso nuevo-cristo rey). *Bol. Invest. Mar. Cost*, 34, 101-119.

[42] Martínez, J. Geomorfología de la *costa* del pacifico *Colombiano* y aspectos relativos a su estabilidad.1992. Available from: http://www.crid.or.cr/cd/CD_El_Nino/pdf/spa/doc9254/doc9254-contenido.pdf.

[43] Díaz, G. (1998). Derrotero del Pacífico Colombiano. Acta Oceanográfica del Pacífico. INOCAR. *Ecuador*., 9, 199-208.

In: Sea Level Rise, Coastal Engineering, Shorelines and Tides ISBN: 978-1-61728-655-1
Editor: Linda L. Wright, pp. 155-170 © 2011 Nova Science Publishers, Inc.

Chapter 7

PROTISTS IN INTERTIDAL SANDY BEACHES

Noriko Okamoto
Department of Botany, University of British Columbia

Abstract

Intertidal sandy beach is full of life. Despite its deserted look and constantly changing conditions, intertidal sandy beach hosts diverse forms of life including meiofaunal or bacterial population, as well as single cell eukaryotes called protists. Either photosynthetic or non-photosynthetic, protists play important roles as the primary producer or the primary predator. The life as a beach protist is different from planktonic one, directly affected by the constant tidal activity that correspondingly changes physical conditions including surface and ground water level, salinity, temperature, light intensity, oxygen availability. As a result, beach protists community is different from the planktonic one even if it's from the same geographic locality. Beach protists community significantly differ within a single beach due to different conditions in microenvironments and zonation across the shore face. Until recently, protists in intertidal sandy beach received little attention, which makes this accessible environment one of the last frontier in exploration of microbial biodiversity. Interestingly, this unique also environment seems to be an experimental field of evolutionary process. The biodiversity of beach protists includes evolutionary intriguing protists, of which the most interesting is *Hatena arenicola*. *Hatena arenocola* is at the intermediate stage of acquiring photosynthetic ability via endosymbiosis of another microalga, demonstrating otherwise in-observable evolutionary process in real time.

1. Introduction

Intertidal sandy beaches are full of life. At first look they may seem deserted, but if you take a pinch of wet sands on a sunny afternoon at low tide and examine it under the microscope, you'll find out millions of tiny life forms restlessly swimming or gliding among the sand grains. These microscopic lives are essentially singe cell eukaryotes called protists. The beach protists are either photosynthetic (microalgae) or non-photosynthetic, and play important roles as primary producers or primary grazers. MacIntyre [1] once referred to the sandy beach as a "secret garden". Of course the garden doesn't have flowers nor hardly any plants. Instead, the rich microalgal populations flourish, enjoying the sun while the heterotrophic protists along with the meiofaunal or macrofaunal grazers harvest those microalgae.

Despite their microscopic body size, protists are key players in beach ecosystem. The microalgae are an important part of the beach ecosystem and include for example the well known diatoms but also euglenids, dinoflagellates and many other groups. Microalgae are a major carbon source of beach ecosystem [1, 2], sometimes forming such a high-density bloom that color the beach during low tides in day time.

In contrast to their photosynthetic allies, the protists that lack photo-synthetic pigments have drawn little attention [2-5]. Those colorless protists are heterotrophic, feeding on bacteria (bacterivor) and /or microalgae (harbivor), thus playing an important role in the microbial loop and the local benthic food web. . In addition to their microscopic body size, the absence of photosynthetic pigments makes them virtually invisible to the naked eyes. This makes the sampling of those colorless protists challenging, and has prevented the deep exploration of the protist diversity in intertidal sandy beaches. Early species surveys revealed the rich protiste fauna [6-8] based solely on light microscopic observations. Although we are still at the stage of alpha taxonomy for beach protists, new tools of phylogenetic analysis based on molecular sequence data are giving us a better picture of their diversity. Not only novel organisms are constantly being discovered, the protists that were once *incertae sedis* are now finding their home in the eukaryotic tree of life (examples in the last 10 years include the following papers [9-21]. We are now realizing the truly rich diversity of beach protists and evolutionarily very interesting stories behind the tiny organisms.

In this review, we will first describe the current picture of the eukaryotic tree of life with focus on beach protists and cover the topics of microalgal adaptation to the beach environments, the discovery of *Hatena arenicola*, a grazer reforming itself to be a new "plant" [11, 13], and cryptic morphological convergence to adapt to the life among the sand grains. The focus of the second part will be the intertidal beach environment as a habitat to the rich protist populations, aiming to benefit the future sampling or survey of the protists in the intertidal beach environment.

2. Beach Protists

What are Protists? This is a simple question, but there is no simple answer. Protists are all eukaryotes except animals, plants and fungi. Most of them are single cell organisms, but their "simple" look does not necessarily reflect their evolutionary history and diversity. They are an assemblage of dozens of distantly related lineages and their entire diversity is larger than the kingdoms sensu Whittaker four decades ago [22] (Figure 1). Currently eukaryotes are classified into six supergroups, namely, Opisthokonta, Amoebozoa, Archaeplastida, Rhizaria, Chromalveolata, and Excavata. Opisthokonta and Archaeplastida both include protists and kingdoms sensu Whittaker, whereas the other supergroups are entirely composed of protists [23-27].

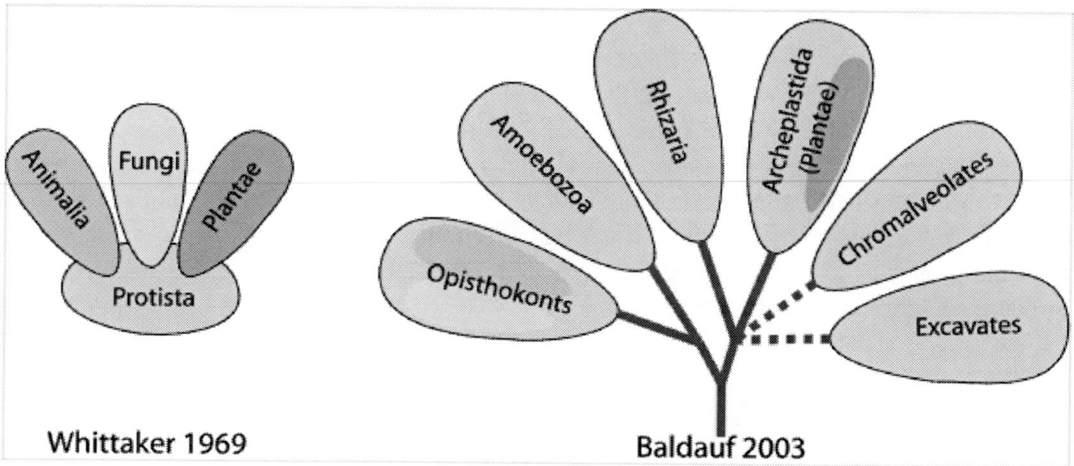

Figure 1. Protists diversity in eukaryotic tree of life.

Opisthokonta includes animals, fungi and several other lifenages of single cell protists. Amoebozoa is a group of lobose amoebae, slime molds and some parasitic amoebae. Opisthokonta and Amoebozoa are grouped together and are referred to as Unikonta that is probably ancestrally uniflagellate [28]. Archaeplastida is the group of the oldest photosynthetic eukaryotes that includes glaucophytes, rhodophytes (red algae) and streptophytes (green algae + plants). Their plastids are directly derived from a symbiosis of a cyanobacterium. Chromalveolata is originally comprised of four photosynthetic lineages of alveolates (dinoflagellates + ciliates + parasitic apicomplexans such as malaria parasites), stramenopiles (including diatoms, brown algae and plants pathogen *Phytophthora* spp.), haptophytes and cryptophytes. Increasing number of non-photosynthetic lineages are assigned to this supergroup [21, 25, 29]. Rhizaria is comprised of cercozoa (amoeboflagellates), radiolaria and foraminifera. Rhizaria is essentially recognized by the molecular phylogeny and includes variety of types of cells, whose unifying morphological character is not known. Two independent cercozoan groups are photosynthetic. Excavata is composed of discicristates (euglenids and kinetoplastids) and amitochondrial excavates (flagellates with a groove on one side of the cell). Of the six supergroup, the unity of Chromalveolata and that of Excavata are yet to be determined (for reference, see [30]).

There's no doubt that the protist diversity is tremendous, though it is unclear how much species are still missing from our knowledge [31-33]. The hidden diversity of protists is either "cryptic" or "missing". Traditionally, protists species have been described based on morphological characters. Some groups, such as the diatoms, have many recognizable morphological characters, and as many species as 200,000 species are estimated to exist, and half of them have been described [34]. However, many protists often lack any conspicuous morphological identity, hence their diversity has largely been underestimated. This cryptic diversity was notably revealed by DNA barcoding projects using short sequence data (for review, [31]). The second type of diversity is simply "missing" from our knowledge. Environmental sequence surveys in various environments brought to light unexpected discoveries of large missing lineages. Some are assigned to known protistan lineages (e.g., *ma*rine *st*ramenopiles, or MASTs to stramenopiles [35]; *m*arine *a*lveolate groups, or MAGs to alveolates [36]), but others are totally enigmatic (e.g., picobiliphytes [37]).

We are still in need of hunting down the tiny organisms from the environment. The majority of studies looking at the hidden protistean diversity has been focused on "extreme" environments such as the open ocean, the deep sea, thermal vents, polar regions etc ([32] and references therein). Despite the familiar look and the proximity to the human activity, the intertidal beach is an extreme environment in its dynamic equilibrium. It undergoes a constant change in the physiochemical and biological conditions [38] and the beach organisms have to cope with the instability in their own ways. These special requirements select some interesting species as well as cradle a premature endosymbiosis of *Hatena arenicola*, an ongoing evolutionary experiment.

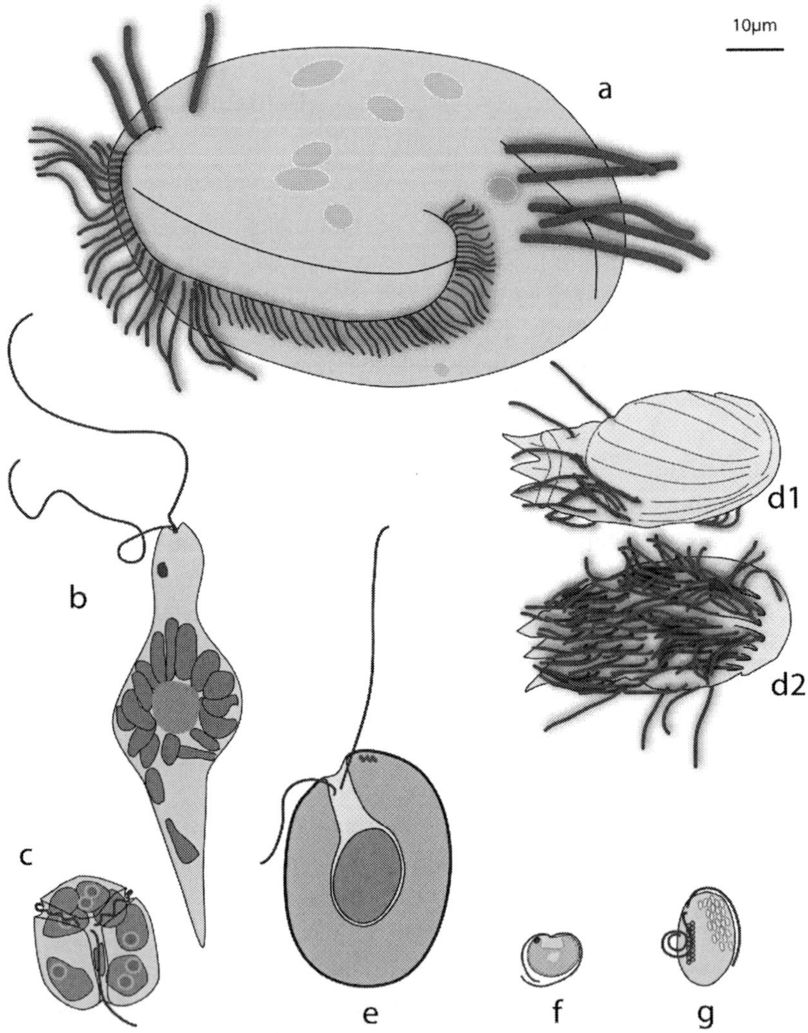

Figure 2. Example of beach protists. a. *Euplotes* sp. (ciliates, Chromalveolata); b. *Eutreptiella* sp. (euglenids, Excavata); c. *Amphidinium latum* (dinoflagellates, Chromalveolata); d. *Stephanopogon minuta* (heterolobosea, Excavata), dorsal (d1) and ventral (d2) view ; e. *Hatena arenicola* (katablepharids, Chromalveolata); f. *Nephroselmis* sp. (prasinophytes, Archaeplastida); g. *Platychilomonas marina* (katablepharids, Chromalveolata).

Microalgae on the Beach

Microalgae are by far the most studied of the beach protists. Diatoms (Stramenopiles) are the most abundant microalgae on the beach, and are a major carbon source in the local food web. A diatom cell is encased in the ridged silica cell wall (frustules), but the intertidal beach species are most often motile, walking on the surface of sediment grains ([39] and references therein). This motility is crucial for their survival in the changing intertidal condition (see below). In addition to diatoms, intertidal beach flora includes various flagellated microalgae such as euglenids, dinoflagellates, cryptophytes, prymnesiophytes, other stramenopiles, and prasinophytes (green algae) (Figure 2).

Microalgae and cyanobacteria on the beach sediments are collectively referred to as microphytobenthos (MPB), which provide a major carbon source in the beach ecosystem [1, 2]. MPB are not only important primary producers, but they also modify physical conditions. When they form a dense bloom on the beach surface in summer months, some MPB, such as diatoms excrete mucus and form a biofilm, which stabilize the sediments from erosion. The biological stabilization of the sediments positively feedback to enhance the further growth of the MPB [40, 41].

One of the widespread adaptations to the intertidal beach environment is the vertical migration. As photoautotrophs, MPB need light for living. The light intensity quickly declines within a few millimeters depth in the sand [38]. To access the light for photosynthesis, they need to be on or near the beach surface. At the same time, they have to avoid the disturbance by the wave during high tide. In response to these factors, the beach MPB migrate upward towards the beach surface during the daytime low tide and migrate back downward during the high tide and night time (e.g. [42, 43]).

Although majority of ecological studies of MPB are based on an assumption that they can be treated as a single unit, MPB is an assembleage of different organisms and, hence, their response to the environmental stress are heterogenous. For instance, diatoms exhibit various protective responses to the high light irradiation. High light irradiation on the beach surface would cause damages to the cell. Some diatoms exhibit downward migration in a microscale [44-47] to avoid the photodamage caused by high light intensity on the very surface, while others physiologically adopt to the high light condition by activating quenching pathway. Each species in a population would have different stress response, or even a single cell of a species could respond differently based on the photohistory (e.g. [48, 49]). With the awareness of their heterogeneity and biodiversity, we can understand a more precise picture of what is going on in nature.

Mixotrophy and Plastid Evolution on the Beach - *Hatena Arenicola*

For the survival in the constantly changing intertidal beach environment, trophic plasticity may be an asset. One organism that may benefit from this unique environment is *Hatena arenicola* that has been shown to be likely halfway in the plastid acquisition process [11, 13, 50]. The same process is thought to have happened millions of years ago to give rise to the extant photosynthetic lineages including diatoms (for a recent review, [51-53]).

Hatena arenicola belongs to the essentially heterotrophic protistan group katablepharids. Unlike the other katablepharids, natural populations of *H. arenicola* possess the green "plastid"-like structure with an eyespot inside (Figure 3).

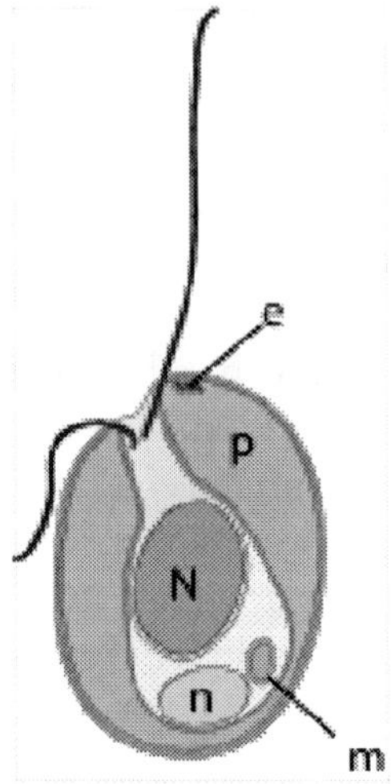

Figure 3. *Hatena arenicola* with the endosymbiont. The endosymbiont retains the plastid, the nucleus and the mitochondrion, but lacks the endomembrane system and the microtoubular cytoskeleton. The eyespot (part of the algal photosensing machinary) in the symbiont's plastid is always placed at the cell apex of the host cell. N: the host's nucleus, n: the symbiont's nucleus, m: the symbiont's mitochondrion, p: the symbiont's plastid, e: the eyespot.

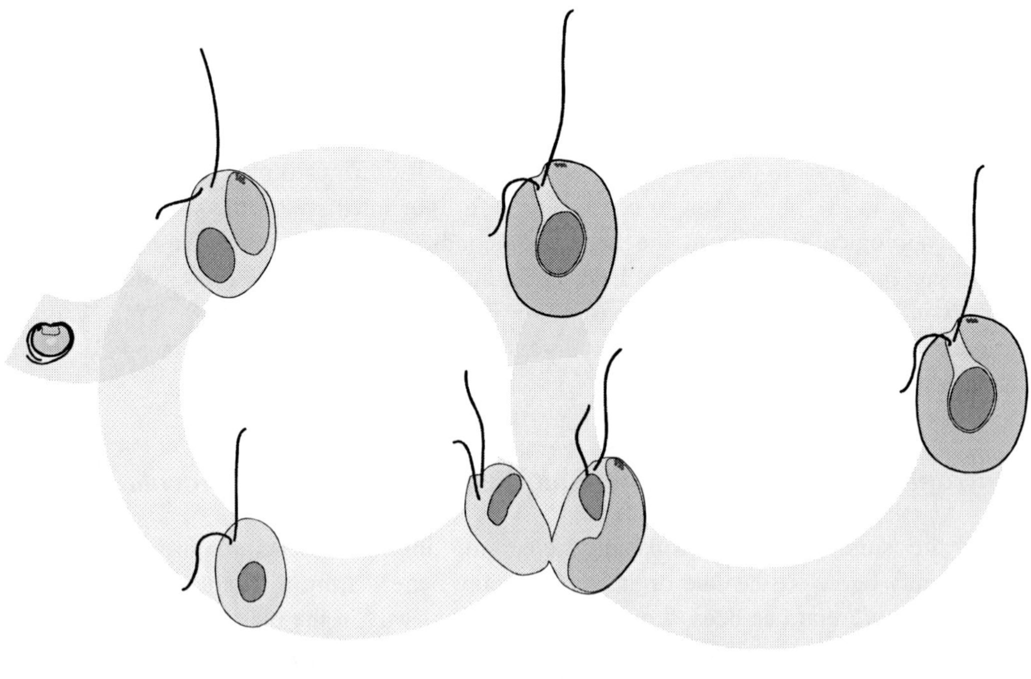

Figure 4. The cell cycle of *H. arenicola*. Majority of the population has the green symbiont. Upon cell division, only one of the daughter cells inherits the symbiont and the other becomes colorless. The colorless daughter cell uptakes a new symbiont cell from the environment.

This seeming plastid is, however, passed on to only one of the daughter cells during cell division. The lopsided inheritance of "plastid" results in one colored and one colorless daughter cell. Because the majority of the population is colored, the colorless daughter cell must uptake the green "plastid" from the environment after every cell division (Figure 4). This green "plastid" is a eukaryotic endosymbiont that belongs to the small green algal genus *Nephroselmis*. Taking a closer look at it revealed that this temporary symbiont still retains its nucleus and mitochondria in addition to the green plastid; but other intracellular structures, such as the endomembrane system (membrane-bound protein expression) and microtubular cytoskeleton, are missing. Paradoxically, the symbiont with reduced cell contents grows more than ten times when uptaken by *H. arenicola,* while it maintains the eyespot (a part of the algal photosensing machinery) within the plastid always positioned at the cell apex of the host cell. Interestingly, *H. arenicola* with the green symbiont shows negative phototaxis., This phototaxis is probably crucial for to the survival of *H. arenicola* cell with the symbiont, as the green *H. arenicola* cells get swollen to death under the high light intensity. It is yet to be confirmed if the symbioint's eyespot contributes to this phototaxis. These pieces of evidence suggest that their relationship is not a casual one, but is more likely to be an early stage of commitment, in which the symbiont will be eventually integrated as a new plastid.

The plastid acquisition through endosymbiosis is widely accepted as the major driving force for the birth and evolution of photosynthetic organism in eukaryotes [51-53]. In the tree of eukaryotes, multiple photosynthetic lineages are patchily distributed across distantly related supergroups. This is better explained by serial endosymbioses of photosynthetic cells (Figure 5). It is a text book acceptance that all extant plastids have originated from a symbiosis with an ancestral cyanobacterium (blue-green alga) and a heterotrophic eukaryote. Through this ancient primary endosymbiosis, the common ancestor of the supergroup Archaeplastida acquired the primary plastid. The primary endosymbiosis gave rise to glaucophytes, rhodophytes (red algae) and green plants (green algae and land plants). Recently, it was revealed that another primary endosymbiosis occurred independently relatively recently in the lineage of testate amoebae *Paulinella* spp. [54-57].

After the primary endosymbiosis, some of the primary algae were uptaken by different heterotrophic protists lineages. Through this process of secondary endosymbioses emerged six other photosynthetic lineages; chlorarachniophytes and photosynthetic euglenids acquired plastids of green algal origin, and dinoflagellates, stramenopiles, cryptophytes and haptophytes acquired plastids of red algal origin. The number of endosymbioses that led to todays' plastid diversity is still under an active debate, in particular for those involving plastid of red algal origin [53].

Plastid acquisition through endosymbiosis is a very interesting phenomenon that requires makeovers in both the host and the symbiont, including harmonization of cell cycles, horizontal gene transfer of the symbiont genes to the host nucleus, and establishment of the transporting machinery for the nucleus-encoded proteins to be transferred back to the symbiont. The degree of plastid integration depends on each photosynthetic lineage. The plastids of cryptophytes and chlorarachniopytes retain the vestigial nucleus of the symbiont, which still encode a minimal amount of genes that keeps genome structure. This nucleus is called nucleomorph (for review, [51]). Despite the evolutionary and cytological importance, the process of host-symbiont integration is hard to tackle, partly because the extant photosynthetic eukaryotes undertook the process million years ago. In contrast, *Hatena arenicola* is most likely currently at the very early stage in this evolutionary process and would offer us an insight to the intermediate state that have lost in the other photosynthetic cells long time ago.

The life as a plant and a predator at the same time (mixotrophy) would have some advantae in the beach environment. There is another variation of temporary plastid known as kleptoplastid (*stolen plastid* in Greek). Beach dinoflagellates *Amphidinium poecilochlorum* and *A. latum* retains plastids of the photosynthetic prey, *Chroomonas* sp. (cryptophytes). Unlike the symbiont of *H. arenicola*, the nucleus of the donor of kleptoplastid is not retained. Although kelptoplastidy is not restricted to the beach environment, these temporary photoautotrophs may hold another key to the plastid evolution [58].

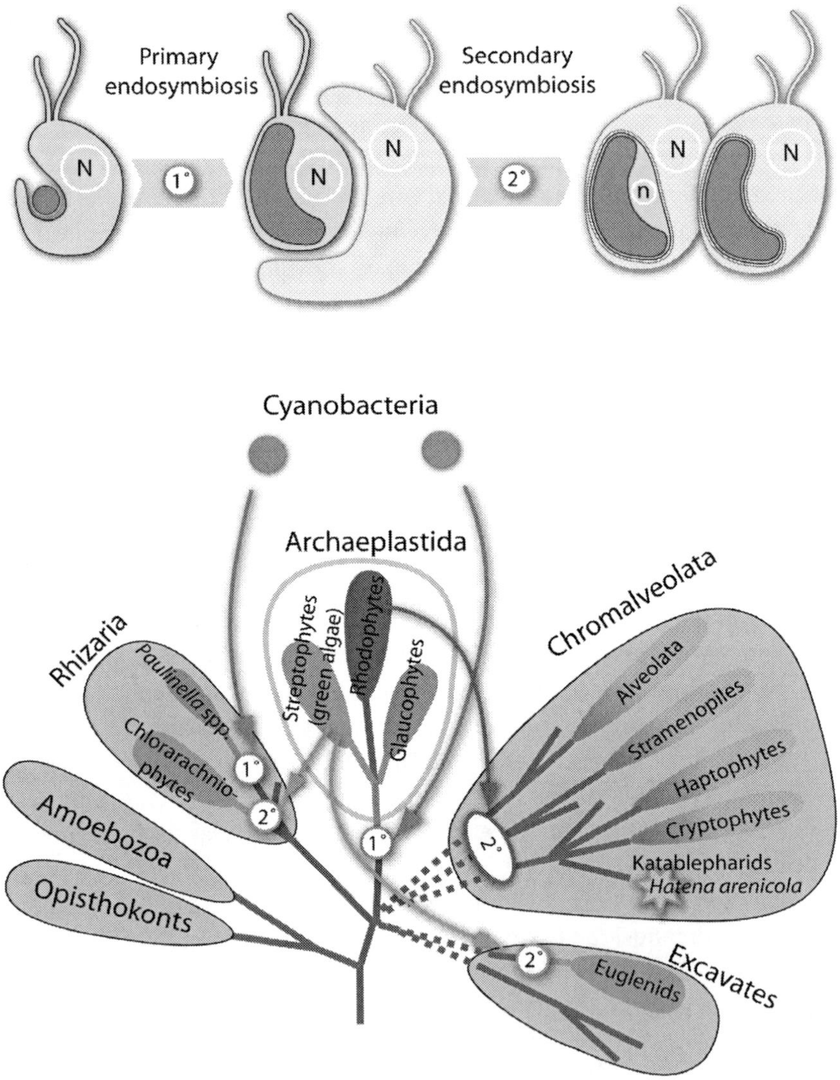

Figure 5. Plastid evolution in the eukaryotic tree of life. All the plastid is originated from the primary endosymbioses of cyanobacterial, either through the primary endosymbiosis or through the secondary endosymbiosis. The enclosed numbers indicate the primary and the secondary endosymbiosis. N: nucleus, n: nucleomorph.

Morphological Convergence in Heterotrophic Protists

Compared to the photosynthetic species, the heterotrophic beach protists have received little attention. As mentioned earlier, there was a general ignorance about the diversity of protists themselves. The morphological convergence in order to adapt to the intertidal beach environment was one of the challenges taxonomists had to face, especially in earlier years before the introduction of molecular phylogeny.

The interstitious environment is a three dimensional structure consisting of narrow labyrinthine void filled with water/air and vast surface area on the sand grains [38]. Many of the beach protists adapted to the environment typically by having a flat body and gliding on the surface of the sediments. Even those with swimming capacity often rest on the bottom surface and require surface area for growth. Interestingly, this gliding behavior seems widespread among the interstitious organisms under a certain size. Recently Rundell and Leander pointed out the convergence in the body shape between meiofaunal organisms such as mesozoans and ciliates [59]. Due to their body size (60 μm - 2 mm), they are adapted to use ciliary gliding to crawl through the narrow void instead of burrowing as commonly observed in macrofaunal animals. Body size of typical protists range from 2 μm up to 500 μm, thus would share the convergence observed in meiofauna to some extent.

An example of convergence between distant lineages is between a very common ciliate genus *Euplotes* and a distant protist genus *Stephanopogon* that belongs to another supergroup Excavata (Figure 2; [19, 60]). Both have a dorsoventrally compressed flat cell, "walk" the surface with cirri (bundles of cilia) or with flagella, possess a cytostome (mouth) and feed on microalgae such as diatoms. Although ultrastructural observations have suggested some differences, it was only after molecular data were available that it is confirmed that they are phylogenetically not related.

3. Intertidal Beach as a Habitat

As a secret garden and a treasure chest for hidden protistan biodiversity, the intertidal beach attracts more attention. However, there has been no empirical study about distribution pattern of protists across the intertidal beach and our understanding is still fragmental. The microalgae and ciliates are relatively well studied, but the only environmental factor that correlates to their distribution or productivity is the grain size of the sediments [3, 48, 61, 62]. This trend gives some clue for protists hunters and primary production researchers in the intertidal beach, but would be far more informative if embedded in a larger picture with the available understanding based on meiofaunal and macrofaunal studies [38]. In the following sections, we'll go through the zonation across the beach and the beach types based on meio- and macrofaunal data, but with implications for the protistan point of view to provide an insight in future studies.

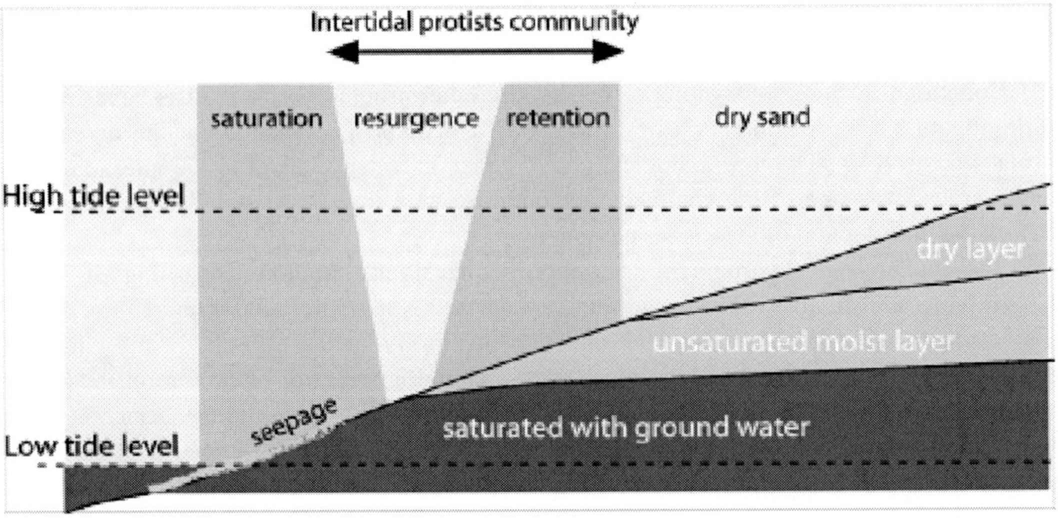

Figure 6. Beach profile and zonation. The beach zonation is dynamically defined by the groundwater. The second and the third strata is where the true intertidal are mostly discovered. Adopted from [38].

Beach Zonation

Zonation on the sandy beach during low tide correlates to the moisture available in the sand. In contrast to the rocky shore zonation, the boundaries of each zones are dynamically changing in response to physical conditions (i.e. waves and tide) as well as migration of organisms. Zonation scheme of sandy beach has been studied and hypothesized by several authors ([38]for review).

One of the classic hypotheses is based on moisture strata originally described by Salvat [63], which recognizes 4 different strata, namely, (1) a stratum of dry sand, (2) a stratum of retention, (3) a stratum of resurgence, and (4) a stratum of saturation (Figure 6). Studies based on beach macrofauna mostly agree with this scheme except for the third stratum, which may be too narrow to be recognizable on exposed beaches with a steep slope, where the majority of the studies have been conducted. In contrast to the other environments, this zonation is not strictly bound to geophysical features of the beach such as mean tidal level, but it is always slightly moving depending on the excursion of the swash zone and the ground water level. The boundaries of the zones are not sharply defined, but it is rather a gradual transition [38].

There seem to be a certain distribution pattern of protists that corresponds to the above scheme to some extent. In general, richest protists population is found on or near the surface, where microalgae form bloom that attract heterotrophic protists. The first stratum is practically devoid of protists, as their existence requires interstitious water. The second and third strata harbor "true" intertidal beach specific species, thus is a hot spot for discoveries of new species as mentioned above. The boundary between the strata of retention and resurgence is rather unclear, but there seems to be a gradient across the two. For example, the cosmopolitan beach protists *Platychilomonas marina* resides more towards the higher end of the second stratum (personal observation). Protists found in the fourth stratum show gradual mix of the intertidal beach species and the surf species. This is probably due to the fact that the groundwater containing protists from the third stratum seepages at the exit point and flow on the surface of the fourth stratum while the surfs run up constantly to bring the surfzone species.

Beach Types

The proportion of the interstitious moisture strata differ between various beach types. There are two extreme types of beaches, i.e., the reflective and the dissipative beaches, and various intermediate beaches [64]. The beach type is defined by the relative tidal range (relative tidal effect to the wave height) and dimensionless fall velocity (how fast the sediments sink in relation to the temporal length of a wave) [65].

As mentioned above, the second and third strata are where there are the "true" intertidal beach protists unique to the habitat. Those strata are less likely to be developed in the reflective beach due to the larger sand grains and steeper slope. In the reflective beach, the available moist is limited to the lower intertidal region, hence would harbor population similar to the surfzone protists.

In contrast, the strata of retention and resurgence are more likely to be developed in more dissipative beaches with more gradient slope. The dissipative beaches tend to have finer sand grains. The combination of finer sands and gentler slope allows less water filtration, making the interstitious environment reductive and often eutrophic due to the deposition of organic detritus. The extreme condition is mud flat, composed of much finer sediments (smaller than 63 μm) and the water exchange rate is extremely low. Such a condition is largely favored by bacteria, but some of cyanobacteria or microalgae may still make a bloom on the surface. Although less abundant than protists bloom in oxic condition, the protists in this environment are unique and a focus of the biodiversity study (e.g., [18, 66, 67]).

One of the interesting habitat for the protists hunters falls somewhere between these two extremes. The beach must be dissipative enough to develop the strata of retention and resurgence, but at the same time reflective enough to allow exchange of ground water so that it can harbor more diverse microalgal population to support the eukaryotic protists. One such intermediate sandy shore is Isonoura beach in Japan, where *Hatena arenicola* was discovered. The microalgae and ciliates in intertidal flat are relatively well studied, and the only environmental factor that correlates to their distribution or productivity is the grain size of the sediments; more biomass or photosynthetic activity is observed in sandy flats than in muddy flats (e.g., [3, 48, 61, 62]).

In addition to those found on the beach surface, there're ones found deep inside the beach (e.g. [68]). Although the biomass seems to quite low, they may contributed to the microbial loop in the deep sediments.

4. Conclusion

Intertidal beach has been a best kept secret of the protists diversity. This dynamic environment harbor rich microalgal flora as well as unique heterotrophic fauna. However the importance of protists in the intertidal beach has been undersampled in terms of ecological role and diversity. Current molecular phylogeny has changed our understanding of protists in the eukaryotic tree of life. Intertidal beach environment is a hot spot for hunting the hidden diversity of protists and Newly discovered species are constantly described each year. Not only the quantitiy but the surprise the new beach protists deliver is unexpectedly important. To fully understand the ecology and diversity of the protists in the intertidal beach, more species survey is needed as well as the empirical study to fill the gap in our understanding on their ecology would be usuful.

Acknowledgment

I am thankful to Dr. Fabien Burki for the constructive and useful comments to the manuscript. I am also thankful to Dr. Rebecca Rundall and Dr. Brian S. Leander for allowance to include the latest publication. This work is supported by TULA foundation.

References

[1] MacIntyre, H; Geider, R; Miller, D. Microphytobenthos: The ecological role of the "secret garden" of unvegetated, shallow-water marine habitats.1. Distribution, abundance and primary production. *Estuaries,* 1996, 19(2A), 186-201.

[2] van Oevelen, D; Soetaert, K; Middelburg, JJ; Herman, PMJ; Moodley, L; Hamels, I; et al. Carbon flows through a benthic food web: Integrating biomass, isotope and tracer data. *J Mar Res,* 2006, 64(3), 453-482.

[3] Herman, P; Middelburg, J; Heip, C. Benthic community structure and sediment processes on an intertidal flat: results from the ECOFLAT project. *Cont Shelf Res,* 2001, 21(18-19), 2055-2071.

[4] Azovsky, A; Saburova, M; Chertoprood, E; Polikarpov, I. Selective feeding of littoral harpacticoids on diatom algae: hungry gourmands? *Mar Biol,* 2005, 148(2), 327-337.

[5] First, MR; Hollibaugh, JT. Protistan bacterivory and benthic microbial biomass in an intertidal creek mudflat. *Mar Ecol-Prog Ser,* 2008, 361, 59-68.

[6] Larsen, J; Patterson, DJ. Some flagellates (Protista) from tropical marine sediments. *J Nat Hist,* 1990, 24(4), 801-937.

[7] Lee, WJ; Patterson, DJ. Heterotrophic flagellates (Protista) from marine sediments of Botany Bay, Australia. *J Nat Hist,* 2000, 34(4), 483-562.

[8] Simpson, AGB; Patterson, DJ. On core jakobids and excavata taxa: the ultrastructure of *Jakoba incarcerata*. *J. Euk. Microbiol,* 2001, 48, 480-492.

[9] Saldarriaga, J; Leander, BS; Taylor, F; Keeling, PJ. *Lessardia elongata* gen. et sp nov (Dinoflagellata, Peridiniales, Podolampaceae) and the taxonomic position of the genus *Roscoffia*. *J Phycol,* 2003, 39(2), 368-378.

[10] Hoppenrath, M; Saldarriaga, J; Schweikert, M; Elbrachter, M; Taylor, F. Description of *Thecadinium mucosum* sp. nov. (Dinophyceae), a new sand-dwelling marine dinoflagellate, and an emended description of *Thecadinium inclinatum* Balech. *J Phycol,* 2004, 40(5), 946-961.

[11] Okamoto, N; Inouye, I. A secondary symbiosis in progress? *Science,* 2005, 310(5746), 287.

[12] Okamoto, N; Inouye, I. The katablepharids are a distant sister group of the Cryptophyta: A proposal for Katablepharidophyta divisio nova/Kathablepharida

phylum novum based on SSU rDNA and beta-tubulin phylogeny. *Protist,* 2005, 156(2), 163-179.

[13] Okamoto, N; Inouye, I. *Hatena arenicola* gen. et sp. nov., a katablepharid undergoing probable plastid acquisition. *Protist,* 2006, 157(4), 401-19.

[14] Tamura, M; Horiguchi, T. *Pileidinium ciceropse* gen. et sp nov (Dinophyceae), a sand-dwelling dinoflagellate from Palau. *European Journal of Phycology,* 2005, 40(3), 281-291.

[15] Tamura, M; Shimada, S; Horiguchi, T. *Galeidiniium rugatum* gen. et sp nov (Dinophyceae), a new coccoid dinoflagellate with a diatom endosymbiont. *J Phycol,* 2005, 41(3), 658-671.

[16] Murray, S; Hoppenrath, M; Larsen, J; Patterson, DJ. *Bysmatrum teres* sp. nov., a new sand-dwelling dinoflagellate from north-western Australia. *Phycologia,* 2006, 45(2), 161-167.

[17] Sparmann, S; Leander, BS; Hoppenrath, M. Comparative Morphology and Molecular Phylogeny of *Apicoporus* n. Gen.: A New Genus of Marine Benthic Dinoflagellates Formerly Classified within *Amphidinium*. *Protist,* 2008.

[18] Chantangsi, C; Esson, HJ; Leander, BS. Morphology and molecular phylogeny of a marine interstitial tetraflagellate with putative endosymbionts: *Auranticordis quadriverberis* n. gen. et sp (Cercozoa). *BMC Microbiol,* 2008, 8, 123.

[19] Yubuki, N; Leander, BS. Ultrastructure and molecular phylogeny of *Stephanopogon minuta*: An enigmatic microeukaryote from marine interstitial environments. *European Journal of Protistology,* 2008, 44(4), 241-253.

[20] Kai, A; Yoshii, Y; Nakayama, T; Inouye, I. Aurearenophyceae classis nova, a new class of heterokontophyta based on a new marine unicellular alga *Aurearena cruciata* gen. et sp nov inhabiting sandy beaches. *Protist,* 2008, 159(3), 435-457.

[21] Okamoto, N; Chantangsi, C; Horák, A; Leander, BS; Keeling, PJ. Molecular phylogeny and description of the novel katablepharid *Roombia truncata* gen. et sp. nov., and establishment of the Hacrobia taxon nov. *PLoS ONE,* 2009, 4(9), e7080.

[22] Whittaker, RH. New concepts of kingdoms of organisms. *Science,* 1969, 163(3863), 150-60.

[23] Baldauf, SL. The deep roots of eukaryotes. *Science,* 2003, 300(5626), 1703-6.

[24] Burki, F; Shalchian-Tabrizi, K; Minge, M; Skjaeveland, A; Nikolaev, SI; Jakobsen, KS; et al. Phylogenomics reshuffles the eukaryotic supergroups. *PLoS ONE* 2007, 2(8), e790.

[25] Burki, F; Shalchian-Tabrizi, K; Pawlowski, J. Phylogenomics reveals a new 'megagroup' including most photosynthetic eukaryotes. *Biol Letters,* 2008, 4(4), 366-369.

[26] Keeling, PJ; Burger, G; Durnford, DG; Lang BF; Lee RW; Pearlman RE; et al. The tree of eukaryotes. *Trends Ecol Evol,* 2005, 20(12), 670-6.

[27] Parfrey, LW; Barbero, E; Lasser, E; Dunthorn, M; Bhattacharya, D; Patterson, DJ; et al. Evaluating support for the current classification of eukaryotic diversity. *PLoS Genet,* 2006, 2(12), e220.

[28] Cavalier-Smith, T. Protist phylogeny and the high-level classification of Protozoa. *Eur J Protistol,* 2003, 39(4), 338-348.

[29] Cavalier-Smith, T. Kingdoms Protozoa and Chromista and the eozoan root of the eukaryotic tree. *Biol Lett,* 2009.

[30] Fehling, J; Stoecker, D; Baldauf, SL. Photosynthesis and the Eukaryote Tree of Life. In, P G Falkowski, A H Knoll editors, *Evolution of Primary Producers in the Sea*. Boston, Elsevier, 2007, 75-107.

[31] Caron, DA. New Accomplishments and Approaches for Assessing Protistan Diversity and Ecology in Natural Ecosystems. *Bioscience,* 2009, 59(4), 287-299.

[32] Epstein, S; Lopez-Garcia, P. "Missing" protists: a molecular prospective. *Biodivers Conserv,* 2008, 17(2), 261-276.

[33] Martiny, JBH; Bohannan, BJM; Brown, JH; Colwell, RK; Fuhrman, JA; Green, JL; et al. Microbial biogeography: putting microorganisms on the map. *Nat Rev Micro,* 2006, 4(2), 102-12.

[34] Mann, D; Droop, S. Biodiversity, biogeography and conservation of diatoms. *Hydrobiologia,* 1996, 336(1-3), 19-32.

[35] Massana, R; Castresana, J; Balague, V; Guillou, L; Romari, K; Groisillier, A; et al. Phylogenetic and ecological analysis of novel marine stramenopiles. *Appl Environ Microbiol,* 2004, 70(6), 3528-3534.

[36] Lopez-Garcia, P; Rodriguez-Valera, F; Pedros-Alio, C; Moreira, D. Unexpected diversity of small eukaryotes in deep-sea Antarctic plankton. *Nature,* 2001, 409(6820), 603-607.

[37] Not, F; Valentin, K; Romari, K; Lovejoy, C; Massana, R; Toebe, K; et al. Picobiliphytes: A marine picoplanktonic algal group with unknown affinities to other eukaryotes. *Science,* 2007, 315(5809), 253-255.

[38] McLachlan, A; Brown, A. *The Ecology of Sandy Shores*. (2nd ed.) Burlington, MA, USA, *Academic Press*, 2006.

[39] Apoya-Horton, M; Yin, L; Underwood, G; Gretz, M. Movement modalities and responses to environmental changes of the mudflat diatom *Cylindrotheca closterium* (Bacillariophyceae). *J Phycol,* 2006, 42(2), 379-390.

[40] Miller, D; Geider, R; MacIntyre, H. Microphytobenthos: The ecological role of the "secret garden" of unvegetated, shallow-water marine habitats.2. Role in sediment stability and shallow-water food webs. *Estuaries,* 1996, 19(2A), 202-212.

[41] Widdows, J; Blauw, A; Heip, C; Herman, P; Lucas, C; Middelburg, J; et al. Role of physical and biological processes in sediment dynamics of a tidal flat in Westerschelde Estuary, SW Netherlands. *Mar Ecol-Prog Ser,* 2004, 274, 41-56.

[42] Du, GY; Son, M; An, S; Chung, IK. Temporal variation in the vertical distribution of microphytobenthos in intertidal flats of the Nakdong River estuary, Korea. *Estuar Coast Shelf S,* 2010, 86(1), 62-70.

[43] Kingston, MB; Gough, JS. Vertical migration of a mixed-species *Euglena* (Euglenophyta) assemblage inhabiting the high-intertidal sands of Nye Beach, Oregon. *J Phycol,* 2009, 45(5), 1021-1029.

[44] Herlory, O; Guarini, J; Richard, P; Blanchard, G. Microstructure of microphytobenthic biofilm and its spatio-temporal dynamics in an intertidal mudflat (Aiguillon Bay, France). *Mar Ecol-Prog Ser,* 2004, 282, 33-44.

[45] Kromkamp, J; Barranguet, C; Peene, J. Determination of microphytobenthos PSII quantum efficiency and photosynthetic activity by means of variable chlorophyll fluorescence. *Mar Ecol-Prog Ser,* 1998, 162, 45-55.

[46] Mouget, JL; Perkins, R; Consalvey, M; Lefebvre, S. Migration or photoacclimation to prevent high irradiance and UV-B damage in marine microphytobenthic communities. *Aquat Microb Ecol,* 2008, 52(3), 223-232.

[47] Serodio, J; Vieira, S; Cruz, S. Photosynthetic activity, photoprotection and photoinhibition in intertidal microphytobenthos as studied in situ using variable chlorophyll fluorescence. *Cont Shelf Res,* 2008, 28(10-11), 1363-1375.

[48] Jesus, B; Brotas, V; Ribeiro, L; Mendes, CR; Cartaxana, P; Paterson, DM. Adaptations of microphytobenthos assemblages to sediment type and tidal position. *Cont Shelf Res,* 2009, 29(13), 1624-1634.

[49] Underwood, G; Perkins, R; Consalvey, M; Hanlon, A; Oxborough, K; Baker, N; et al. Patterns in microphytobenthic primary productivity: Species-specific variation in migratory rhythms and photosynthetic efficiency in mixed-species biofilms. *Limnol and Oceanogr,* 2005, 50(3), 755-767.

[50] Okamoto, N; Inouye, I. Intertidal sandy beaches as a habitat where plastid acquisition processes are ongoing. In, J Seckbach editor, *Algae and cyanobacteria in extreme environments.* Dordrecht, Springer Netherlands, 2007, 225-237.

[51] Gould, SB; Waller, RF; McFadden, GI. Plastid evolution. *Annu Rev Plant Biol,* 2008, 59, 491-517.

[52] Keeling, PJ. The endosymbiotic origin, diversification and fate of plastids. *Philos T R Soc B,* 2010, 365(1541), 729-748.

[53] Lane, CE; Archibald, JM. The eukaryotic tree of life: endosymbiosis takes its TOL. *Trends Ecol Evol,* 2008, 23(5), 268-275.

[54] Nakayama, T; Ishida, KI. Another acquisition of a primary photosynthetic organelle is underway in *Paulinella* chromatophora. *Curr Biol,* 2009, 19(7), R284-R285.

[55] Nowack, ECM; Melkonian, M; Gloeckner, G. Chromatophore genome sequence of *Paulinella* sheds light on acquisition of photosynthesis by eukaryotes. *Curr Biol,* 2008, 18(6), 410-418.

[56] Yoon, HS; Nakayama, T; Reyes-Prieto, A; Andersen, RA; Boo, SM; Ishida, KI; et al. A single origin of the photosynthetic organelle in different *Paulinella* lineages. *BMC Evol Biol,* 2009, 9, 98.

[57] Yoon, HS; Reyes-Prieto, A; Melkonian, M; Bhattacharya, D. Minimal plastid genome evolution in the *Paulinella* endosymbiont. *Curr Biol,* 2006, 16(17), R670-R672.

[58] Stoecker, DK; Johnson, MD; de Vargas, C; Not, F. Acquired phototrophy in aquatic protists. *Aquat Microb Ecol,* 2009, 57(3), 279-310.

[59] Rundell, R; Leander, BS. Masters of miniaturization: convergent evolution among interstitial eukaryotes. *Bioessays,* 2010, 1-25.

[60] Cavalier-Smith, T; Nikolaev, S. The Zooflagellates *Stephanopogon* and *Percolomonas* are a Clade (Class Percolatea: Phylum Percolozoa). *J Euk Microbiol,* 2008, 55(6), 501-509.

[61] Billerbeck, M; Roy, H; Bosselmann, K; Huettel, M. Benthic photosynthesis in submerged Wadden Sea intertidal flats. *Estuar Coast Shelf S,* 2007, 71(3-4), 704-716.

[62] Hamels, I; Muylaert, K; Sabbe, K; Vyverman, W. Contrasting dynamics of ciliate communities in sandy and silty sediments of an estuarine intertidal flat. *Eur J Protistol,* 2005, 41(4), 241-250.

[63] Salvat, B. Les conditions hydrodynamiques interstitielles des sediments meubles intertidaux et la repartition verticale de la fauna dndogee. *C R Acad Sci Paris,* 1964, 259, 1576-1579.

[64] Short, AD. Beaches. In: Short AD; editor. *Handbook of Beach and Shoreface Morphodynamics*. Hoboken, NJ; USA: John Wiley & Sons, LTD, 2001, 3-20.

[65] Masselink, G; Turner, IL. The effect of tides on beach morphodynamics. In: Short AD, editor. *Handbook of Beach and Shoreface Morphodynamics*. Hoboken, NJ, USA: John Wiley & Sons, LTD, 2001.

[66] Berninger, U; Epstein, S. Vertical distribution of benthic ciliates in response to the oxygen concentration in an intertidal North Sea sediment. *Aquat Microb Ecol,* 1995, 9(3), 229-236.

[67] Patterson, DJ; Simpson, AGB; Weerakoon, N. Free-living flagellates from anoxic habitats and the assembly of the eukaryotic cell. *Biol Bull,* 1999, 196(3), 381-383.

[68] Horiguchi, T; Fumiyasu, K. *Roscoffia minor* sp. nov. (Peridiniales, Dinophyceae): A new, sand-dwelling. armored dinoflagellate from Hokkaido, Japan. *Phycol Res,* 1997, 45, 65-69.

In: Sea Level Rise, Coastal Engineering, Shorelines and Tides ISBN: 978-1-61728-655-1
Editor: Linda L. Wright, pp. 171-190 © 2011 Nova Science Publishers, Inc.

Chapter 8

APPLICATION OF THE WORK-ENERGY THEOREM FOR COMPUTING INUNDATION FROM LONG GRAVITY WAVES

***G. Muraleedharan*[1,*], *C. Guedes Soares*[1], *T.S. Murty*[2], *Indu Jain*[3], *A.D. Rao*[3] *and S.K. Dube*[3]**

[1]Centre for Marine Technology and Engineering (CENTEC), Technical University of Lisbon Instituto Superior Tecnico, Lisboa, Portugal

[2] Department of Civil Engineering, University of Ottawa, Canada

[3] Centre for Atmospheric Sciences, Indian Institute of Technology Delhi, Hauz Khas, New Delhi, India

Abstract

An expression developed for beach run-up heights for long gravity waves based on work energy theorem concept of classical physics is validated against real historical tsunami data including the 26th December 2004 Indian Ocean tsunami. It can also be applied for pilot estimation of storm surge inundation approximately as tsunamis and storm surges differ mainly in their generation mechanisms but behave in a similar way in shallow waters (long-period gravity waves). The time required by a wave to travel from shallow depths (d_1 m) to 0 m depth and shore slopes show significant nonlinear functional relationships (power series, $R^2 > 0.99$). The average deceleration ($\bar{a}$) estimated by Monte-Carlo method (Composite mid-point rule) rather than constant deceleration (a), when a tsunami wave travels from 1 m to 0 m depth seems to be more effective in estimating near shore tsunami heights for steep shore slopes (>0.3). Storm surge beach run-up height depends on the height near coastline, on the shore slopes and shallow water bottom topography over which the wave travels (as that determines the greater momentum of terminal speed) and slopes on land. Inundation is estimated for 61 cyclonic events in the Bay of Bengal during the period, 1972 - 2006.

Keywords: Momentum of Terminal Speed; Shore Slope; Pilot Estimation

* E-mail address: muralg@mar.ist.utl.pt. (Corresponding author)

1. Introduction

Many coastal locations around the world are in risk of being flooded as a result of storm surges induced by different phenomena. There can be little doubt that the number of casualties would be considerably lower if the flooding could be predicted say, 24 hours in advance, thus allowing effective warnings in the threatened areas. The accurate prediction of inland inundation is an utmost pre-requisite for predicting storm surge effect and to initiate evacuation of people along the coast during the occurrence of storm surge flooding.

Even though there is a reasonable amount of data available for cyclone tracks and storm surge heights at the coast, estimation of storm surge inundations at various coastal locations are still seems to be difficult. For instance, the floods along the coasts of New Orleans and Gulf of Mexico caused by hurricanes Katrina and Rita. The case is not different for Indian Ocean tropical cyclones. Even though several numerical storm surge inundation models are now available, their validation is mainly limited to hypothetical and laboratory conditions. The limited application of numerical storm surge inundation models in real time situations and the time lapsed during inundation estimations before the storm surge stroked the coasts were the reasons for serious property damage and loss of life.

Natural hazards have always been a challenge to the modelers and tsunamis and storm surges are at the top of the list. Various techniques such as empirical, statistical, analytical and numerical are employed for the prediction of coastal flooding around the world. In general, empirical and statistical methods require several years of observations on vast shelf areas which are difficult to obtain. In particular, a simplistic horizontal inland extrapolation of the predicted surge height at the fixed side wall in order to estimate the distance of inland penetration does not in itself take account of the dynamics of the intrusion process. Several authors developed numerical models that allow the flow of the water across the initial position of the model coastline. This has been accomplished by arranging for the advancing water front to move discontinuously from one grid point to another according to a prescribed criterion (Reid and Bodine, 1968; Flather and Heaps, 1975; Flather, 1994; Jelesianski and Chen, 1979).

However, to obtain acceptable accuracy this procedure requires an extremely fine grid spacing which may not be otherwise necessary. A finite element simulation of shallow water flow in continuously deforming lateral boundary has been proposed by Lynch and Gray (1979), which involves the continuous motion of the shoreline. However, finite element procedures may make the calculations expensive. Johns et al. (1982) employed a finite difference method which models a continuously moving lateral boundary. They applied their model to entire east coast of India in which inland inundation is confined to few grid points along the Andhra coast in India and simulated inland inundation associated with the Divi cyclone of 1977. Even though numerical ocean circulation models such as the POM-WAD scheme are applied to estimate tidal inundation under laboratory conditions (Lie-Yauw Oey, 2005), they are not validated for bottom topography and basin geometry. Peng et.al. (2004) made use of an integrated storm surge and inundation modeling system to simulate storm surge height and inundation for 10 hypothetical hurricanes of category 2 and 3 and found that slower translation speed is associated with higher storm surge height and more inundation.

Storm surge and heavy downpours are major causes for coastal flooding in the countries bordering the Bay of Bengal. Hence, the disasters due to inundations along the east coast of India, Bangladesh and Myanmar associated with these cyclones are very severe.

Long gravity waves such as storm surges behave in a similar manner to tsunamis in shallow water where the speed of travel is approximately proportional to the square root of water depth. The surges take the form of long waves and approach the coast with considerable phase speed approximately defined by $\sqrt{(g.h)}$, where g is the acceleration due to gravity and h is the ocean depth. Since the surges are long wave phenomena, their amplitude increases as they reach the coastal shallow waters (Roy et.al., 1999).

An expression developed from the concept of work-energy theorem of classical physics for beach run up from long gravity waves (Muraleedharan et. al., 2006a, 2006b) is dimensionally valid and it is also validated for various tsunami inundations that had occurred in Atlantic and Pacific Oceans including the 26th December 2004 Indian Ocean Tsunami (Muraleedharan et. al., 2006c). Nirupama (2006) considered that this work has a great practical utility in the sense that data such as this is very relevant for possible evacuation purposes during real tsunami events, as well as for long-term planning of coastal development for protection from tsunamis.

The focus of this study is the computation of storm surge inundation with the work energy theorem model, which has been calibrated with tsunami data. A few available observed severe storm surge inundations when compared with estimated ones provide confidence on the work energy theorem approach in estimating inundation from long gravity waves.

2. Methodology

The study area (Figure 1) shows the locations which are highly vulnerable to storm surge inundation along the border of the Bay of Bengal. The storm surge heights are collected from various sources, e.g. IMD's (India Meteorological Department) atlases, SMRC (SAARC Meteorological Research Centre in Dhaka, Bangladesh,1998), Unisys Weather Information Services (1945 - 2007), and from several research publications. The data source of bathymetry is General Bathymetric Chart of the Oceans **(GEBCO) 1-minute global bathymetric grid.** A bilinear cubic spline interpolation is used to obtain the bathymetry data field to the required resolution. With this procedure sufficiently accurate and realistic bathymetry is generated.

The long wave inundation (or tsunami inundation) equation given by Muraleedharan et. al., (2006b) is developed based on the concept of Work-Energy Theorem of classical physics. The theorem states that the sum of initial potential energy (PE_i), initial kinetic energy (KE_i) and external work done is equal to the sum of final potential energy (PE_f) and final kinetic energy (KE_f). This is the conceptual aspect of this theorem. When a storm surge enters shallow water its kinetic energy decreases (KE_i) due to decrease in speed and potential energy increases (due to increase in surge height). ie. a decrease in kinetic energy (ΔKE) is associated with an increase in potential energy (ΔPE). At the shore since the water depth is approximately equal to zero, the celerity of the wave is also expected to be zero. However the momentum and terminal velocity are quite considerable at this stage and this is what propels the tsunami or the surge on to the land, thus causing inundation.

In addition to this increase in potential energy (increase in tsunami height or surge height) associated with a decrease in kinetic energy due to decrease in water depth (Muraleedharan et.al., 2006b), the storm surge height is also influenced by remote and local wind fields, river runoff and tide. Example, during a high tide, the 0m depth (landfall point) will be further shifted towards onshore (point "O" in Figure 2). Since storm surges are long waves, the energy loss due to friction and dissipation (inversely proportional to wave length) are not as significant as for short period wind waves.

Work is being done by the wave (or storm surge) as it climbs the beach and the potential energy of the wave is gradually dissipated. The initial potential energy (PE_i) (due to surge height or flow depth) is maximum at the point O and final potential energy (PE_f) is zero (surge height or flow depth is zero) at the point S (limit of inundation). The reference level for zero potential energy due to flow depth is the line segment $\overline{OS}$ (inundation distance, Figure 2). The work done serves to reduce the depth of the water column as the wave continues climbing the land, the depth of the water column get reduced until at a certain stage the flow depth is zero, which is the limit of inundation.

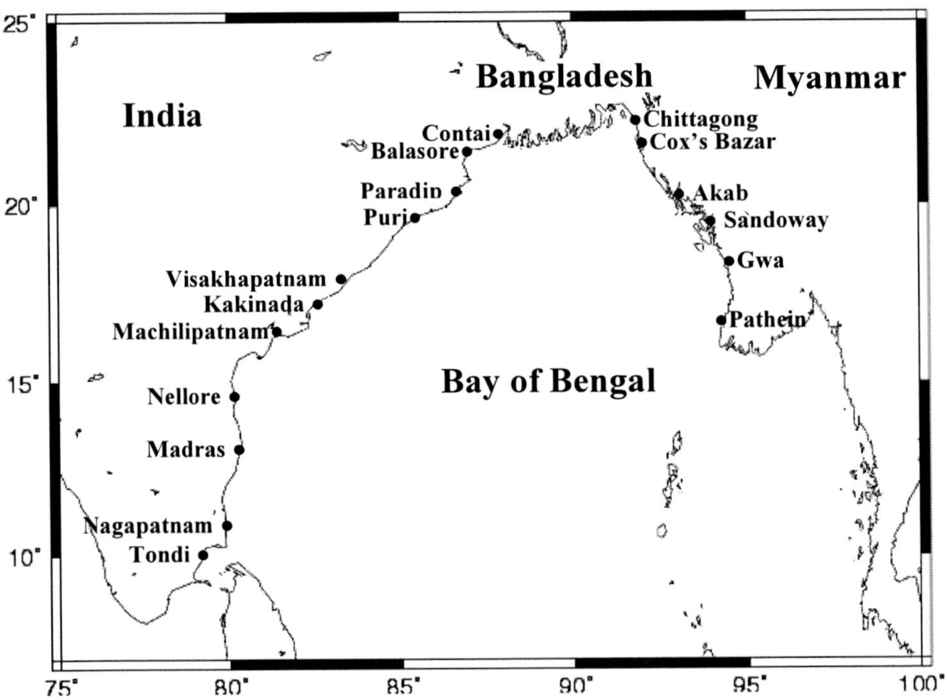

Figure 1. The rim coastal locations of Bay of Bengal vulnerable to tropical cyclones.

Our intuition in this area can be further clarified through the relationship between the rate of change of PE with respect to displacement and the force a system exerts on a particle. Consider a conservative system, one in which no energy is transferred across the system boundary. This means that the change in kinetic energy (**ΔKE** = KE_i - KE_f), plus the change in potential energy (ΔPE = PE_i - PE_f) equals zero. At shore (point O), surge wave height (H_s) and potential energy (PE_i) is maximum, and zero (H_s and PE_f) at point (S) (Figure 2). The celerity of the surge is a minimum (depth $\approx$ 0) at shore (point O). But **ΔKE** + ΔPE = **0** or ΔPE= **−ΔKE**= −work done by the system= − Fdr, where dr is the horizontal displacement and F is the force exerted on a particle by the system (Figure 2).

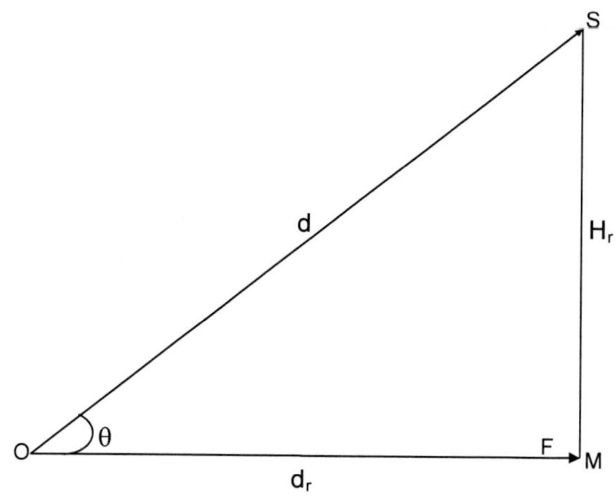

Figure 2. Schematic representation showing work done by a tsunami wave height (H_s) from the point O to point S against a beach slope (tanθ).

The significance of the minus sign is that when F and d_r are in the same direction, then PE decreases. When they are opposed, PE increases. Since Fcosθ, the component of force F along $\overline{OS}$ (inundation distance) and the displacement d are in the same direction (Figure 2) PE decreases as the wave climbs the beach ($\overline{OS}$).

In systems of many particles, the centre of mass moves exactly like a single particle, so the whole set of results already developed for one particle can be adopted to describe the translational motion of the whole system.

2.1. Calibration of the Model with Tsunami Data

The expression for tsunami beach run-up heights (H_r) developed from work-energy theorem is dimensionally valid and is given by Muraleedharan et.al., (2006b) as

$$H_r = \sqrt{\left(\frac{g}{d_1}\right)}\, H_s\, t \tan\theta \qquad (1)$$

where

d_l- shallow water depth (m)
H_s- tsunami height near shore (m)
t- time required by a tsunami wave to travel from d_lm to 0m depth.
tanθ- slope on land

The inundation distance is given by (Figure 2) as:

$$d = \frac{H_r}{\sin\theta} \tag{2}$$

The various error criteria employed to validate the performance of the models are relative-root-mean-square error (ε_{rrms}), root-mean-square-error (ε_{rms}) and mean error or relative- bias (ε_{mean}) given by

$$\varepsilon_{rrms} = \frac{\sqrt{\frac{1}{N}\Sigma[P-C]^2}}{\frac{1}{N}\Sigma C} \tag{3}$$

$$\varepsilon_{rms} = \sqrt{\frac{1}{N}\Sigma[P-C]^2} \tag{4}$$

and

$$\varepsilon_{mean} = \frac{\Sigma(P-C)}{\Sigma C} \tag{5}$$

where,

P-predicted statistics,
C-computed statistics,
N-total number of data.

Muraleedharan et al., (2006b) estimated the time tuning coefficients t (s) from equation (1) using known wave height (5.9m), beach slopes ($\tan\theta$) (beach slopes $\tan\theta$ and near shore slopes $\tan\theta_1$, $\theta_1<90$ are equal), beach run-up height (H_r) of the numerical experiment (Marchuk and Anisimov, 2001) and for d_1=1m as

$$t = \frac{H_r}{\sqrt{g}\ H_s \tan\theta} \tag{6}$$

Hence the estimated t(s) is the time required by the tsunami wave to travel from 1m depth (d_1=1m) to 0m depth (Table 1,col.2). To ensure a shoal free shallow water bottom topography up to shoreline is the purpose of choosing a small depth, d_1=1m. Also it will aid in practical application of the model (6). Except in a few cases, the assumption of a linear near shore bottom topography (Marchuk and Anisimov, 2001; Figures 3 and 4) is applicable in many coastal regions.

Table 1. Computed (C) (from 6) and predicted (P) (from 7) Time tuning coefficients (t) for a wave (5.9 m) to travel from 1m depth to 0m depth for different shore slopes ($\tan\theta_1$) (Muraleedharan et. al., 2006b)

tanθ1	Time tuning coefficient (t)				
	C (s)	P (s)	ε_{rrms}	ε_{rms}	ε_{mean}
0.1	7.60	7.81	0.034	0.082	0.004
0.2	3.80	3.74			
0.3	2.50	2.43			
0.4	1.82	1.79			
0.5	1.43	1.42			
0.6	1.16	1.17			
0.7	0.97	0.99			
0.8	0.85	0.86			
0.9	0.76	0.76			
1.0	0.68	0.68			

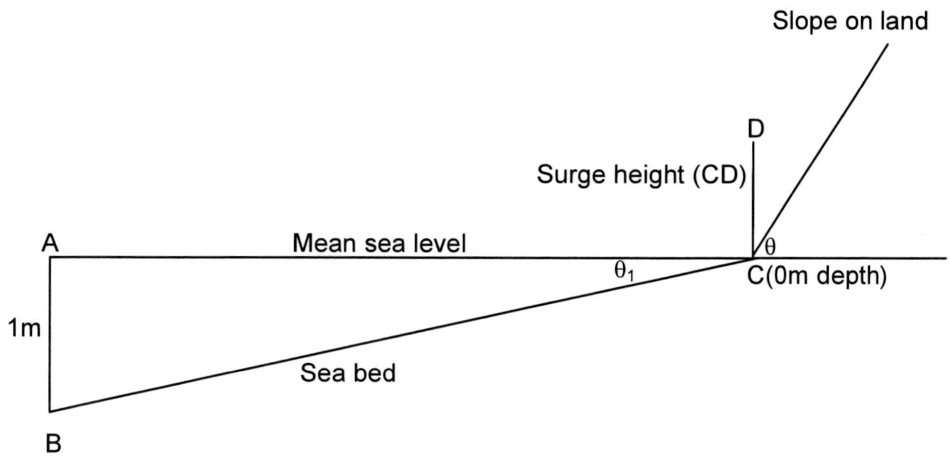

Figure 3. Schematic representation showing the distance between 1m depth and 0m depth is large for small offshore slopes.

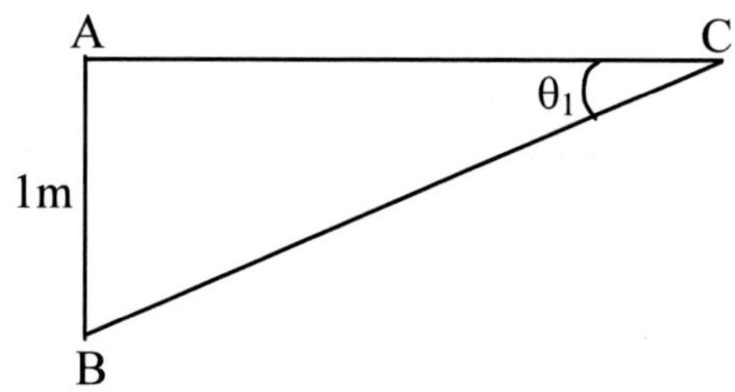

Figure 4. Schematic representation showing the distance between 1m depth and 0m depth is small for large offshore slopes.

The time tuning coefficients (t) decrease as the shore slopes increase. This seems to be reasonable since the horizontal distance from d_l m depth (small depth) to 0 m depth is more in the case of shallow shore slopes compared to steep slopes (Figures 3 and 4) and hence time tuning coefficients are large. Also, the deceleration of wave celerity (from d_lm to 0m) and hence terminal speed will be comparatively small for shallow shore slopes and vise-versa.

The time tuning coefficient (t) and the shore slope ((tan (θ_1)) exhibits a significant nonlinear functional relationship (power series) and hence the nonlinear regression equation fitted is given (Figure 5) as

$$t = 0.6791\left[\tan\left(\theta_1\right)\right]^{-1.0606} \tag{7}$$

with a coefficient of determination 0.9995 revealing that there exists significant goodness of fit. The fitting procedure is usually iterative and based on maximum likelihood methods (Wilks, 2006). The predicted time tuning coefficients (Table.1, col.3) using expression (7) is validated against the computed values from (6) (Table.1, col.2) for emphasizing reliability on expression (7). The values of the error criteria employed are small.

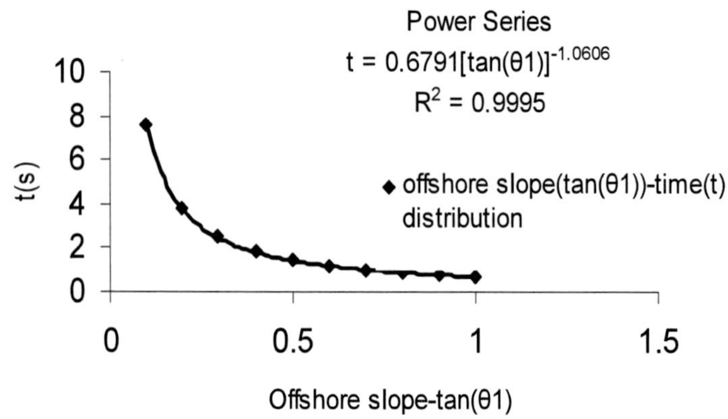

Figure 5. Scatter diagram showing the nonlinear relationship of offshore slopes and time required by a tsunami surge to travel from 1m depth to 0m depth.

Table 2. Non-linear functional relationships between time tuning coefficients (t) and shore slopes ($\tan\theta_1$) for a tsunami wave to travel from various shallow water depths(d_lm) to shore(0m)

Depth(d_lm)	Non-linear functional relationships between time tuning coefficients (t) and shore slopes ($\tan\theta_1$)	R^2
2.0	$t=0.9591(\tan\theta_1)^{-1.0613}$	0.9994
5.0	$t=1.5178(\tan\theta_1)^{-1.0607}$	0.9995
10.0	$t=2.1466(\tan\theta_1)^{-1.0607}$	0.9995
20.0	$t=3.0366(\tan\theta_1)^{-1.0603}$	0.9995
25.0	$t=3.3962(\tan\theta_1)^{-1.0603}$	0.9994
50.0	$t=4.8027(\tan\theta_1)^{-1.0602}$	0.9994
100.0	$t=6.79(\tan\theta_1)^{-1.0604}$	0.9995

Non-linear functional relationships between the time tuning coefficients (t) required by the tsunami wave to travel from various shallow water depths (d_1m) to 0 m depth and shore slopes ($\tan\theta_1$) are given in table 2. The coefficients of determinations (R^2) are more than 99% in all cases. The errors between the computed (7) and predicted are insignificant.

The non-linear functional relationships between the time (t) required to travel from small depths (d_i m) ($0 < d_i \text{ m} \leq 1$ m) to 0m depth estimated by work-energy theorem model (6) for various shore slopes and the associated depths (d_i m) are given in Table 3. The associated average decelerations (estimated by Monte-Carlo method, Composite mid-point rule of integration and by Reimann integral) are given in Table 3. The average deceleration ($\bar{a}$) is given by

$$\bar{a} = -\frac{1}{d_i t_i} \int_0^{d_i} \sqrt{gd}\; d(d) \qquad (8)$$

where t_i is the time required by the tsunami wave height to travel from d_i m depth to 0 m depth for a given shore slope ($\tan\theta_1$) estimated by the expressions given in Table 3.

Table 3. Non-linear functional relationships between time tuning coefficients (time required by a wave to travel from d_i m to 0 m) and shore depths (d_i m) for various shore slopes ($\tan\theta_1$) and the deceleration (a) and average deceleration ($\bar{a}$) (by Reimann and Monte-Carlo integration) when a tsunami wave travels from 1m depth to 0 m depth

$\tan(\theta_1)$	Functional relationship between time tuning coefficient (t) and depth (d_i m) for a tsunami wave to travel from d_i m($0 < d_i \leq 1$ m) to 0 m for various shore slopes	$a = -\left(\frac{\sqrt{g}}{t_i}\right)$ m/s²	$\bar{a}$ (by Riemann Integration) m/s²	$\bar{a}$ (by Monte-Carlo Integration) m/s²	R^2
0.1	$t_i = 7.5772(d_i)^{0.4996}$	0.412	0.276	0.413	1.0000
0.2	$t_i = 3.7993(d_i)^{0.5003}$	1.041	0.550	0.825	1.0000
0.3	$t_i = 2.5134(d_i)^{0.5016}$	1.583	0.831	1.248	1.0000
0.4	$t_i = 1.8231(d_i)^{0.4979}$	2.174	1.145	1.714	1.0000
0.5	$t_i = 1.426(d_i)^{0.5004}$	2.767	1.464	2.197	0.9999
0.6	$t_i = 1.1588(d_i)^{0.4967}$	3.411	1.802	2.694	0.9999
0.7	$t_i = 0.9688(d_i)^{0.4987}$	4.079	2.155	3.229	0.9998
0.8	$t_i = 0.8507(d_i)^{0.4983}$	4.655	2.455	3.676	0.9999
0.9	$t_i = 0.759(d_i)^{0.5009}$	5.206	2.751	4.130	0.9998
1.0	$t_i = 0.68(d_i)^{0.4976}$	5.818	3.071	4.595	0.9993

Alternatively by Monte-Carlo Method (Composite mid-point rule), for $d_1 = 1$ m

$$\bar{a} \approx -\frac{1}{n}\sum_{i=1}^{n}\frac{\sqrt{gd_i}}{t_i}$$

$$d_i = \left(i - \frac{1}{2}\right)\Delta d \quad \text{for } i = 1, 2, \ldots\ldots\ldots n$$

$$\Delta d = 0.001$$

$$n = \frac{1}{\Delta d} \tag{9}$$

t_i , for a given shore slope ($\tan\theta_1$), is the time required by the tsunami to travel from d_i m ($0 < d_i \leq 1$ m) to 0 m depth (Table 3).

Alternatively by solving the Reimann integral

$$\bar{a} = -\frac{2\sqrt{g}\left(d_i\right)^{\frac{3}{2}}}{3\,d_i t_i} \tag{10}$$

There exists significant non-linear functional relationship (R^2=0.9994 and 0.9995) between average deceleration ($\bar{a}$) and shore slope ($\tan\theta_1$) given by the power series

$$\bar{a} = 4.606\left(\tan\theta_1\right)^{1.0597} \quad \text{(by Monte-Carlo Integral)} \tag{11}$$

and

$$\bar{a} = 3.0741\left(\tan\theta_1\right)^{1.0598} \quad \text{(by Reimann Integral)} \tag{12}$$

Then the beach run-up height (1) can be given by

$$H_r = \frac{g\,H_S \tan\theta}{\bar{a}} \tag{13}$$

For known shore slopes, the constant (a) and average decelerations ($\bar{a}$) (by Reimann and Monte Carlo Integration) of a tsunami as it travels from 1 m depth to 0 m depth is given in table 3. The discrepancies between the two decelerations are significant for steep shore slopes (>0.3).

The predicted tsunami heights (from 1) and time tuning coefficients t (from 7 for constant deceleration, a, as shore slopes are <0.3) for the historical tsunami events (NGDC data) in the Pacific and Atlantic Oceans are given in table 4. The beach run-up heights can be more than twice the tsunami heights due to grater momentum of the terminal speed.

Table 4. Predicted tsunami heights (H_s) and time tuning coefficients (t- to travel from 1 m depth to 0 m depth for constant deceleration concept) for given historical tsunami beach run-up heights (H_r) and maximum inundation events (d) in the Pacific and Atlantic Oceans

Year	Country	Location	Beach Run-up Ht (H_r) (m)	Max. Inund. Dist.(d) (m)	Beach Slope tan(θ)	Predicted t (s)	Predicted tsunami ht. at 0 m depth (H_s) (m)
1835	CHILE	MAULE RIVER	3.5	140	0.025	34.01	1.32
1835	CHILE	TALCAHUANO	9	4000	0.003	436.59	2.92
1867	SAINT VINCENT AND THE GRENADINES	BEQUIA ISLAND: ADMIRALTY BAY	1.8	146	0.012	71.86	0.65
1867	USA TERRITORY	VIRGIN ISLANDS: FREDERIKSTED	7.6	76	0.101	7.76	3.11
1867	BRITISH VIRGIN ISLANDS	TORTOLA ISLAND: ROAD TOWN	1.5	9	0.169	4.47	0.63
1867	USA	PUERTO RICO: ARROYO	0.9	40	0.023	37.96	0.34
1868		BEQUIA ISLAND: ADMIRALTY BAY	0.9	21	0.043	19.15	0.35
1878	USA	WAIALUA, OAHU, HI	3	150	0.020	43.01	1.11
1899	USA	YAKUTAT BAY, WEST SHORE, AK	9	500	0.018	48.10	3.32
1918	USA	PUERTO RICO: CAJA DE MUERTOS	1.5	15	0.101	7.76	0.61

Table 4. (Continued)

Year	Country	Location	Beach Run-up Ht (H_r) (m)	Max. Inund. Dist.(d) (m)	Beach Slope tan(θ)	Predicted t (s)	Predicted tsunami ht. at 0 m depth (H_s) (m)
1918	USA	PUERTO RICO: PUNTA BORINQUEN	4.5	100	0.045	18.18	1.75
1944	JAPAN	TEMMA	5	200	0.025	33.94	1.88
1946	USA	PAKALA POINT, HI	10	227	0.044	18.60	3.89
1964	USA	KLAWOCK, AK	4.6	15.2	0.318	2.29	2.02
1969	RUSSIA	OK'KHOVAYA RIVER	15	500	0.030	27.97	5.70
1975	USA	PUNALUU BAY, HAWAII, HI	7.6 m	137	0.056	14.56	3.00
1985	MEXICO	PLAYA AZUL	2.5 m	150	0.017	52.19	0.92
1985	MEXICO	ZIHUATANEJO	2.5 m	200	0.013	70.82	0.90
1991	COSTA RICA	CAHUITO-PUERTO VIEJO	2 m	300	0.007	137.93	0.69
1992	NICARAGUA	MASACHAPA	6 m	100	0.060	13.39	2.38
1994	PAPUA NEW GUINEA	SIMPSON HARBOUR, NEW BRITAIN	1.2 m	200	0.006	154.24	0.41
2001	PERU	CAMANA	8 m	1000	0.008	113.68	2.81

Table 5. Predicted time (t) required to travel from 1m depth to 0m depth and tsunami height (H_s) (for constant and average decelerations) at shore for 26th December 2004, Indian Ocean Tsunami for given tsunami beach run-up heights (H_r) and inundation distances along Tamil Nadu Coast, India

Location	Beach Run-up height (m) (H_r)	Inundation distance (m) (d)	Beach slope-tan(θ)	t in secs (assumed $\theta = \theta_1$)	H_s using constant deceleration (a)	H_s using $\bar{a}$ by Monte Carlo method	H_s using $\bar{a}$ by Reimann Integral
Ovari	19	150	0.1277	6.02	7.89	7.89	5.26
Idinthakarai	18	175	0.1034	7.53	7.38	7.38	4.92
Perumanal	17	200	0.085	9.24	6.88	6.89	4.60
Navaladi	16	200	0.0803	9.85	6.46	6.46	4.31
Vattakottai	15	300	0.0500	16.26	5.88	5.89	3.93
Lakshmipuram	16	250	0.0641	12.50	6.37	6.38	4.25
Chinna Muttam	17	350	0.0486	16.77	6.65	6.66	4.45
Kanyakumari	21	300	0.0702	11.36	8.41	8.41	5.61
Keelamanakudi	9	600	0.0150	58.36	3.28	3.29	2.19
Pallam	14	400	0.0350	23.75	5.37	5.38	3.59
Rajakkamangalam	16	150	0.1073	7.24	6.57	6.58	4.39
Muttom	11	200	0.0551	14.69	4.34	4.34	2.90
Kadiapatanam	14	100	0.14	5.41	5.84	5.85	3.90
Mandakadu	16	100	0.16	4.68	6.73	6.73	4.49
Colachel	12	750	0.0160	54.50	4.39	4.40	2.94
Vaniakudy	16	100	0.16	4.68	6.73	6.73	4.49
Midalam	17	300	0.0567	14.23	6.72	6.73	4.49
Enayam	15	130	0.1162	6.66	6.19	6.19	4.13
Taingapatnam	14	200	0.07	11.36	5.61	5.61	3.74

The predicted tsunami heights (from 13) near Tamil Nadu coast using average deceleration ($\bar{a}$, by Monte-Carlo integration)) of the tsunami wave as it travels from 1 m to 0 m depth (11) are equal to the values by using constant deceleration (using 7 for estimating t) (Table 5). This is because the discrepancies between the decelerations (constant and average decelerations) are small for shallow near shore slopes (<0.3, Table 3). The heights are under predicted using average deceleration by Reimann integral (12).

Numerical modeling of tsunami wave run-up heights was carried out for real shore profile near Minehama village in Akita prefecture, Japan (Marchuk and Anisimov, 2001). Run-up height by the tsunami of May 26, 1983 was 14.08 m. To reach such a height, the tsunami wave height should be 8 m at the shore.

This complicated beach is having 11 different slopes on land. Hence the one-third significant beach angle is computed to be 51°. From expressions 1 and 7 (using constant deceleration approach from 1 m depth to 0 m depth), the tsunami wave height at 0 m depth will be 8 m if the near shore slope is 55° and time tuning coefficient is 0.46s (Muraleedharan et.al., 2006b). For a near shore slope ($\tan\theta_1$) of the above magnitude, by average deceleration concept (11), the tsunami height (from 13) will be 7.81 m. Usually a 10m tsunami wave near shore could have a run-up height inland which may be much higher than 10m (and even as much as double) because of the topography and due to greater momentum of the terminal speed.

The expressions for estimating beach run-up heights are derived for a plane beach slope with no obstacles. Hence the run-up height estimated by these expressions give the maximum beach run-up height limit from the mean sea level to which a particular tsunami wave height will climb.

3. Results and Discussions

3.1. Adaptation of the Calibrated Model for Storm Surge Inundation

The storm surge inundation estimations using (1), (2) and (7) for the tropical cyclones that has occurred in the period (1972-2006) in Bay of Bengal are given in table 6. Since the discrepancies between the constant and average decelerations are small for shallow near shore slopes, only constant deceleration concept is considered for storm surge inundation estimations as near shore slopes are very shallow (<<0.1) for all cyclonic events. . A few observed details of storm surge inundations along the coasts of Bay of Bengal are compared with estimated values to emphasize reliability on the latter.

The Bay of Bengal, is a very low-lying and densely populated area, consisting of extensive mangrove swamps and many river deltas including those of the large Ganges, Meghna, and Brahmaputra rivers. Its funnel-shape and extensive shallow bathymetry help contribute to the potential for large storm surges and associated inundations.

Table 6. Estimated storm surge inundation distance (d) (for constant deceleration concept) along the coasts of rim nations of the Bay of Bengal for the period from 1972 to 2006

Year	Near shore water depth location		Nearshore slope($\tan\theta 1$)	Hs (m)	Beach slope ($\tan\theta$)	Predicted Inundation dist (d) m
	E	N				
15-23 NOV 1972	80.2584	13.7126	0.0007	1.00	0.0038	4682.91
14-18 NOV 1973	90.50	22.0208	0.0024	1.00	0.00027	1288.55
23-28 Nov. 1974	90.4958	22.1416	0.0024	1.37	0.00073	1758.52
4-8 MAY 1975	94.4542	17.0667	0.0004	1.00	0.0034	13741.94
8-20 SEP 1976	87.7958	21.6916	0.0011	1.00	0.0053	2927.80
14-20 NOV 1977	80.7941	15.8205	0.0008	3.20	0.0041	13839.64
27 OCT-4 NOV 1977	80.0874	14.9042	0.0018	1.0	0.0077	1747.34
5-13 MAY 1979	80.0874	14.8874	0.0023	1.37	0.0037	1857.79
5-10 DEC 1981	88.1333	21.8916	0.0009	1.0	0.00054	3673.52
30 APR-5 MAY 1982	94.5584	17.4042	0.0015	3.2	0.0105	6596.73
30 MAY-4 JUN 1982	86.5292	20.1708	0.0024	1.0	0.00056	1305.42
13-16 OCT 1982	82.1458	16.5042	0.0014	1.0	0.0014	2256.89
17-19 OCT 1982	80.2458	13.8416	0.0053	1.0	0.00003	554.06
1-4 OCT 1983	83.5667	17.9708	0.0046	1.0	0.0043	647.45
10-14 OCT 1984	86.8626	21.2374	0.0009	1.0	0.0016	3746.74
9-15 NOV 1984	80.1667	14.00	0.0044	1.37	0.0014	912.04
27 NOV-8 DEC 1984	79.7458	11.60	0.0020	1.37	0.00049	2100.54
22-25 MAY 1985	91.2833	22.6084	0.0030	1.0	0.0044	1020.62
8-11 OCT 1985	83.425	17.8208	0.0081	1.0	0.0250	353.52
15-16 OCT 1985	86.8626	20.5833	0.0025	1.0	0.000002	1243.85

Table 6. (Continued)

Year	Near shore water depth location		Nearshore slope($\tan\theta_1$)	Hs (m)	Beach slope ($\tan\theta$)	Predicted Inundation dist (d) m
	E	N				
13-18 NOV 1985	80.1374	14.7042	0.0010	1.0	0.0040	3070.56
9-14 DEC 1985	80.1584	14.30	0.0017	1.0	0.0015	1799.16
30 MAY-5 JUN 1987	90.8542	22.2874	0.0022	1.0	0.0048	1417.61
14-16 OCT 1987	81.2042	16.1626	0.0002	1.0	0.002	21911.71
30 OCT-3 NOV 1987	80.0916	14.8042	0.0019	1.0	0.0016	1623.14
8-13 NOV 1987	80.3833	15.7626	0.0001	1.0	0.0022	39076.08
21-30 NOV 1988	88.8416	21.55	0.0001	3.2	0.002388	91388.81
23-27 MAY 1989	87.2084	21.4958	0.0018	1.0	0.00016	1736.66
3-11 MAY 1990	80.7916	15.7542	0.0009	3.2	0.0036	12046.87
13-19 DEC 1990	92.1292	21.0667	0.0011	1.0	0.0089	2900.07
22-30 APR 1991	91.8626	22.0416	0.0014	4.72	0.000022	10299.39
30 MAY-3 JUN 1991	91.4084	22.75	0.0004	1.0	0.0018	7617.75
9-16 NOV 1991	79.8584	11.0042	0.0028	1.0	0.00018	1084.87
14-18 JUN 1992	86.7874	20.4667	0.0020	1.0	0.00083	1513.62
24-28 JUL 1992	86.575	20.1958	0.0018	1.0	0.00067	1710.41
4-9 OCT 1992	81.1958	16.1458	0.0023	1.0	0.00118	1312.86
27 NOV-5 DEC 1993	79.8458	11.20	0.0009	1.37	0.00164	5187.80
26 APR-3 MAY 1994	92.1792	21.0042	0.0003	3.2	0.0171	33713.22
28-31 OCT 1994	80.2458	12.6667	0.0037	1.0	0.00073	797.69
12-20 SEP 1995	87.5374	21.5958	0.003	1.0	0.0021	12028.13
5-10 NOV 1995	83.5667	17.9708	0.0046	1.37	0.0043	887.01
18-25 NOV 1995	92.05	21.2958	0.0089	2.1	0.00529	669.32
1-8 MAY 1996	92.05	21.2874	0.0034	1.0	0.0104	890.49

Table 6. (Continued)

Year	Near shore water depth location		Nearshore slope($\tan\theta1$)	Hs (m)	Beach slope ($\tan\theta$)	Predicted Inundation dist (d) m
	E	N				
11-18 JUN 1996	83.5708	17.9958	0.0066	1.0	0.008	435.46
21-29 OCT 1996	88.7958	21.5916	0.0022	1.0	0.0005	1379.41
1-7 NOV 1996	82.3	16.5626	0.0011	4.72	0.00096	14241.70
26 NOV-7 DEC 1996	80.25	12.8458	0.0064	1.0	0.00091	454.04
13-20 MAY 1997	91.7584	22.3	0.0008	4.72	0.00136	19412.00
19-27 SEP 1997	91.2833	22.6084	0.0030	1.37	0.00445	1398..25
13-20 MAY 1998	91.7626	22.225	0.0012	1.37	0.00014	3678.55
13-16 NOV 1998	83.0791	17.5042	0.0056	1.37	0.00031	710.49
16-23 NOV 1998	89.8708	21.8916	0.0018	1.0	0.00002	1759.34
8-11 JUN 1999	87.9626	21.7916	0.0023	1.0	0.0011	1356.78
15-18 OCT 1999	84.6333	18.8958	0.0052	4.72	0.00236	2627.59
25 OCT-3 NOV 1999	86.20	19.8874	0.0008	1.0	0.000023	25407.92
25-28 OCT 2000	88.8208	21.5833	0.0022	4.72	0.00031	1404.25
26 NOV-6 DEC 2000	79.7458	11.60	0.0020	1.37	0.00049	2100.54
23-28 DEC 2000	77.9416	8.30	0.0018	1.37	0.00266	1739.33
11-16 DEC 2003	81.1458	15.9792	0.0013	1.37	0.00031	2455.78
28-29 SEP 2006	84.8626	19.1958	0.0009	1.0	0.0075	3742.89
24-29 APR 2006	94.5292	17.7042	0.0150	4.72	0.0086	862.03

Table 7. Reported and predicted inundations on the rim nations bordering the Bay of Bengal

Year	Reported inundation (Km)	Predicted inundation (Km)
May,1975,Myanmar	–	13.7
21-30 November 1988, Bangladesh	90.0	91.4
22-30 April 1991, Bangladesh	10.0	10.3
26 April-03 May 1994 Myanmar	–	33.7
14-20 Nov, 1977, Andhra	12.0	13.8
3-11 May, 1990, Andhra	11.3	12.0
9-16 November 1991,Tamil Nadu	200-250 m	1085 m
1999 Orissa cyclone	26.0	25.4

The validation with few reported storm surge inundation instances support the view point that the inundation equation provides an immediate approximation sufficient enough for pilot estimations. (Table 7). The low lying regions bordering the Bay of Bengal are severely vulnerable to storm surge inundations. Hence the inundation data from these locations are valuable asset for model validations as Bay of Bengal is notorious for severe cyclones.

It is to be concluded that the bathymetry from the General Bathymetric Chart of the Oceans (GEBCO) is good enough for applying the model developed from work energy theorem for pilot estimation of storm surge inundations. This is really an attraction as it is readily available in the websites. The methodology (bilinear cubic spline interpolation) used seem to be appropriately estimated the bathymetry with required precision.

The equation is very sensitive to shore slopes. The inland inundation depends upon the shore bathymetry which in turn determines the surge height (along with wind, river run-up, and tide) and greater momentum of the terminal speed and slopes on land. A moderate surge height in favourable environment can penetrate inland considerably and cause great devastations. In short, the height to which a surge will reach on land (H_r) will depend on the surge height(H_s) at 0m depth (landfall), on the shore slope ($\tan\theta_1$) over which the surge travels (as that determines the momentum of terminal speed), the slopes on land ($\tan\theta$) and the land topography (trees, buildings, obstacles which may create friction).

Even though numerical storm surge models can estimate surge inundations more accurately, it may not be available in many cases before the storm stroked the coasts. This is one of the main reasons for great loss of life and property. Hence a pilot estimation of surge inundation approximately at the time of warning (cyclone category, its landfall point, storm surge height etc) will be of great assistance for evacuation purposes.

4. Conclusion

The tsunami surge inundation equation developed from the concept of work-energy theorem of classical physics also seems to be effective for storm surge flood estimations (pilot estimates) provided it is validated against observations by calibrating with landfall bathymetry. There exists a significant nonlinear functional relationship (power series) between the time required by a wave to travel from small shallow water depths, d_1 m to 0 m depth and respective shore slopes. The decrease of the time tuning coefficient (t) for the surge (or tsunami) to travel from 1m depth to 0 m depth, with increase in shore slopes is in agreement with reality as the horizontal distance between these depths decreases. There also exists a significant non-linear functional relationship (power series) between the average deceleration ($\bar{a}$) of a tsunami wave as it travels from 1 m to 0 m depth and shore slopes. The predicted tsunami heights using average deceleration (from 13) and constant deceleration (from 1 and using 7 for estimating t) are equal because the discrepancies between the two decelerations are small for shallow shore slopes (<0.3). One of the attractions is that the GEBCO, 1-minute bathymetric grid (readily available in websites) with required precision is applicable for storm surge inundation evaluations. The height to which a surge will reach on land will depend on the surge height at 0m depth (landfall), on the shore slope over which the surge travels (as that determines the greater momentum of terminal speed), the slopes on land and the land topography (trees, buildings, obstacles which may create friction). For different estimated surge heights at landfall, beach angle and time tuning coefficients (t) for known shore slopes, land inundation tables could be constructed for various coastal locations which are vulnerable to storm surge flooding.

Acknowledgments

The authors wish to thank IMD, SMRC, BODC and Unisys Corporation for providing the data used here. Indu Jain acknowledge with pleasure for granting the financial support by Integrated Coastal Marine and Area Management, Govt. of India, to carry out this work.

References

Flather, RA; Heaps, NS. Geophy. *J. Royal Astro. Soc.,* 1975, 42, 489-517.
Flather, RA. *J. Phy. Oceanogr*., 1994, 24, 172-190.
Jelesnianski, CP; Chen, J. SLOSH (Sea Lake and Overland Surges from Hurricanes). *Report of TDL, NWS, Silver Spring:* MD, 1979.
Johns, B; Dube, SK; Sinha, PC; Mohanty, UC; Rao, AD. *Comp*., Fl. 1982, 10, 105-116.
Lie-Yauw Oey. *Ocean Mod*., 2005, 9, 133-150.
Lynch, DR; Gray, WG. *Comp*., Fl. 1979, 7, 207-228.
Marchuk, AG; Anisimov, AA. ITS Proc. *Session,* 7. 2001, No.7-27, 933-940.
Muraleedharan, G; Mourani, S; Rao, AD; Murty, TS. Internl Con. Mesoscale Processess in *Atmos. Ocean Environ.Sys.* (IMPA2006). 2006a. Indian Institute of Technology Delhi.
Muraleedharan, G; Mourani, S; Rao, A.D; Murty, T.S. *Mar. Geod.* 2006b, 29(3), 223-231.
Muraleedharan, G; Rao, AD; Murty, TS; Mourani, S. The Indian Ocean Tsunami; Murty,T.S; Aswathanarayana, U; Nirupama, N; *Taylor and Francis,* NL, 2006c, 131-142.
Nirupama, N. The Indian Ocean Tsunami; Murty, T.S., Aswathanarayana, U., Nirupama, N., Taylor and Francis: NL, 2006, 209-211.
Peng, M; Lian Xie; Leonard.J; Pietrafesa. *Coast. Shelf Sci*., 2004, 59, 121-137.
Reid, RO; Bodine, BR. *J. Waterways Har. Div. ASCE*., 1968, 94, 33-57.
Roy, GD; Humayun Kabir, ABM; Mandal, MM; Haque, MZ. Dyns. *Atmos. Oceans*., 1999, 29 (2-4), 397- 413.
SMRC. SAARC Meteorological Research Centre publication No.1, Dhaka, 1998.

Unisys Weather Information Services, (http://weather.unisys.com/hurricane/n_ indian/index.html).

Wilks, DS. Statistical Methods in the Atmospheric Sciences; *International Geophysics Series*, Elsevier: US, 2006, 59, 627.

In: Sea Level Rise, Coastal Engineering, Shorelines and Tides ISBN: 978-1-61728-655-1
Editor: Linda L. Wright, pp. 191-204 © 2011 Nova Science Publishers, Inc.

Chapter 9

DYNAMICAL MODE DECOMPOSITION OF TIDAL CURRENTS OVER THE MAIN SILLS OF THE STRAIT OF GIBRALTAR

A. Sanchez-Roman, F. Criado-Aldeanueva, J. Garcia-Lafuente, J.C. Sanchez-Garrido, J. Soto-Navarro, C. Naranjo and C. Calero
Physical Oceanography Group, University of Malaga, Spain

Abstract

ADCP velocity data collected in the two main sills of the Strait of Gibraltar (Camarinal and Espartel sills) have been used for analysing the vertical structure of main tidal constituents (M_2, S_2, O_1 and K_1) currents in this area. Two different periods (winter and summer) were considered in correspondence to seasonal variations in density profiles. Amplitudes and phases of the various tidal constituents have been compared for both periods and locations. Barotropic and baroclinic parts of the tidal currents have been extracted using the dynamical mode decomposition technique and the relative importance of each mode has been established in terms on the energy associated. In Espartel sill, the barotropic mode is more energetic in wintertime for all constituents except for K_1. Baroclinic modes have smaller contribution to total energy. Over Camarinal sill, barotropic mode accounts for more than 90% of total energy in all the tidal constituents, the highest value (97%) observed for M_2.

Keywords: Tidal currents, dynamical mode decomposition, harmonic analysis, seasonal variability, velocity profilers, Strait of Gibraltar.

1. Introduction

The Strait of Gibraltar connects the Mediterranean Sea and the Atlantic Ocean through a rather complicated system of sills and narrows. It has a length of nearly 60 km and a mean width of 20 km. Its shallower depth, less than 300m, is found in the main sill of Camarinal, CS (west of Tarifa) and its minimum width of around 14 km coincides with the contraction of

Tarifa Narrows (see Figure 1). The Strait is the scenario of a well-studied baroclinic exchange between two basins with different densities.

The tidal signal in the Strait has been extensively studied and analysed in the past. García-Lafuente et al. (1990) described the structure of the barotropic (vertical) tide in the Strait and Candela et al. (1990) extended the analysis of the barotropic tide to M_2 tidal currents. The tidal transport through the Strait was studied by Bryden et al. (1994) and García-Lafuente et al. (2000). Bruno et al. (2000) have described the vertical structure of the semidiurnal tide at CS and, in accordance with previous estimations (Candela et al., 1989; Mañanes et al., 1998), have reported that the semidiurnal variance at CS is mainly due to the barotropic mode, with less than 10% accounted by baroclinic modes. Tsimplis (2000) has performed a similar analysis in CS from 100 days (18/01/97-29/04/97) series of ADCP data. Wang (1993) studied the tidal flows, internal tide and fortnightly modulation by mean of a numerical model and Tsimplis and Bryden (2000) have estimated the water transports through the strait; García Lafuente et al. (2000) have analysed the tide in the eastern section of the strait and Baschek et al. (2001) have estimated the transport with a tidal inverse model. Recently, Sannino et al. (2002, 2004) have implemented a numerical model for analysing the mean exchange through the Strait and the semidiurnal tidal exchange respectively. In contrast to CS, there are no studies on the vertical structure of tidal currents over Espartel sill (ES), which also plays an important role in the exchange. This work describes the vertical profile of the main diurnal (O_1 and K_1) and semidiurnal (M_2 and S_2) tidal constituents at ES from observed data in two different periods of the year. For comparison purposes, winter data from CS have also been analysed.

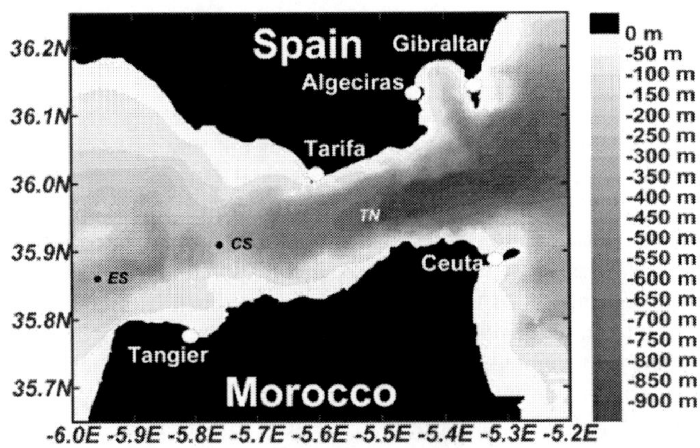

Figure 1. Map of the Strait of Gibraltar showing the main topographic features. TN corresponds to the minimum width section of Tarifa narrows. ES and CS indicates the locations of Camarinal and Espartel sills, respectively, and also the mooring lines location of the observed data.

Tidal velocities can be separated into barotropic and baroclinic components by a dynamical mode decomposition, DMD (Kundu et al., 1975; Schott, 1977; Müller and Siedler, 1992; Bruno et al., 2000, Tsimplis, 2000). With DMD, a set of vertical dynamical modes of pressure (or another flow variable) is obtained from the hydrodynamic equations for given density profiles. A least-square estimation procedure is then used to determine the modal amplitudes and phases from the observed values of tidal velocities at different depths.

2. Data and Methods

ADCP data were collected in the two main sills of the Strait of Gibraltar (ES and CS, see Figure 1) at its western approach. The ES data comes from a monitoring station located at 35 ° 51.70'N, 05° 58.60'W installed in September 2004. It consists of an up-looking moored ADCP at 340 m depth (20 m above the seafloor) that resolves 40 bins, 8-meter thick each one, and provided horizontal velocity at 40 levels every 30 min. This work analyses the data collected in 2005. CS data come from a monitoring station installed at 35° 54.80'N, 05° 44.70'W within the frame of a special programme of Woods Hole Oceanographic Institution. The station consisted of an up-looking mooring ADCP that measured the water column velocity between 54 and 274 m depth with sampling interval of 60 minutes. The bin-size of the ADCP was 10 m and 23 bins were resolved. The time series in this sill spans from October 1995 to April 1996, covering the winter season.

MEDATLAS database provided historical Conductivity-Temperature-Depth (CTD) profiles in the area of CS and ES in order to determine mean density profiles in these points. The area within 35° 53.6'N – 35° 55.1 'N / 05° 43.1'W – 05° 45.3'W has been considered to be representative for CS and the area within 35° 48.6'N – 35° 53.9'N / 05° 56.7'W – 06° 00.8'W, for ES. In order to account for seasonal variations of the density profiles, two representative periods have been selected: "winter" period between 01/02 and 30/04 and "summer" period between 01/06 and 31/10. This selection relies on the vertical density stratification of the upper layer of the water column. From February to April, there is no signature of the seasonal thermocline that develops in summertime, whereas from June to October it is clearly established.

3. Results and Discussion

3.1. Semidiurnal Constituents

ADCP time series were subjected to the standard harmonic analysis (Foreman, 1978; Pawlowicz, et at., 2002). The vertical pattern of the along-strait velocity of M_2 and S_2 in ES is shown in Figure 2. The amplitude of M_2 (panel A) reaches a maximum of ~58 $cm \cdot s^{-1}$ at ~240 m, 50m below the mean depth of the interface, inside of the Mediterranean layer. Then, the amplitude decreases quickly towards the bottom. Only slight variations between winter and summer periods are found, one of them being the relative minimum of ~47 $cm \cdot s^{-1}$ at 190m observed in summertime that is not evident in winter. Amplitude of S_2 (panel C) remains almost constant at ~16 $cm \cdot s^{-1}$ for summertime until ~240 m, where it starts decreasing with depth. In wintertime, the amplitude slowly increases from ~18 $cm \cdot s^{-1}$ near the surface until ~26 $cm \cdot s^{-1}$ at ~220m, then decreasing with depth. Discrepancies between both periods are evident, with winter values 30% higher on average and up to 50% higher at ~220m, level of the maximum difference.

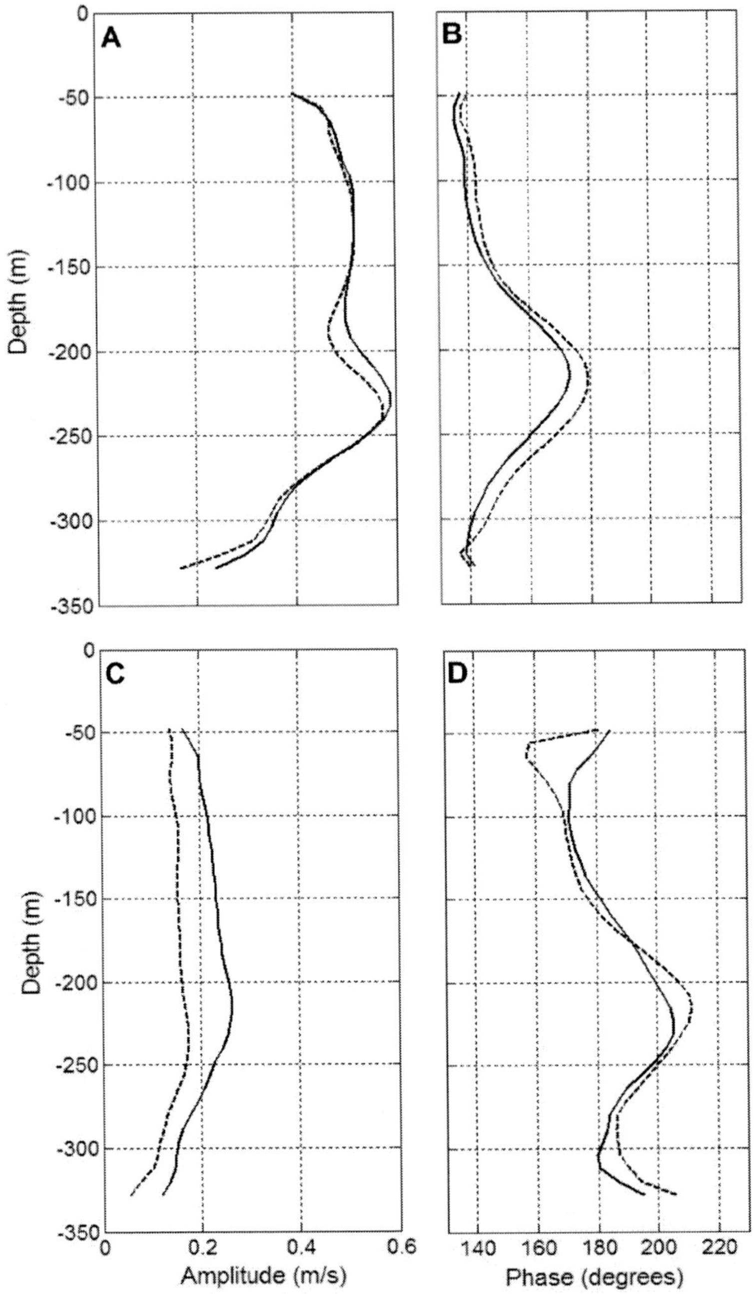

Figure 2. Vertical structure of the semidiurnal tidal constituents M_2 (amplitude, panel A; phase panel B) and S_2 (amplitude panel C; phase panel D) both for winter (solid line) and summer (dashed line) periods at ES.

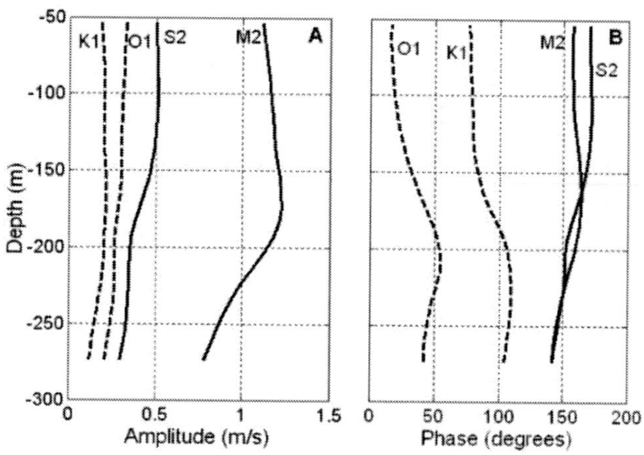

Figure 3. Vertical structure (amplitude, panel A; phase, panel B) of the main semidiurnal (solid line) and diurnal (dashed line) tidal constituents for winter period at CS.

The phase of M_2 is maximum nearly, but slightly above, the depth of the maximum amplitude. It is more pronounced in summer, a fact that could be related to the different stratification of the upper water column. The dependence of the phase with depth implies downwards phase propagation above the depth of local maximum and upwards propagation below it. According to the linear theory of internal waves in a continuously stratified flow, the group velocity (that is, the energy) has a vertical propagation of different sign that phase, that is upwards (downwards) from the depth of maximum phase, which means a local divergence of energy towards the surface and the bottom. The maximum of M_2 amplitude found below this depth as well as the secondary local maximum around 150 m above this depth could be related to this vertical energy propagation.

S_2 constituent shows a similar vertical pattern, although the relative position in the water column of the amplitude and phase maxima in winter is different from the pattern of M_2 and also from that of S_2 in summer. The nearly constant phase difference of 30° between S_2 and M_2 (S_2 leading M_2) indicates an age of the tide of 1.2 days, that is, spring tide happens one day later than new or full moon. The increase of S_2 phase near the bottom implies greater age of the tide in this area. The greater S_2 amplitude in winter implies stronger (weaker) spring (neap) tides in this season, a feature enhanced between 170 and 230 m (around the depth of the mean interface) because of the increase of M_2 amplitude within this depth range in winter.

The amplitude of M_2 in CS slightly increases from the first bins to ~175m depth (40m below the mean depth of the interface, Figure 3), where a maximum over 120 $cm \cdot s^{-1}$ is reached. The maximum is shallower and more than twice greater than in ES. Below 175m, the amplitude decreases moderately. This pattern agrees with that of Tsimplis (2000) from mid-January to April and of Candela et al. (1990) spanning different periods. S_2 amplitude remains almost constant at ~50 $cm \cdot s^{-1}$ until ~150m, then gradually decreasing with depth. The values are again greater than in ES.

The phase of M_2 in CS is maximum some tens meters below the surface of zero mean velocity, at around 175 m, the same depth in which the amplitude also reaches maximum. It is less pronounced than in ES and, as in this place, it implies phase propagation from above and below and, therefore, energy divergence upwards and downwards from this depth. Phase of S_2 is greater than phase of M_2 above 160 m in the upper layer but it is less or equal in the

Mediterranean layer. It is interesting to emphasise the good agreement of these results and those reported by Tsimplis (2000) with data of the same season but different year. According to the values in Figure 3B, spring tides in CS delay 0.5 to 1 day the new/full moon in the upper layer while they happen simultaneously with or even shortly before new/full moon in the lower layer. The diminution of S_2 phase in the Mediterranean layer in CS contrasts with the increase in ES, where it reaches a noticeably relative maximum well below the interface (Figure 2D). The depth averaged phase of M_2 in CS (156°) is quite similar to that in ES (150°) while the depth averaged phase of S_2 in ES (186°) is greater than in CS (160°).

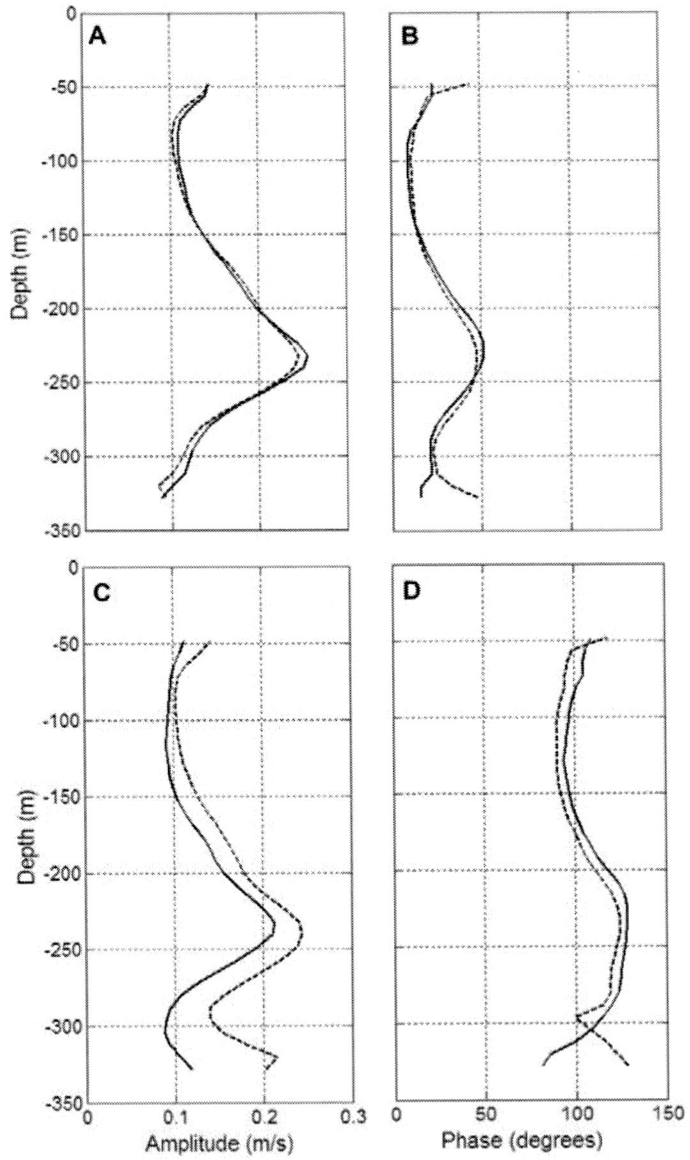

Figure 4. Vertical structure of the diurnal tidal constituents O_1 (amplitude, panel A; phase panel B) and K_1 (amplitude panel C; phase panel D) both for winter (solid line) and summer (dashed line) periods at ES.

3.2. Diurnal Constituents

The vertical distribution of K_1 and O_1 for winter and summer is shown in Figure 4. Amplitude of O_1 (panel A) reaches a maximum of 25 $cm \cdot s^{-1}$ at 240 m, the same depth of M_2 maximum. There are no differences between summer and winter periods. In contrast, K_1 amplitude is 2 to 4 $cm \cdot s^{-1}$ greater in summer increasing to 6-7 $cm \cdot s^{-1}$ in the lower bins (panel C). Similarly to O_1, maximum value is observed at ~240 m in both periods. Phases of both constituents (Figure 4B-D) exhibit a similar behaviour with maxima at ~230 m and ~220 m, respectively. The phase of K_1 in summer leads by ~8° the phase in winter until the lower bins, where they diverge.

Amplitude of diurnal constituents is nearly depth-independent in CS (Figure 3). Rather constant values of ~30 $cm \cdot s^{-1}$ (O_1) and ~20 $cm \cdot s^{-1}$ (K_1) remains until ~200m, then decreasing slightly to the bottom. The values are somewhat higher than in ES and match reasonably well those reported by Tsimplis (2000). The vertical structure of the phases is very similar, with a rather constant difference of 50° and greater values in the lower layer that implies a delay in the occurrence of maximum diurnal tidal currents in this layer.

3.3. Normal Mode Decomposition

In order to separate tidal velocities into barotropic and baroclinic components, dynamical mode decomposition (DMD) technique has been used. DMD uses the linearized hydrodynamics equations with the Boussinesq approximation and the assumptions that dissipative and nonlinear advective effects are negligible.

DMD assumes that the solutions in all variables can be expressed as the sum of various vertical orthogonal normal modes. The vertical structure of each mode is described by $\psi_n(z)$ and the solutions of the velocity components (u, v, w) and pressure and density perturbations (p, ρ) are then written in the form:

$$w(x,y,z,t) = \sum_{n=0}^{\infty} w_n(x,y,t)\psi_n(z) \qquad [1]$$

$$\{u(x,y,z,t), v(x,y,z,t), p(x,y,z,t)/\rho_0\} = \sum_{n=0}^{\infty} \{u_n(x,y,t), v_n(x,y,t), p_n(x,y,t)\} \frac{d\psi_n(z)}{dz} \qquad [2]$$

$$\rho(x,y,z,t) = \sum_{n=0}^{\infty} \rho_n(x,y,t) \frac{d^2\psi_n(z)}{dz^2} \qquad [3]$$

Using the orthogonallity condition of the different modes (Birkhoff and Rota, 1962; Gill, 1982)

$$\int_{-H}^{0} N^2 \psi_n(z)\psi_m(z)dz = 0 \qquad [4]$$

and substituting equations [1]-[3] into the equation of motion leads to the classical eigenvalue equation:

$$\frac{d^2\psi_n(z)}{dz^2}+\frac{N^2(z)}{c_n^2}\psi_n(z)=0 \qquad [5]$$

where c_n is the phase speed of mode n and N^2 is the buoyancy frequency. Equation [5] is solved using boundary conditions of flat bottom (w_{-H} =0) and rigid lid free surface (w_0 =0), which in terms of ψ_n *(z)* are:

$$\psi_n(z)=0 \qquad \text{for } z=0 \text{ and } z=-H \qquad [6]$$

Equations [5]-[6] constitute a Sturm-Liouvuille problem for eigenvectors ψ_n *(z)* with eigenvalues $1/c_n^2$.

3.4. Normal Modes Solutions for Espartel and Camarinal Stratification

Figure 5A shows the density profiles from MEDATLAS for all the stations selected for ES. The *N(z)* profiles are shown in Figure 5B-D. The seasonal variation of density is evident in ES, where the winter profile has a sole peak at ~210m whereas the summer profile also peaks at ~50 m. In CS, the maximum in Brunt-Vaisalla frequency is founded at ~140m.

The solution of equation [5] using these *N(z)* profiles with boundary conditions [6] provides the vertical shape of the normal modes ψ_n *(z)* for the vertical component of velocity. First derivative of ψ_n *(z)* provides the vertical shape of the normal modes for the horizontal components $\varphi_n(z)$, according to equation [2]. Figure 6 shows the vertical shape of the first four baroclinic modes, $\varphi_n(z)$ (n = 1,…,4) for winter (panel A) and summer (panel B) periods in ES and for winter in CS (panel C) normalised to maximum amplitude. However, the normalisation

$$\int_{-H}^{0}\varphi_n^2(z)dz=1 \qquad [7]$$

turns out to be more convenient and will be used henceforth as it allows for a direct interpretation of the relative importance of each mode in terms of the energy associated to it.

Let us consider along-strait tidal currents of the form:

$$u_l=a_l\cos(\omega t-\theta_l) \qquad [8]$$

where a_l, θ_l (l = 1,2,…L, number of bins sampled) are, respectively, the computed current amplitude and phase (harmonic constants) of the constituent of frequency ω. The above expression can be written:

$$u_l=A_l\cos(\omega t)+B_l\sin(\omega t) \qquad [9]$$

with

$$A_l=a_l\cos\theta_l \qquad ;B_l=a_l\sin\theta_l \qquad [10]$$

The next step is to reproduce the coefficients A_l and B_l through a linear combination of the theoretical modes:

$$A_l = \sum_{n=1}^{K} C_n \cdot \varphi_n + noise \quad ; \quad B_l = \sum_{n=1}^{K} S_n \cdot \varphi_n + noise \qquad [11]$$

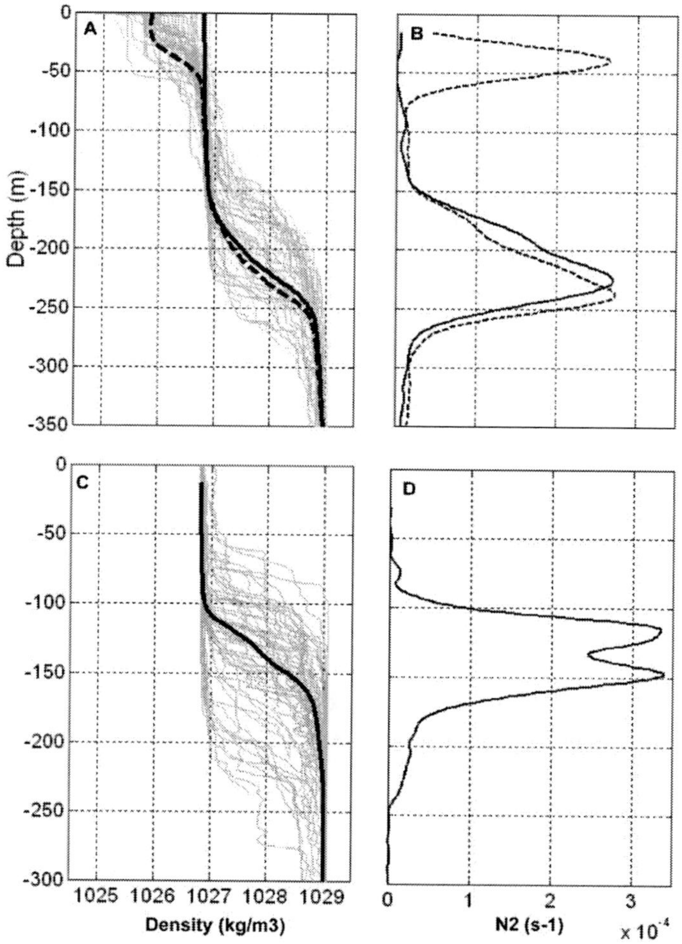

Figure 5. A) Density profiles for all the stations selected for ES (thin grey lines). Mean density profile for winter (black solid line) and summer (black dashed line) periods have been superposed. B) Brunt-Vaisalla (N^2) mean profiles for winter (solid line) and summer (dashed line) periods at ES. C) Density profiles for all the stations selected for CS (thin grey lines). Mean density profile for winter period have been superposed. D) Brunt-Vaisalla (N^2) mean profile for wintertime at CS.

Coefficients C_n and S_n are determined by minimisation of the noise in [11] through least-squares fitting. The number of modes in the summation K cannot exceed the number of bins M and must be selected taking into account the number of degrees of freedom (NDF) in the fitting. Due to the negligible contribution of higher modes, K was fixed to 10 so that NDF is 30 for ES and 13 for CS. The vertical structure of each mode is obtained through the products

$(C_n, S_n) \cdot \varphi_n(z)$, where the coefficients are independent of depth. This vertical structure is often described in terms of amplitude a_n and phase θ_n of each mode, computed according to

$$a_n^2 = C_n^2 + S_n^2 \quad ; \quad \theta_n = \tan^{-1}\left(\frac{S_n}{C_n}\right) \qquad [12]$$

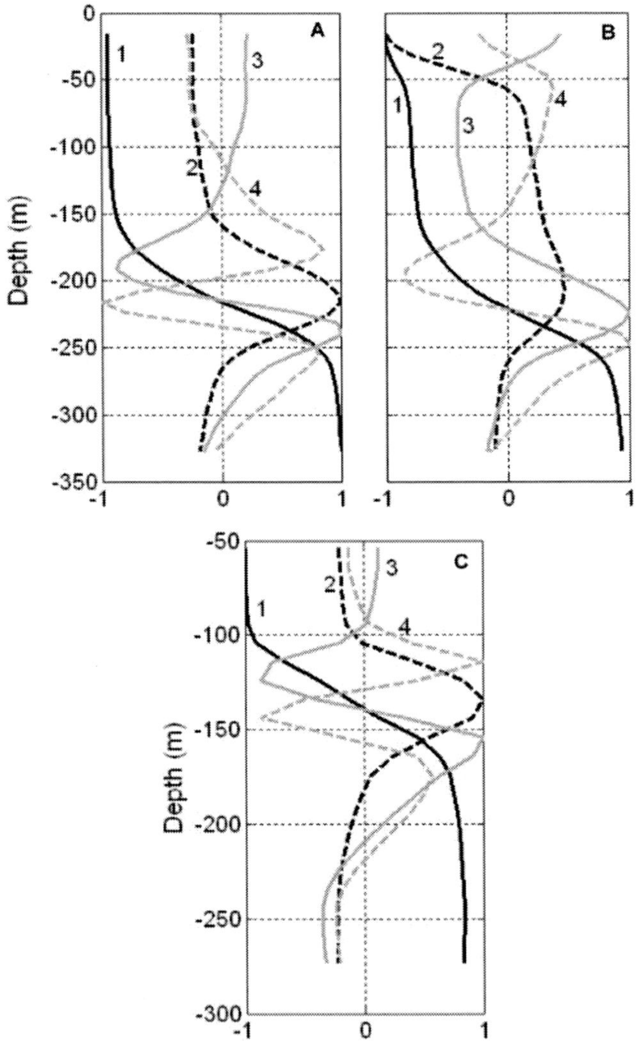

Figure 6. First four baroclinic "horizontal" normal modes (#1, black solid line; #2, black dashed line; #3, grey solid line; #4, grey dashed line), $\varphi_n(z)$ for winter (panel A) and summer (panel B) periods at ES and for wintertime at CS (panel C) normalised to maximum amplitude. An identificative number for each mode has also been labelled. This normalisation to maximum amplitude has only been used for this Figure for a better visualisation of the vertical shape. For calculations the more convenient (7) is preferred.

In all cases, the combination of the barotropic and the first five baroclinic modes (normalised to [7]) reproduces accurately the amplitude and phase profiles of the tidal constituents, as shown in Figure 7 for semidiurnal constituents. Relative importance of each mode can be established in terms of the energy associated. Squaring equation [2] and then integrating between bottom and the surface, yields:

$$\int_{z=-H}^{z=0} u^2(x,y,z,t)dz = \sum_{n=0}^{\infty} u_n^2(x,y,t) \int_{z=-H}^{z=0} \varphi_n^2(z)dz = \sum_{n=0}^{\infty} u_n^2(x,y,t) \qquad [13]$$

where the normalisation [7] has been used. The LHS is proportional to kinetic energy associated to horizontal movement. The coefficients u_n in [13] are the amplitudes a_n computed from [12] and each term on the summation of the RHS, which in practice extends from n=0 to n=K, indicates the amount of energy accounted by each mode. The quotient:

$$E_n = \frac{a_n^2(x_A, y_A)}{\sum_{n=0}^{K} a_n^2(x_A, y_A)} \qquad [14]$$

represents the fraction of total energy explained by mode n.

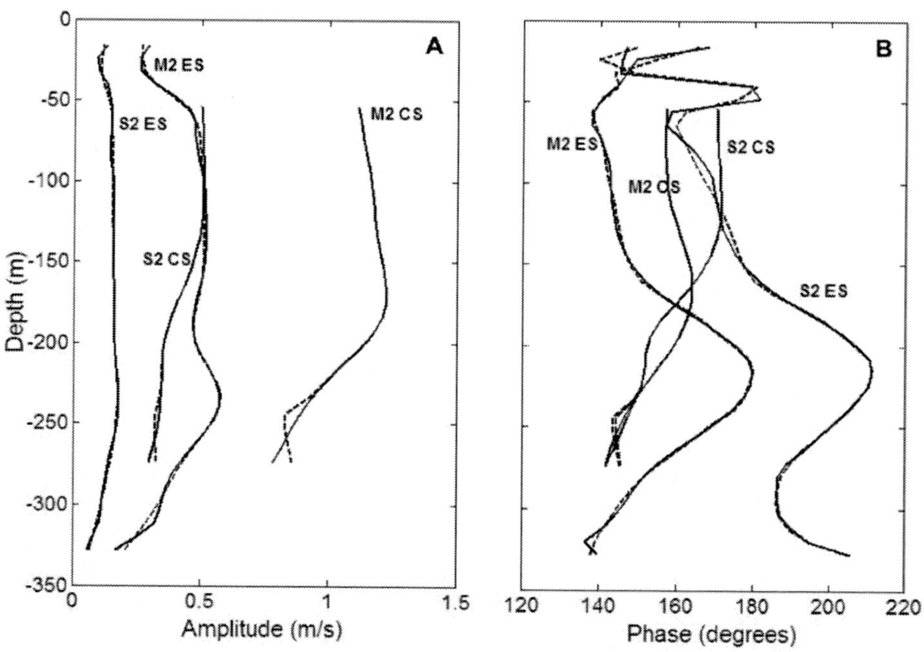

Figure 7. Reproduction of amplitude (panel A) and phase (panel B) profiles of the semidiurnal tidal constituents (dashed line) from the combination of the barotropic and the first five baroclinic modes (normalised to (7)). For comparison purposes, the real profiles have also been displayed (solid line).

Barotropic mode is by far the most energetic in all cases. In ES, it is more energetic in winter than in summer for constituents M_2 (above 85% of total energy in winter and 77% in summer), S_2 (87.5% in winter and ~73% in summer) and O_1 (above 78% in summer and ~70% in summer). K_1 is more energetic in summer (almost 80%) than in winter (around 70%). In CS the barotropic mode explains more than 93% for all constituents, a result previously reported by Candela et al. (1990), Mañanes et al. (1998) and Tsimplis (2000). The small fraction of energy not explained by the barotropic mode is accounted by the first baroclinic one for all constituents analysed except M_2. Both modes explain around 99% of the total energy. For M_2 constituent, modes 2 and 3 are more important than mode 1. The barotropic mode in ES explains a considerable less percentage of total energy, particularly in summer. For M_2 constituent, the second baroclinic mode is more important than the first one in winter and comparable to the first one during summer. Summer stratification favours a first-mode internal oscillation but also, and particularly, the third mode. It is not a surprising fact if we consider the vertical structure of amplitude in Figure 2A with two well differentiated relative maxima in summer. The second shallower maximum is smoothed in winter and, consequently, the existence of a relevant third mode is not as necessary as in summer. The richer high mode structure in ES is thus consequence of the more complex vertical structure of the harmonic constants.

4. Conclusion

The observations of the exchange at the two main sills of the Strait of Gibraltar (ES and CS) analysed in this work have evidenced the noticeable spatial variability of the tidal flow as well as some minor seasonal differences. Over ES, only slight variations between winter and summer periods are observed for M_2 tidal current, especially through the interface layer. For S_2 and K_1 discrepancies between both periods are clearly evident. Amplitude in CS reaches maximum at a shallower depth than in ES and its value is higher for all tidal constituents (more than twice for M_2). Phases of diurnal constituents near the bottom are delayed with regards to phases near the surface, in contrast to what happened to the semidiurnal constituents.

DMD has been carried out to separate tidal velocities into barotropic and baroclinic components and relative importance of each mode has been determined in terms of the energy associated. For ES, barotropic mode is more energetic during wintertime for M_2, S_2 and O_1 constituents, K_1 being more energetic during summertime. In any case, total energy of barotropic mode is always above 70% (above 85% for semidiurnal constituents in winter) and baroclinic modes have smaller contributions. During wintertime, second baroclinic mode is more energetic than the first one and relatively high percentage of energy is accounted by third baroclinic mode for M_2 (8%) and O_1 (9%) constituents. Over CS, barotropic mode accounts for more than 90% of total energy in all the tidal constituents, the highest value (97%) observed for M_2.

Acknowledgments

This work has been carried out in the frame of the Spanish Junta de Andalucia-funded project P07-RNM-02938.

References

Baschek, B., Send, U., García Lafuente, J. & Candela, J. (2001). Transport estimates in the Strait of Gibraltar with a tidal inverse model. *Journal of Geophysical Research*, 106, 31, 033-31, 044.

Birkhoff, G. & Rota, G. C. (1962). *Ordinary differential equations*. Ginn, Boston, Massachusetts.

Bruno, M., Mañanes, R., Alonso, J. J., Izquierdo, A., Tejedor, L. & Kagan, B. (2000). Vertical structure of the semidiurnal tidal currents at Camarinal Sill, the Strait of Gibraltar. *Oceanologica Acta*, 23, 15-24.

Bryden, H. L., Candela, J. & Kinder, T. H. (1994). Exchange through the Strait of Gibraltar. *Oceanologica Acta*, 33, 201-248.

Candela, J., Winant, C. & Bryden, H. L. (1989). Meteorologically forced subinertial flows through the Strait of Gibraltar. *Journal of Geophysical Research*, 94, 12667-12674.

Candela, J., Winant, C. & Ruiz, A. (1990). Tides in the Strait of Gibraltar. *Progress in Oceanography*, 33, 7313-7335.

Foreman, M. G. G. (1978, revised 2004). Manual for tidal currents analysis and prediction. Pacific Marine Science Report 78-6, Institute of Ocean Sciences, *Patricia Bay*, 57.

García Lafuente, J., Almazán, J.L., Castillejo, F., Khribeche, A. & Hakimi, A. (1990). Sea level in the Strait of Gibraltar: Tides, *Int. Hydrog. Rev*., 47, 111-130.

García Lafuente, J., Vargas, J. M., Plaza, F., Sarhan, T., Candela, J. & Baschek, B. (2000). Tide at the eastern section of the Strait of Gibraltar. *Journal of Geophysical Research*, 105, 14197-14213.

Gill, A. E. (1982). *Atmosphere-ocean dynamics*. Academic Press, San Diego, California, 662.

Kundu, P. K., Allen, J. S. & Smith, R. L. (1975). Modal decomposition of the velocity field near the Oregon coast. *Journal of Physical Oceanography*, 5, 638-704.

Mañanes, R., Bruno, M., Alonso, J., Fraguela, B. & Tejedor, L. (1998). The nonlinear interaction between tidal and subinertial flows in the Strait of Gibraltar. *Oceanologica Acta*, 21, 33-46.

Müller, T. J. & Siedler, G. (1992). Multi-year current time series in the eastern North Atlantic Ocean. *Journal of Marine Research*, 50, 63-98.

Pawlowicz, R., Beardsley, B. & Lentz, S. (2002). *Classical Tidal Harmonic Analysis Including Error Estimates in MATLAB using T_TIDE*", Computers and Geosciences.

Sannino, G., Bargagli, A. & Artale, V. (2002). Numerical modelling of the mean exchange through the Strait of Gibraltar, *Journal of Geophysical Research*, 179, C08,3094, doi:10.1029/2001JC000929.

Sannino, G., Bargagli, A. & Artale, V. (2004). Numerical modelling of the semidiurnal tidal exchange through the Strait of Gibraltar, *Journal of Geophysical Research*, 109, C05011, doi:10.1029/2003JC002057.

Schott, F. (1977). On the energetics of baroclinic tides in the North Atlantic. *Ann. Geophys.* 33, 41-62.

Tsimplis, M. N. (2000). Vertical structure of the tidal currents over the Camarinal Sill at the Strait of Gibraltar. *Journal of Geophysical Research*, 104, 13,541-13,557.

Tsimplis, M. N. & Bryden, H. L. (2000). Estimation of the transport through the Strait of Gibraltar. *Deep Sea Research, Part I*, 47, 2219-2242.

Wang, D. P. (1993). The Strait of Gibraltar Model: Internal tide, diurnal inequality and fortnightly modulations. *Deep Sea Research*, Part I, 40, 1187-1203.

In: Sea Level Rise, Coastal Engineering, Shorelines and Tides ISBN: 978-1-61728-655-1
Editor: Linda L. Wright, pp. 205-234 © 2011 Nova Science Publishers, Inc.

Chapter 10

THE TELLURIC FIELD INDUCED BY TIDAL MOTION: A REVIEW OF THE PORTUGUESE EXPERIENCE

Fernando A. Monteiro Santos*[*,1]*, Rita Nolasco*[2]*,
Martinho Marta-Almeida*[2]*, António Soares*[1]*, Ivo Bernardo*[1]*,
***Rafael Luzio*[1]*, Jesus Dubert*[3] *and João M. Dias*[3]**
[1]Instituto Don Luíz-Laboratório Associado,
Universidade de Lisboa,Campo Grande, Lisboa, Portugal
[2]CESAM, Universidade de Aveiro, Portugal.
[3]CESAM, Departamento de Física, Universidade de Aveiro, Portugal

Abstract

Electromagnetic fluctuations in the ocean have external sources like ionospheric–magnetospheric current systems, and purely internal oceanic sources associated with interaction between water velocity fields and the geomagnetic field. The oscillations of the telluric field originated by tides were first predicted by Faraday in 1832. The physical phenomena that explain those oscillations are connected to the tidal water movement in the earth's geomagnetic field. This phenomenon has been used to estimate average water mass transport in straits, channels, throats or even in the open ocean, using submarine cables. However, the telluric field induced by tides that spreads out far inland could be used to characterize tidal phenomenon. This result opened the possibility to estimate mass transport alongshore associated with tidal flow using onshore measurement of the telluric field.

A review of the fundaments, processing and interpretation of telluric oscillations with tide origin is presented and discussed in this article. The paper presents results obtained from the analysis of data collected in two different systems located in Portugal: 1) a submarine cable crossing the channel at the entrance of a lagoon (Aveiro, Portugal) and, 2) two antennas installed close to the coast line (Portuguese west coast). Spectral analysis of the data revealed that measured voltages are dominated by semidiurnal M_2, S_2/K_2 periods. Values of 720 $m^3\ s^{-1}\ mV^{-1}$ and of 3.0×10^4 and 4.25×10^3 were estimated for the coefficient relating voltage and water transport at Aveiro channel and at the two different sites in the coast line, respectively. The results show that it is possible to indirectly measure the water transport by tidal flow

* E-mail address: fasantos@fc.ul.pt. Instituto Don Luíz-Laboratório Associado, Universidade de Lisboa,Campo Grande, Ed. C8, 1749-016 Lisboa,Portugal

measuring onshore differences of electrical potential. The method shows great potential for low cost accurate long-term monitoring of integral water transport.

Introduction

Sea water flow through the Earth's magnetic field induces secondary electric fields, electric currents and secondary magnetic fields that have a potential for the remote sensing of ocean flow variability, as noted by several authors (see, e.g., Palshin et al. 1996, Tyler et al. 1999, Maus and Kuvshinov 2004). Magnetic field variations, particularly those originated by periodic tidal motion, have been observable in land observatories and satellites, generally with magnitudes of a few nT (see, e.g., Malin 1970, McKnight, 1995, Tyler et al. 1997, 2003). Secondary electric fields generated by tidal movements have been observed in the sea (e.g., Duffus and Fowler 1974, Larsen 1985, Fujii and Utada 2000, Monteiro Santos et al. 2002) and on land and has a potential for electromagnetic soundings of the Earth (Rooney 1938, Harvey et al. 1977, Junge 1988, Monteiro Santos et al 2007).

The oscillations of the electric field originated by tides were first predicted by Faraday in 1832 and observed by Egedal in 1937. The physical phenomena that explain those oscillations are connected to the tidal water movement in the earth's geomagnetic field (Longuet-Higgens 1949, Sanford 1971). This phenomenon has been used to estimate average water mass transport in straits, channels, throats or even in the open ocean, using submarine cables (Harvey et al. 1977, Flosadottir and Taire 1997, Palshin et al. 1999, 2001, Monteiro Santos et al. 2002, 2007, Nolasco et al. 2006). In 1988, Junge demonstrated that the telluric field induced by tides that spreads out far inland can be used to characterize tidal phenomenon. This result opened the possibility to investigate tidal characteristics and to estimate mass transport alongshore associated to tidal flow using onshore measurement of the telluric field.

A review of the theoretical fundamentals, processing and interpretation of telluric oscillations with tide origin measured in narrow channels using short-submarine cables and on land, is presented in this article. The paper presents results obtained from the analysis of data collected in two different systems: 1) in a channel at the entrance of a lagoon (Aveiro, Portugal) and, 2) in two antennas installed close to the coast line (Portuguese west coast). The results show that it is possible to indirectly measure the water transport by tidal flow measuring on shore differences of electrical potential.

Theoretical Fundamentals of the Motionally Induced Voltage (MIV)

Electromagnetic fluctuations in the ocean have external sources and internal, purely oceanic sources, associated with interaction between water velocity fields and the earth's magnetic field. Motionally induced voltage (MIV) phenomenon is generated by movements of free electrical charges through the earth magnetic field. The charged particles in the water having a movement component perpendicular to the vertical component of the earth's magnetic field are deviated by the Lorenz force. The phenomenon was first studied in 1832 by Michael Faraday who also quantified the electrical field $\vec{E}'$, induced by the charge q+ (or q-) moving through the magnetic field $\vec{B}$ with a velocity $\vec{v}$:

$$\vec{E}' = \vec{v} \times \vec{B} \tag{1}$$

Positively and negatively charged particles having a movement component perpendicular to the vertical component of the earth's magnetic field are deviated in opposite direction generating electrical fields that spread far inland. It is thus possible, at least theoretically, by measuring $\vec{E}'$ and $\vec{B}$, to deduce the velocity of water currents, since these in general carry electrical charges in the form of dissolved ions. The characteristics of this field depend on: 1) the water flow and the electric properties of the water and 2) the electric structure of the earth in the area. This phenomenon has been used to estimate average water mass transport in straits, channels, throats or even in the open ocean, using submarine cables (Larsen 1992, Harvey et al. 1977, Flosadottir and Taire 1997, Palshin et al. 2001, Palshin et al. 2002, Monteiro Santos et al. 2002, Fristedt et al. 2002, Nolasco et al. 2006).

The theory of motional induction applicable to the ocean has been developed by many authors, (i.e., Larsen 1968, 1971, Sanford 1971, Filloux 1987, Chave and Luther 1990, Lanzerotti et al. 1992). There are two principal modes for electric currents generated by ocean movements: the toroidal mode, with the currents restricted to the vertical plane and the poloidal mode that includes the currents that are in the horizontal plane. The toroidal mode is the most important one and describes the electric currents in the surface layer forced by surface flow and has weaker returning currents in the deep ocean (Szuts 2008). The poloidal mode exists where gradients in the downstream direction exist. This mode will not be described in this chapter.

Sanford (1971) presented a general theory for MIV assuming: 1) a horizontal ocean bottom (H) with small topography perturbations (h/H << 1), 2) width scales (L) larger than the bottom depth (H/L << 1), 3) predominance of horizontal oceanic movements ($\vec{v} = \vec{v}_h$), 4) distant lateral boundaries, 5) a sedimentary layer with uniform electrical conductivity, and 6) negligible magnetic self-induction and mutual induction. As pointed out by Sanford (1971), the induction parameter $\mu\omega\sigma L^2$ (μ is the magnetic permeability, σ is the electrical conductivity and ω the frequency) must be less than 1 for mutual induction to be negligible. It means that this phenomenon can be discarded for periods longer than 10 hours (Chave and Luther 1990).

The electrical field in a reference frame fixed to the sea floor, is (Sanford, 1971):

$$\vec{E} = \frac{\vec{J}}{\sigma} - \vec{v} \times \vec{B} \tag{2}$$

where $\vec{v} \times \vec{B}$ is the motional field associated with the velocity field $\vec{v}$, $\vec{J}$ is the electric current density and $-\vec{J}/\sigma$ is the resistive or ohmic force opposed by the environment to electric current flow. The form for horizontal electric current divided by conductivity is (Sanford 1971, Szuts 2008),

$$\frac{\vec{J}_h}{\sigma} = (\vec{v} - \vec{v}^*) \times B_z \hat{k} - \nabla_h \left[\frac{1}{H(1+\lambda)} \int_{-H}^{0} \hat{k} \cdot (\vec{v} \times \vec{B}) z' \, dz' + \int_{-H}^{z} \hat{k} \cdot (\vec{v} \times \vec{B}) \, dz' \right] \quad (3)$$

with

$$\vec{v}^* = \frac{\int_{-H+h}^{\varsigma} \sigma \vec{v} \, dz'}{\int_{-H_s}^{\varsigma} \sigma \, dz'} \quad (4)$$

where ς is the sea surface, $\hat{k}$ is the unitary vector in Z direction, $-H$ is the mean depth of the seafloor, h is the perturbation of the seafloor, λ is the sediment conductivity factor and $-H_s$ is the bottom of the conductive sediment. $\vec{v}^*$ is the conductivity-weighted vertically-average velocity and it is linearly related to the vertically averaged velocity. λ is the sediment conductance factor and is given by

$$\lambda = \frac{\int_{-H_s}^{-H+h} \sigma(z) \, dz}{\int_{-H+h}^{\varsigma} \sigma(z) \, dz} \quad (5)$$

From equation (3) the electric current density divided by σ has two terms. The first one is equal to the difference between the current density driven by the vertically uniform electric field and that generated by local horizontal water motion, $(\vec{v}(z) - \vec{v}^*) \times B_z \hat{k}$ (Szuts 2008). Considering baroclinic flows and that the width L of the current is much larger than the depth H, the resulting horizontal electric field is vertically uniform and is given by $\vec{v}^* B_z$. For a non-stratified water $\vec{v}^*$ becomes $\vec{v}^* = \vec{v} / \Sigma$, where $\Sigma = 1 + \lambda_s / \lambda_w$. λ_w and λ_s are, respectively, the conductances of sea-water layer and the ocean bottom sediments and $\vec{v}$ is the mean velocity (Sanford 1971).

The second term of (3) depends on horizontal gradients of topography and of the velocity field. Horizontal gradients of H, λ and $\vec{v}$ can make this term a non-zero term and, frequently, the assumption of small horizontal gradients is invalid. Therefore, high order terms must be took into account in specific areas of the ocean where those condition are verified (Szuts 2008).

Absolute electric fields can be measure using stationary instrumentation such as submarine cables. In this case the local velocity ($\vec{v} \times \vec{B}$) cancels part of the first term of (3) leaving only the electric field caused by v^* , e.g., $\vec{E}_h = \vec{v}^* B_z$. The difference of electric potential between the terminals of a cable (with a distance between end electrodes of L) placed perpendicularly to the water flow depends basically on: i) the telluric field (electric), generated by the variations of the geomagnetic field; ii) the water flow rate and the electric properties of the fluid and, iii) the electric structure of the terrain in the site (e.g., Sanford 1971, Filloux 1987). In this way, the monitoring of that potential will provide an indirect knowledge of the flow rate and of the mass transport associated with this flux (e.g., Lanzerotti et al. 1992). The induced voltage should be proportional to the average transport ($t = \overline{v}LH$):

$$\frac{t}{\Delta V} = \frac{H\Sigma}{B_z} \tag{6}$$

Therefore, the recording of the variation of the electrical potential between the terminals of the cable will allow us to obtain a measure of the variations of the mass transport with time, and hence, its seasonal and even interannual variations.

Data Processing

The electrical potential difference (*dop*) between a pair of electrodes must be processed before any interpretation. The data are processed with several objectives. For example, moving average of the *dop* data is commonly used to suppress outliers in the time-series. The amplitude spectrum of the averaged electrical potential allows identifying the main spectral components in the recorded signal. The spectrum of the voltages measured in a cable is dominated by semidiurnal M_2, S_2/K_2, P_1/K_1, O_1 and M_4 components (Figure 1). The constituents M_2, N_2, O_1 and M_4 are mainly of oceanic tidal motional origin. The constituents S_2/K_2 and P_1/K_1 are a combination of oceanic tidal motion and geomagnetic origin.

If the main objective is the use of MIV for tidal studies, a calibration coefficient relating the motionally induced voltage of lunar origin (MIVL) and the water transport of the lunar constituents of tides must be estimated. Therefore, the focus will be on the four (M_2, N_2, O_1, M_4) constituents of lunar origin. Let us to consider the signal measured in a cable crossing a narrow channel. Firstly, the different components of the measured signal must be separated. For such the *dop* measured time-series is approximated by a set of lunar harmonic constituents using the least square fitting (LSF) method (Pawlowicz et al. 2002). Sea surface elevation (SSE) measured at a tide gauge can be used to calibrate the cable. In this case, harmonic constituents of lunar origin (M_2, N_2, O_1 and M_4) should be calculated from SSE using the same least square fitting method and the water flow for each component must be estimated. The water flow with lunar origin is then correlated with MIVL measurements allowing the estimation of a calibration factor. The correlation of MIVL can also be done with independent flow measurements or with water flow values obtained from a previously calibrated hydrodynamic model (Nolasco et al. 2006).

Differences of potential measured on shore can be processed and analyzed in a similar way (Marta-Almeida et al. 2010). The separation of the signals in very low frequency and high frequency components can be done by Singular Spectrum Analysis (Marques et al 2006). High frequency component is usually analyzed using Fourier analysis in order to study the components contained in the *dop* signal. Specific analysis of tides components can be done using harmonic fitting (Pawlowicz et al. 2002), as exposed before. Also, the possible variation of tidal parameters can be investigated making harmonic fit using 60 days moving window (Marta-Almeida et al. 2010).

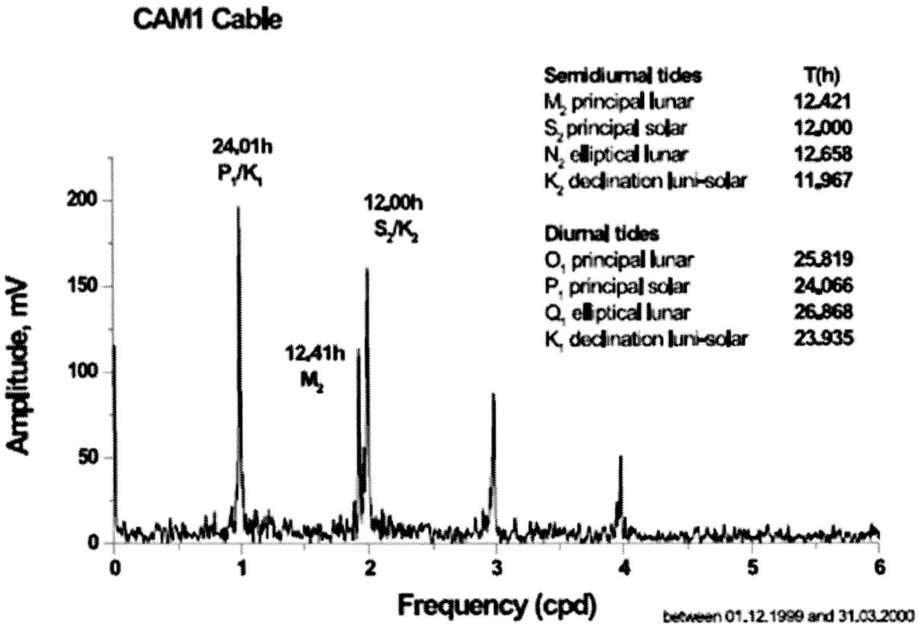

Figure 1. Amplitude spectrum from data collected in the submarine cable CAM-1 between Sesimbra and Madeira Island. From Monteiro Santos et al. (2002).

Case Studies

MIV Measurements in Estuarine Environments Using Submarine Cables

Coastal and estuarine environments are nowadays recognized as significant for bio-diversity evolution. The biotas in these environments are extremely sensitive to changes in water properties. This perception has originated a large research activity in order to understand the parameters controlling the dynamics of those systems. Water flow into the ecosystem is one of the significant physical parameters that need to be monitored and MIV can be used for that.

The Ria de Aveiro (Northwest Portugal) is an example of this kind of system. It is a shallow water lagoon with a very complex geometry, connected with the Atlantic Ocean through a single artificial channel, the Barra channel of 300 m width and from 3 to 25 m depth (Figure 2). All of its water exchange with the ocean takes place by input/output across this narrow entrance. The current velocity varies from 0 to 3 m s^{-1} with typical values of about 1–1.5 m s^{-1} (Dias et al. 2003). The input of fresh water comes from several small rivers distributed by different channels of the lagoon. These characteristics make Ria de Aveiro an ideal place to implement and test innovative sensing systems dedicated to monitoring estuaries and coastal zones (Nolasco et al. 2006).

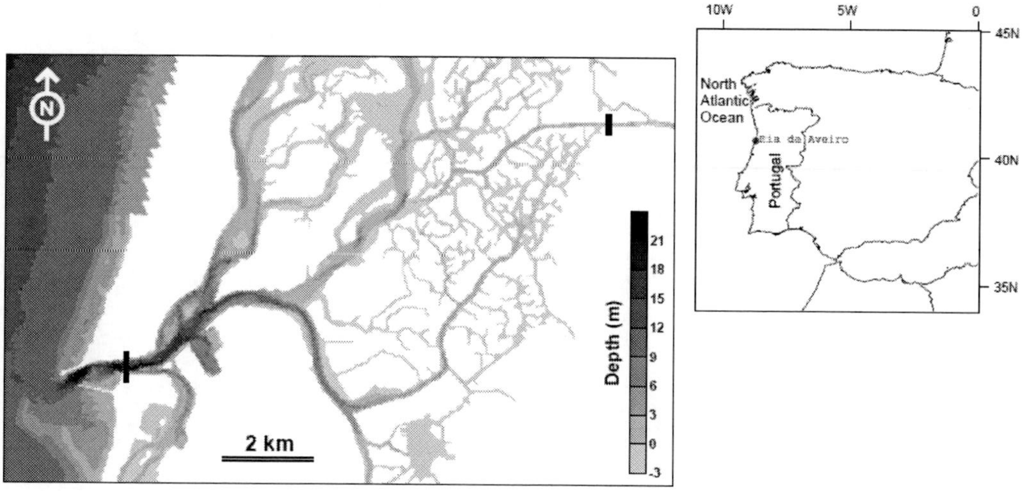

Figure 2. Location of the cable at the Ria de Aveiro. From Nolasco et al. (2006).

A system composed of two silver-silver-chloride (Ag/AgCl) nonpolarizable electrodes, connected to a voltmeter by submarine cables, was installed at the Barra channel (entrance of the lagoon), to allow the estimation of the water transport/flow at this site. The electrodes were specially designed for sea applications and the distance between them was 280 m. The electrical potential difference between the two electrodes was measured using a 16-bit datalogger. The sampling rate of the datalogger was 10 s, but only the average of six values over 60 s was recorded. The main objective was to test the possibility of using the MIV method at Ria de Aveiro to calculate the water transport (by tidal) through Barra channel from the differences of electrical potential measured across the channel.

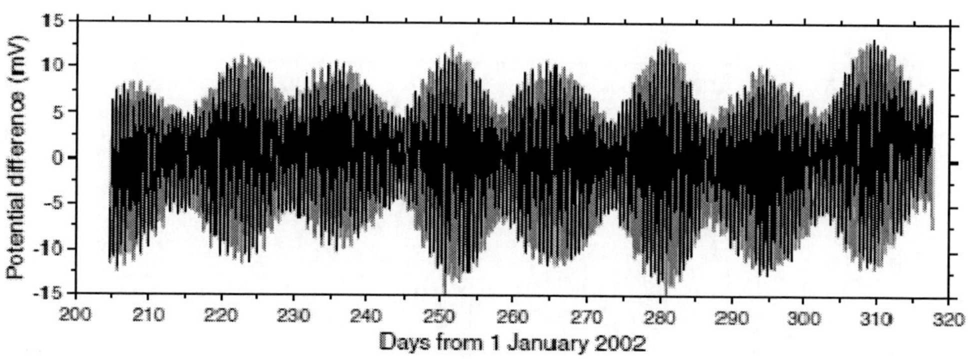

Figure 3. Voltage no cabo em Aveiro (hourly mean values). From Nolasco et al. (2006).

Figure 3 shows the electric potential difference (hourly means) recorded during approximately four months in 2002, across the channel at Barra. The potential difference varies between about −13 and 13 mV. Hourly mean values have been computed to suppress outliers in the time-series. The amplitude spectrum of the hourly mean electrical potential differences is plotted in Figure 4. The spectrum of the measured voltages is dominated by semidiurnal M_2, S_2/K_2 frequencies. P_1/K_1, O_1 and M_4 frequencies are also present, with smaller amplitudes. The M2 constituent amplitude is higher than the S_2/K_2 or any other frequency in the spectrum. This means that the tidal, M_2, induced electric field is stronger than other constituents of the electric field. The time series also provides the visualization of the spring-neap tidal cycle and the diurnal-inequality characteristic of mixed tidal regime (Figure 3).

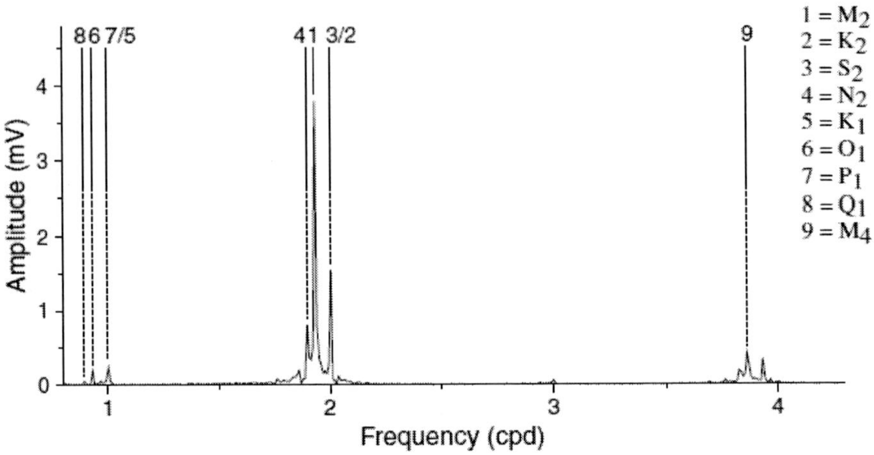

Figure 4. Amplitude spectrum calculated from *dop* recorded at Barra. From Nolasco et al. (2006).

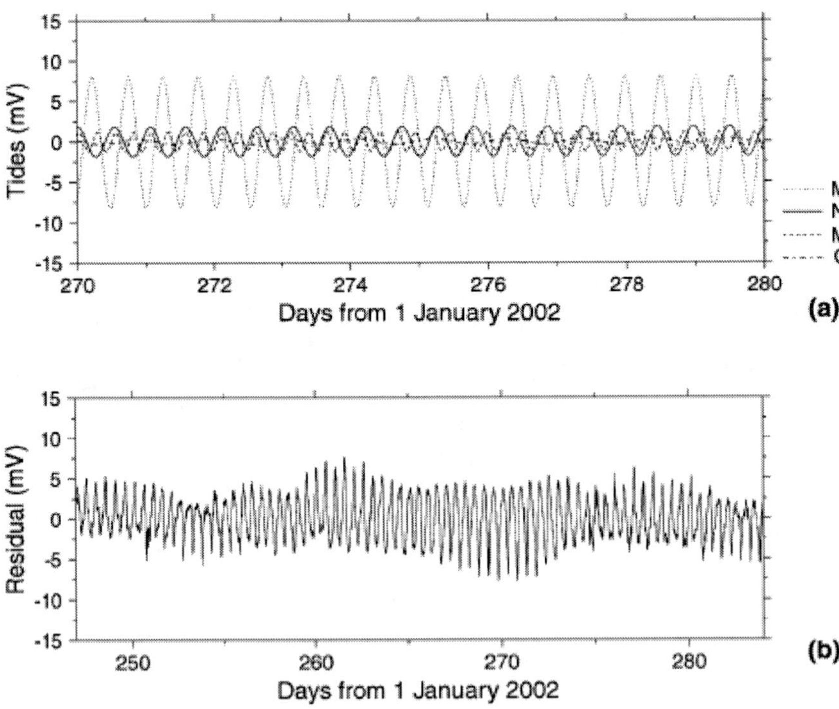

Figure 5. Results of the LSF analysis of the hourly mean dop data collect at Barra. (a) The selected four components with lunar origin (MIVL). (b) Residual time-series obtained after removing the sum of the MIVL components. From Nolasco et al. (2006).

The measured time-series were processed according to the method detailed before: e.g., the series were approximated by a set of lunar harmonic constituents using LSF method (Nolasco et al. 2006). The results of LSF are presented in Figure 5 where Figure 5(a) shows the four lunar constituents with 95 per cent confidence interval and Figure 5(b) presents the residual time-series calculated as the difference between recorded data and MIVL.

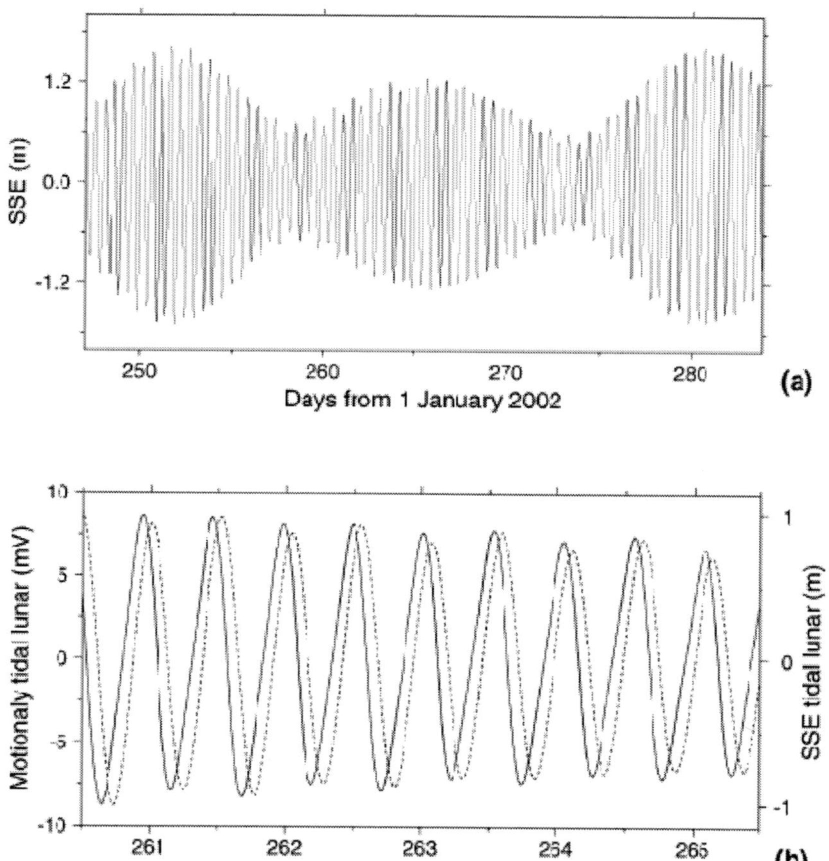

Figure 6. (a) Sea surface elevation (SSE) measured at the tide gauge. (b) MIV*L* in solid line, and the resulting of the sum of the significant harmonics of the SSE in dashed line. The time lag between the two series is about 0.088 days. From Nolasco et al. (2006).

Sea surface elevation from a tide gauge located at the southern side of the channel were digitized, using a sample rate of 1 hr (Figure 6(a) and processed in the same way in order to obtain the harmonic constituents of lunar origin (M_2, N_2, O_1 and M_4). Figure 6(b) shows the sum of the constituents with lunar origin (dashed line). A difference of 0.088 days between the maximum values of electric potential difference and SSE (sums of the lunar constituents) can be noted. Dias (2001) analyzed the delay of the local current inversion relative to high and low water at the lagoon mouth, for both neap and spring tide conditions, based on numerical modeling results. Dias has observed that tidal elevations and velocities are usually not in phase and that high water occurs before high slack tide and low water before low slack tide. The values found are not the same for the high water and the low water cases, ranging from about 70 min (high water of neap tide) to about 120 min (low water of spring tide). Therefore, the phase shift determined from MIVL and SSE analyses (0.088 days or about 127 min) is consistent with the numerical results. The phase shift is originated by the balance between the pressure force due to the slope of the water surface, and the retarding force resulting from bottom friction (Dias 2001).

As our objective is to correlate water flow/mass transport with the *dop* measured in the cable, we need to have an independent estimation of the water flow/mass transport. In the absence of water fluxes measurements at the mouth of the lagoon in the Barra-Atlantic Ocean channel, results from a 2-D vertically integrated hydrodynamic numerical model (Dias 2001) were used. The program was properly calibrated for Ria de Aveiro lagoon through comparison between model results and measurements of sea surface elevation and current velocity, for several stations distributed along the main lagoon channels (Dias 2001; Dias and Lopes 2006). The model results describes the depth-averaged circulation in a tidal basin, and allows the determination of the sea surface elevation, current velocity and water fluxes in any area of the computational domain.

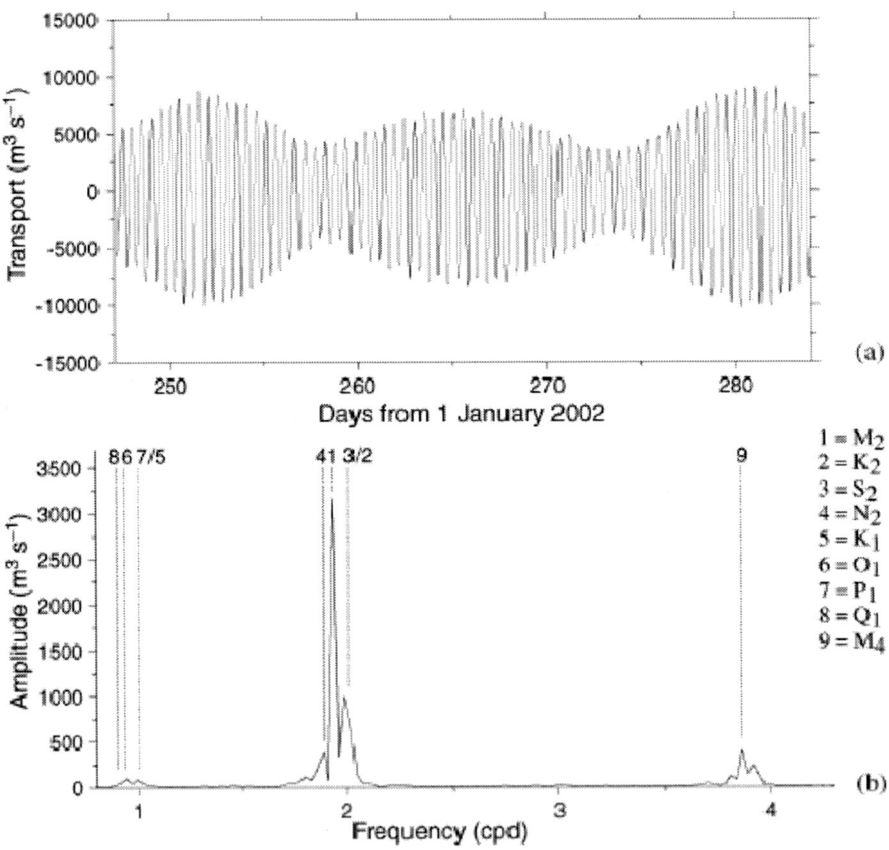

Figure 7. Water transport computed by the hydrodynamic numerical model (a) and its amplitude spectrum (b). From Nolasco et al. (2006).

The water transport time-series computed by the numerical model for the transversal section corresponding to the cable location and the water transport amplitude spectrum are displayed in Figures 7(a) and (b), respectively. As expected, different tidal constituents, with oceanic moon and sun origin and respective combinations, are identified in the water transport spectrum. In order to obtain the water transport with tidal lunar origin (M_2, N_2, O_1 and M_4), the model time-series was approximated by a set of lunar harmonics using the statistical least square fitting (LSF) method (Pawlowicz et al. 2002). Again the same data processing procedure used regarding the electric potential difference and the SSE data was used. Figures 8 shows the relation between the sum of the four oceanic constituents of the electric potential difference (MIVL) and the sum of the four constituents of the water transport, for the same time period. The best-fit line, in Figure 8, was obtained by least-square method. The correlation between the two time-series allows an estimation of a water transport of 720 m3 s^{-1} for each 1 mV potential difference between the ends of the cable. This coefficient found between both variables allows the calibration of the electrical cable.

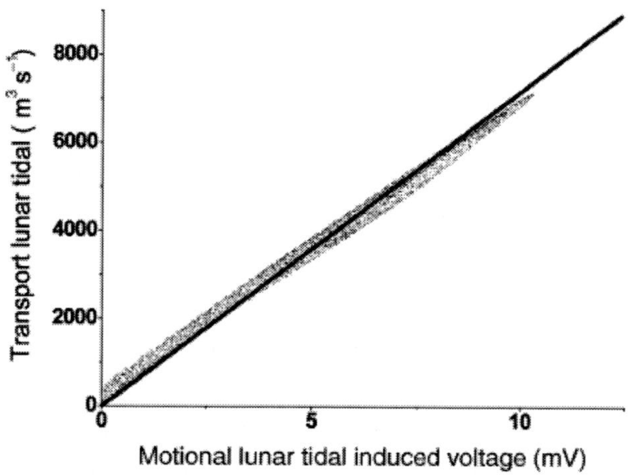

Figure 8. Calibration between motional field due to moon constituents of tides and the correspondent water transport. From Nolasco et al. (2006).

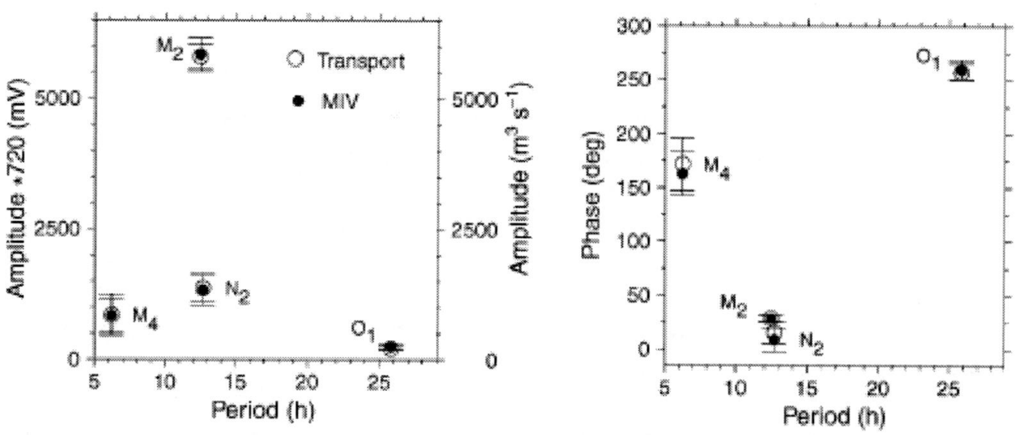

Figure 9. Amplitude and phase of the lunar tidal constituents of the water transport determined by the numerical model and from the MIVL, results of LSF approximation of the data with 95 per cent confidence interval. From Nolasco et al. (2006).

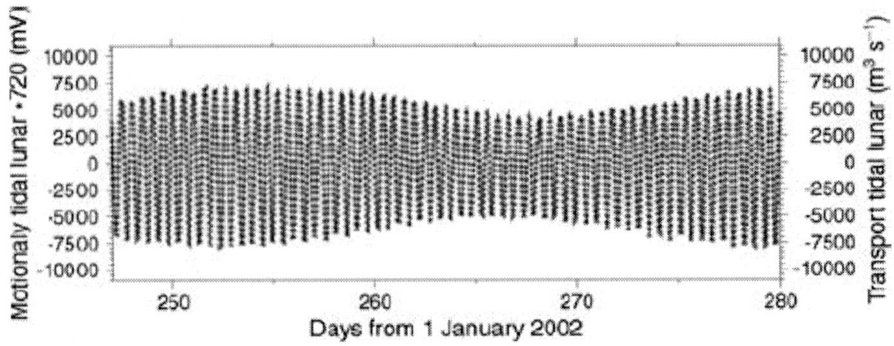

Figure 10. Sums of the lunar tidal constituents of the water transport determined by the numerical model and from the MIVL. From Nolasco et al. (2006).

MIVL values are converted into water transport of lunar origin multiplying them by the calibration factor. Amplitude and phase lag for lunar origin tidal constituents for water transports both numerically computed and determined from the MIVL (with the errors of the LSF for a significance level of 95 per cent), are plotted in Figure 9. The adjustment between both series as regard amplitudes and phases is very good, and small differences between them are included in error bar. As an example of good adjustment of both series, the sums of the lunar constituents for water transports (both numerically computed and determined from the MIVL) are compared in Figure 10.

MIV Measurements on Share around the Ria De Aveiro

An electromagnetic antenna constituted by two pairs of nonpolarizable electrodes (PbPbCl) was installed in the north side of the channel, onshore and close to the Barra. The electrodes were installed in the N–S (approximately N9°E and 360 m apart) and E–W direction (68 m apart) to investigate the possibility to estimate the water transport/ flux in the channel, using onshore electrical potential measurements. They have been placed in salted bentonite in 1.0 to 1.5 m deep holes. The electric potential was sampled at a rate of 10 s, but only the average of six values over 60 s was recorded. Figure 11 shows hourly means of the electric field recorded between days 247 and 317 since January 2002 in the two onshore pair of electrodes. The variations observed in the onshore dipoles are lesser than those observed across the channel in the submarine cable, mainly in the E–W dipole. The electric variations registered on onshore dipoles show a predominantly N–S polarized electric field in the vicinity of the Barra. Since the orientation of the water flow in the Barra channel is predominantly E–W, the N–S electrodes are the most sensible pair of the onshore antennas to the electric signals originated by water transport in the channel.

The amplitude spectrums of the hourly mean electric field are shown in Figure 12. The spectrums are dominated by the same frequencies that are present in the submarine cable. However, the small importance of the M_2, S_2/K_2 frequencies should be noted. The 3 cpd (8 h) tidal component is present in the onshore measured fields but is not revealed in the cable data. Moreover, the 4 cpc (6 h) tidal component, corresponding to shallow water harmonics, is shown in the N–S cable spectrum and is absent in the E–W and cable measured fields. The spectrum for the E–W dipole shows that this component is noisier than the N–S one.

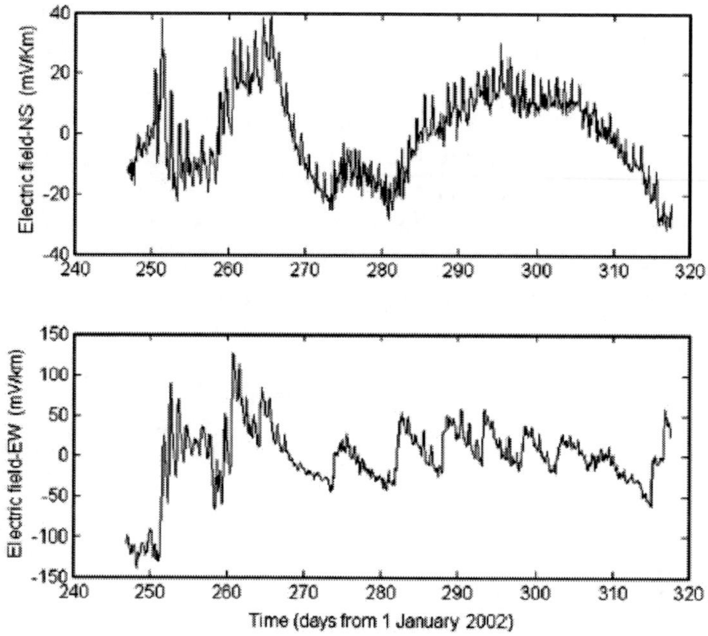

Figure 11. Hourly mean of the electric field measured in the N–S onshore dipole (top) and with the E–W onshore dipole (bottom). From Monteiro Santos et al. (2007).

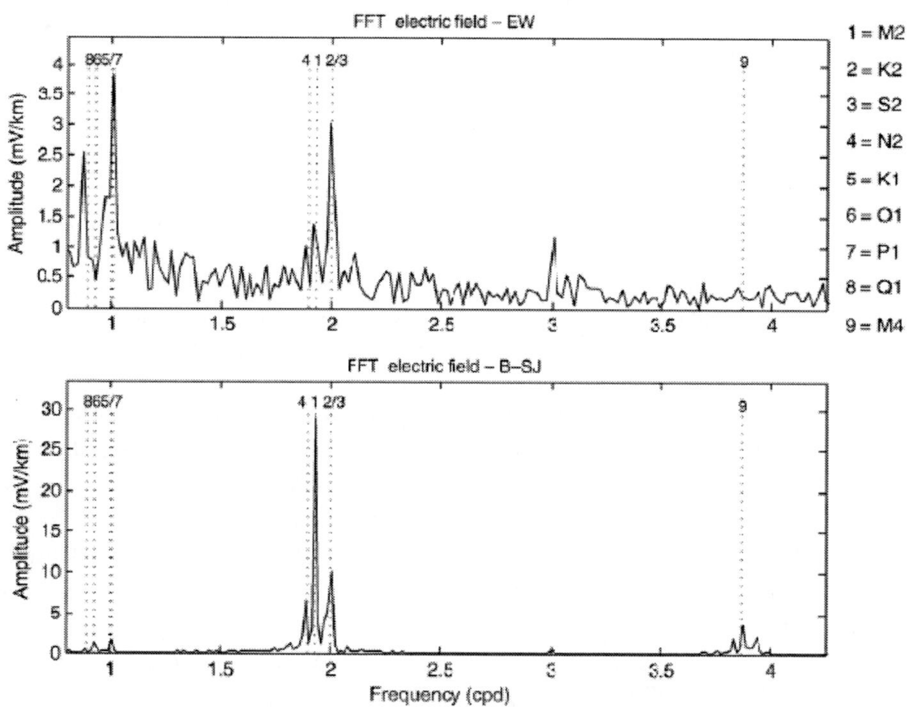

Figure 12 . Amplitude spectrum obtained from data recorded with the N–S dipole (top) and the E–W dipole (bottom). From Monteiro Santos et al. (2007).

The main objective is to use onshore electric potential measurements in the vicinity of a channel to evaluate the mass transport for the water flow in the channel. Therefore, let us concentrate our attention into the M2 component in the antenna. In order to separate this component from the total signal, the measured time series were approximated by a set of lunar harmonics constituents using the least square fitting method, as explained before. A detailed explanation of the method is presented in Nolasco et al. (2006). The results obtained are presented in Figure 13 where amplitude and phase lag for tidal components for water transport, electric field in the submarine cable and in the onshore dipoles (with the errors of the least square fitting) are plotted. The adjustment between the series regarding amplitude and phases is very good, except for the E–W onshore dipole where a significant error is observed.

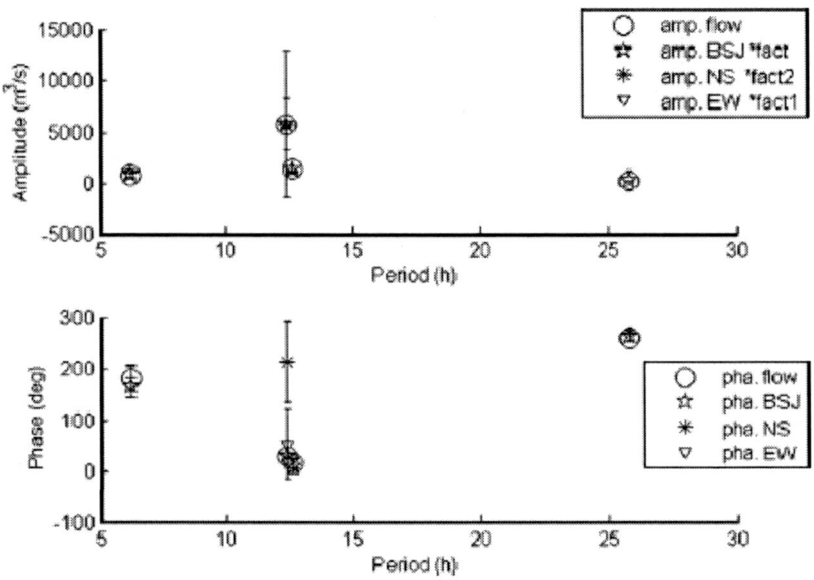

Figure 13. Amplitude and phase of the lunar tidal constituents for water transport, electric field in the submarine cable and on-shore dipoles determined by least square fitting approximation of the data with 95% confidence interval. From Monteiro Santos et al. (2007).

As the *dop* measurements in the onshore antennas and in the submarine cable were carried out simultaneously the results from both systems can be analyzed together in order to estimate the calibration coefficient for the onshore antennas (Monteiro Santos et al. 2007). Figure 14 shows the M_2 harmonics of the electric field measured in the submarine cable and in the N–S dipole together with the water flux/submarine cable calibration. A delay of 3.62 h between the maximum values of the water flux and the electric field measured in the N–S dipole can be noted.

Comparing the M_2 component of the electric field measured in the submarine cable with that measured in the N–S dipole, one notes the relation $E_{cable} = 36.46\ E_{N\text{-}S}$ between their amplitudes. A calibration factor for the onshore antenna can be estimated using the correlation between the water transport in the channel (or their equivalent electric potential in the cable) and the electric potential differences with tidal origin measured in the N–S dipole.

A value of 7350m^3 s^{-1} for each mV/km in the N–S onshore dipole and of 5476.9m^3 s^{-1} for each mV/km in the E–W dipole was estimated for water transport.

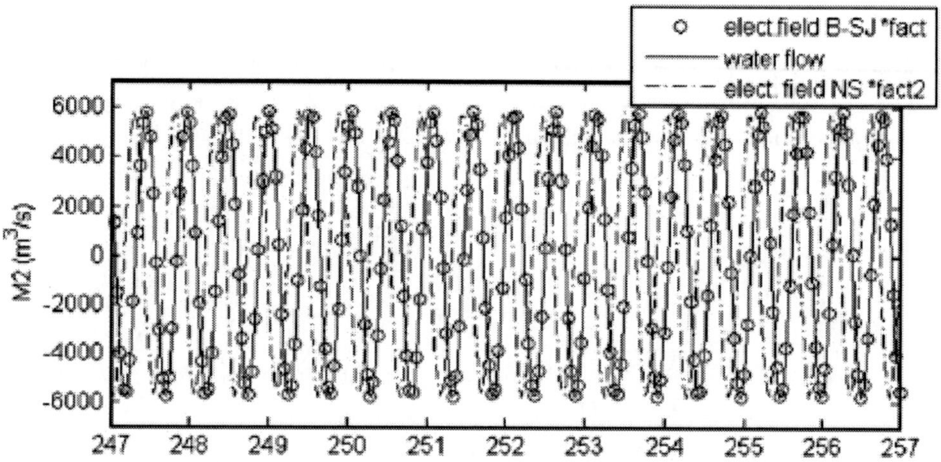

Figure 14. Comparison between the M2 components of the electric field measured with the submarine cable and the N–S dipole with the water transport. The cable electric field values were multiplied by the factor 201.6 m^3 s^{-1} mV^{-1} km, whereas the N–S field values were multiplied by the factor 7350 m^3 s^{-1} mV^{-1} km. From Monteiro Santos et al. (2007).

Modelling

We adopt the model proposed by Junge (2001) in order to explain some of our onshore observations around Ria de Aveiro. Figure 15 gives a schema of the current at the Barra channel with tidal origin for a fixed time t_0. As the ions in the water moves perpendicularly to the vertical component of the earth magnetic field ($\vec{B}$), the charges separation takes place. These charges will cause electric currents in vertical planes in and beneath the water. The scale length of the horizontal water current, characterised by its wavelength λ_y, exceeds the thickness of the uppermost conducting earth layer with an integrated conductivity τ_L. The vertically integrated electric current density $\vec{I}$ is:

$$\vec{I}(x,y,t) = \int_{z_1}^{z_2} \vec{J}(x,y,z,t)dz \qquad (7)$$

where z_1 and z_2 are the lower and upper boundary of the conducting layer and $\vec{J}$ is the current density which is given by (Sanford 1971), $\vec{J} = \sigma(\vec{E} + \vec{v} \times \vec{B})$. Integrating (7) we obtain

$$I_L - (\tau_w + \tau_L^{'})E_x = -\tau_w B_z v_y \qquad (8)$$

where τ_w and τ_L' are the integrated conductivities of seawater and conducting sedimentary layer beneath the channel. The right side of the equation (8) is the current source and the left hand side represents the different parts of the resulting currents in the vertical and horizontal planes.

Considering that the current along the "Ria de Aveiro" I_L varies harmonically with the period T_{M2} (= 12.4206 h) and the Greenwich phase φ_{M2} (≈ 28°), the potential $V(P,t)$ at $P(x,y)$ distant of r_P from the current source at y_0, is (Junge 2001)

$$V(x,y,t) = \int_{y_A}^{y_B} V^*(x,y,t;y_0)dy_0 \tag{9}$$

where $V^*(x,y,t;y_0) = \dfrac{I_L(t)}{\pi\tau_L}\ln(\dfrac{1}{r_P})$ represents the logarithm potential at P originated by the current I_L at y_0. Here, $r_P = \left[x^2 + (y-y_0)^2\right]^{1/2}$ and

$$I_L(t,y_0) = I_{L0}\sin\left[2\pi(\frac{y_0}{\lambda_y} - \frac{t}{T_{M2}}) + \varphi_{M2}\right] \tag{10}$$

where

$$I_{L0} = h_w\, v\, \sigma_w\, B_z) \tag{11}$$

is the mean maximum amplitude of the integrated electric current density. Therefore, the components of the electric field with motionally origin is (Junge 2001):

$$E_x(x,y,t) = \frac{I_{L0}}{2\pi\tau_L}x\int_{y_A}^{y_B}\frac{1}{x^2+(y-y_0)^2}\sin\left[2\pi(\frac{y_0}{\lambda_y} - \frac{t}{T_{M2}}) + \varphi_{M2}\right]dy_0 \tag{12i}$$

$$E_y(x,y,t) = \frac{I_{L0}}{2\pi\tau_L}\int_{y_A}^{y_B}\frac{y-y_0}{x^2+(y-y_0)^2}\sin\left[2\pi(\frac{y_0}{\lambda_y} - \frac{t}{T_{M2}}) + \varphi_{M2}\right]dy_0 \tag{12ii}$$

y_A and y_B are the limits of integration in the direction of the channel flow. It should be noted that this model does not consider the contribution of the currents parallel to the coast line. In this study and for the Aveiro particular case, it was assumed that these contributions are negligible.

The mean water depth in Barra is 16 m, and its wavelength is about 40 km. The vertical water flow will be assumed as negligible compared to the horizontal one, which will be v_y =1 m/s. The conductivity of the sea water will be considered uniform (3.33 S/m), as well as the water depth and water velocity. The vertical component of the mean magnetic field at the time of the observations was 36200 nT. Then, the mean maximum amplitude of the integrated electric current density is $I_{L0} = 1.9 \times 10^{-3}$ A/m. The water conductance is $\tau_w = 53.3$ S.

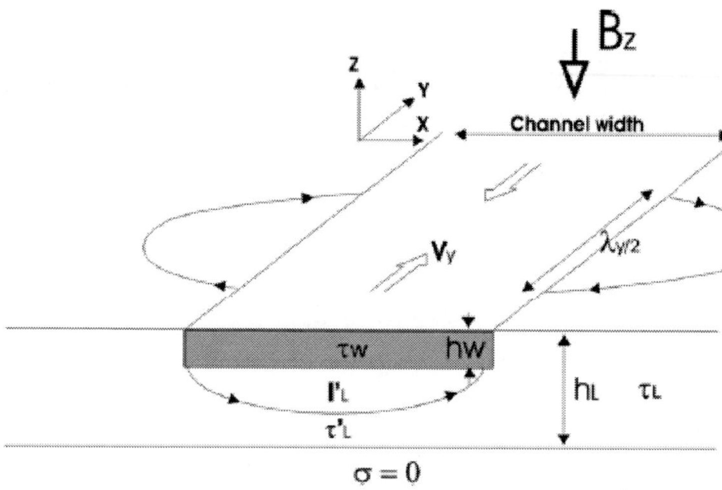

Figure 15. Schematic view of the tidal current in a channel and induced telluric currents at a fixed time (adapted from Junge 2001).

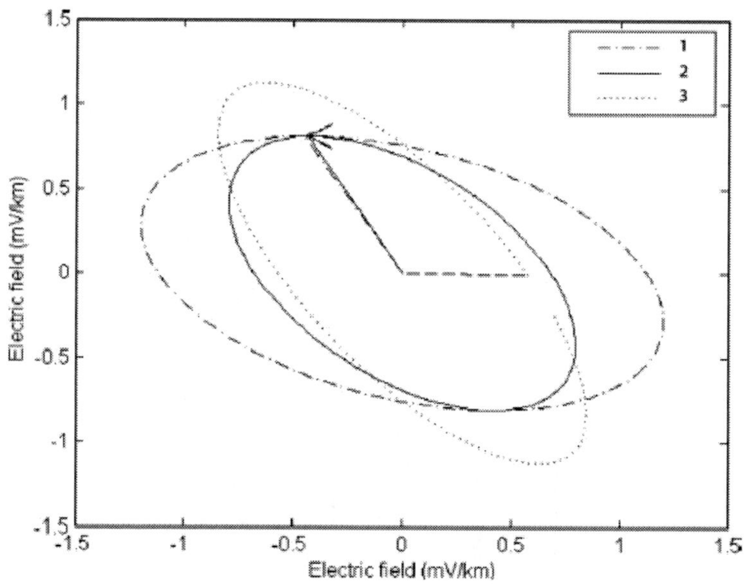

Figure 16. Comparison between observed (solid and dash-point lines) and theoretical (pointed) telluric polarization ellipses. Lines 1 and 2 represent the ellipses calculated using the extreme values of the lunar tides constituents of the E-W electric field component. From Monteiro Santos et al. (2007).

It was observed that the electric field measured in the N-S dipole is about 1/36.46 lesser than the electric field measured in the submarine cable. Our model does not allows for the modelling of the electric field in the cable, therefore we compare the theoretical electric field calculated at the N-S dipole site with that one calculated at a point 5 m apart from the source current located at northern edge of the channel. The value of τ_L is unknown. For this reason, we proceeded by trial-and-error. A value of 1/32.1 for the ratio $E_{N\text{-}S}/E_A$ was obtained with τ_L = 1200 S. In this case the calculated ratio is closer to the observed value and the conductance value was assumed for the next step.

Figure 16 shows the comparison between the measured and theoretical telluric ellipse calculated at a location corresponding to the middle of the N-S dipole. The figure shows two measured ellipses calculated from the extreme values of the lunar tides constituents of the E-W electric field component. The theoretical ellipse has the major axes approximately in the NNW-SSE direction. The agreement of the theoretical and experimental ellipses can be considered good, taking into account our simplified model and the errors in the E-W electric field component.

MIV Measurements in the West Portuguese Coast (Onshore)

MIV measurements carried out onshore close to the coast line can also be used for tides characterization and alongshore transport estimation. The principle is exactly the same as exposed before: MIV values recorded onshore using two electric dipoles are correlated with estimated/measured tidal transport and a calibration coefficient is calculated. In our experiment the pairs of electrical dipoles were installed close to the sea at two locations along the western coast of Portugal in N-S and E-W directions. The sites, São Jacinto and Sines are indicated in Figure 17. At Sines the electrodes are installed 200 m apart roughly in the direction perpendicular and parallel to the coastline. At São Jacinto the distance between electrodes are 360 m and 68 m for the N-S (9°E) and E-W dipoles, respectively.

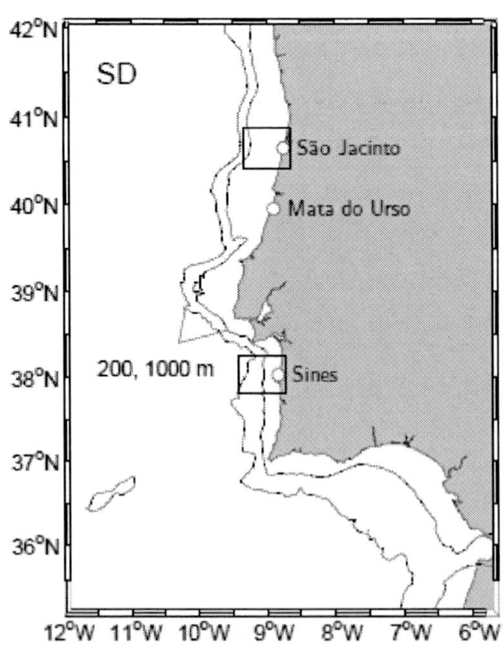

Figure 17. Location of the antennas installed in the West Portuguese coast.

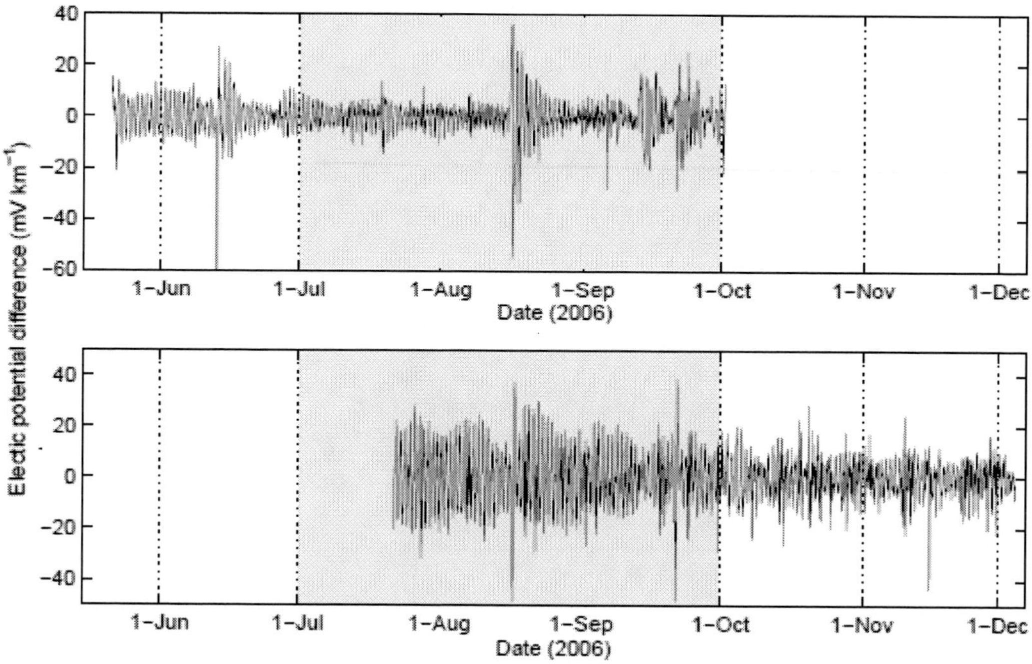

Figure 18. Data from São Jacinto (top) and Sines (bottom) E-W antennas. The shaded regions (1-Jul to 1-Oct) represents the ocean simulations period.

Since the coastline is mainly N-S oriented the electric signal is predominantly sampled by E-W dipoles and the signal observed in N-S dipoles was not used in the present investigation. All the electrodes have been placed in salted bentonite into 1.0 to 1.5 m deep holes. The electric potential difference between electrodes was sampled at a rate of 10 s, but only the average of six values over 60 s was recorded. The distance from dipoles sites and the coastline was about 500 m. The measurements used in this study took place during the Summer and Autumn of 2006.

Figure 18 shows the difference of potential measured in the E-W dipoles located in São Jacinto and Sines after the background long period on the signals has been removed using a Singular Spectrum Analysis algorithm (Marques et al. 2006). The signals have typical amplitudes between 10 and 20 mV km^{-1} for São Jacinto and arround 20 mV km^{-1} for Sines. The signals were explored with Fourier analysis (Figure 19). Both signals have very strong components around the diurnal (stronger) and semidiurnal periodicities. Since the main tidal components in the western Portuguese region are M2 and S2 (Marta-Almeida and Dubert 2006), i.e., semidiurnal, it can be realised that much of the *dop* must have geomagnetic and not oceanic origin. Aiming to quantify the importance of tides in the *dop* signal, the signals were examined with harmonic fit using the tidal analysis package T TIDE (Pawlowicz et al. 2002). The amplitudes and phases (and its estimated errors) of the main lunar components (M2, N2, O1 and Q1) are shown in Table 1.

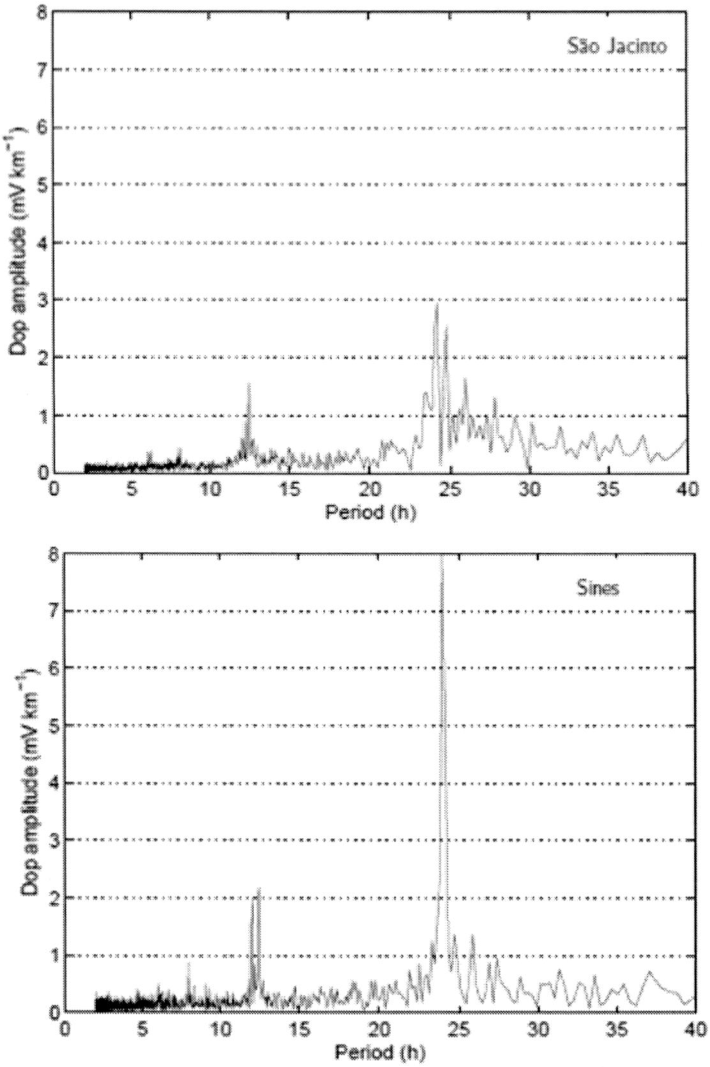

Figure 19. Amplitude spectrum of the dop signals measured in the E-W dipoles installed in São Jacinto and Sines.

Table 1. Tidal analysis of the *dop* signals. Amplitude (amp), amplitude error (eamp), phase (pha) and phase error (epha) of the main lunar constituents for the region, M_2, N_2, O_1 and Q_1. The units are mV km^{-1} for amplitudes and degree for phases.

	São Jacinto				Sines			
	amp	**eamp**	**pha**	**epha**	**amp**	**eamp**	**pha**	**epha**
M_2	2.06	0.52	148.48	15.46	3.39	0.94	61.18	15.68
N_2	0.66	0.56	133.91	50.79	0.73	0.77	17.10	69.96
O_1	0.55	0.99	120.05	121.31	1.22	1.39	344.39	57.23
Q_1	0.72	0.94	59.74	110.72	0.88	1.16	251.85	85.54

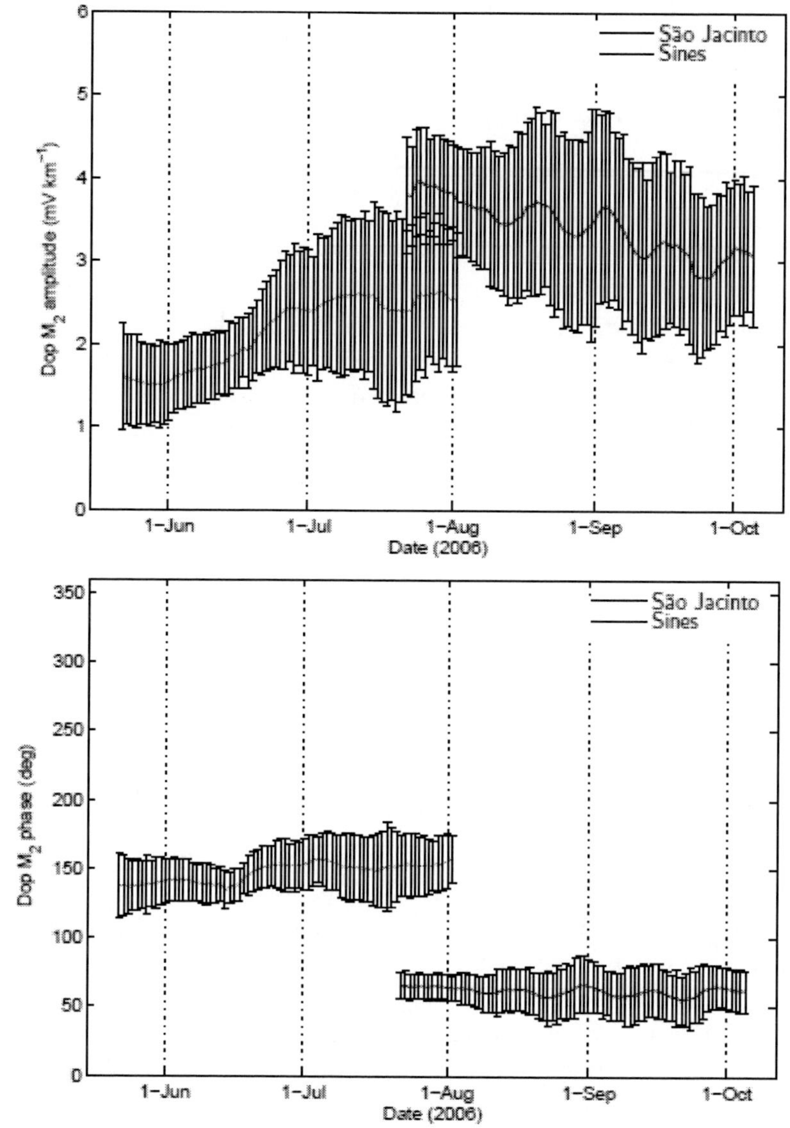

Figure 20. Tidal analysis of the São Jacinto and Sines *dop* signal using a moving 60 days period. The amplitudes (a) and phases (b) refer to the frequencies of the tidal component M_2.

The tidal harmonic fit was repeated using a 60 days moving window, instead of the entire dataset available, with daily increments and starting from the beginning of the *dop* data available for each of the two dipoles (Figure 20) in order to investigate the possible variation of tidal parameters with time. For São Jacinto the M2 amplitude ranges from 1.5 and 2.5 mV km^{-1} with errors around 0.75 mV km^{-1}. Phase ranges between 140° and 160° with errors of about 20°. Amplitude for Sines lays between 3 and 4 mV km^{-1} with errors around 1 mV km^{-1} and the phase is between 55° and 65° with errors of about 15°◦.

In order to establish the relationship between the *dop* measured with electrical dipoles and the oceanic current field, numerical modelling techniques that allows to forecast the tidal and subtidal currents in the coastal zone offshore the dipoles were used. The ocean model used was the Regional Ocean Modeling System (Shchepetkin and McWilliams 2003, 2005) with embedded nesting capabilities, ROMS-AGRIF (Adaptative Grid Refinement in Fortran, Penven et al. (2006)), a 3D free-surface s-coordinate terrain following primitive equation model with hydrostatic approximation configurable for fully realistic regional applications. ROMS-AGRIF enables the use of online and offline nesting thus permitting regional applications to be built based on large-scale configurations covering a wide range of time and space scales. Previous works of ROMS-AGRIF applications to Western Iberia are reported in Peliz et al. (2007, 2009); Teles-Machado et al. (2007); Marta-Almeida et al. (2008).

With the objective to resolve the shelf circulation of the west coast of the Iberian Peninsula, a nested grids configuration has been implemented for the Summer 2006 (July 1st to October 1st). The nested model was forced with climatological surface fluxes, the surface momentum and heat fluxes data were obtained from simulations with the widely used atmospheric model Weather Research and Forecast, WRF (Skamarock et al. 2005). Surface winds, humidity, pressure, air temperature and radiative data were generated by WRF model and passed to the oceanic model to calculate the air-sea fluxes. Tidal circulation was achieved by imposing a tidal forcing at the boundaries. This forcing consisted in the superposition at the lateral boundaries of tidal parameters for elevations (amplitude and the Greenwich phase) and currents (barotropic tidal ellipses in the form of semi-major and semi-minor axis, inclination and Greenwich phase) from the TPXO database for the North Atlantic (Erofeeva and Egbert 2002). This database provides information for the main diurnal and semidiurnal constituents, M_2 S_2, N_2, K_2, K_1, O_1, P_1 and Q_1, with a resolution of 1/6°. The details and modeling techniques of the tidal dynamics for the region, using a similar approach to the present one, has already been done for the homogeneous case (Marta-Almeida and Dubert 2006).

With the objective of analyse the alongshore (N-S) transports, perpendicular to the installed E-W dipoles, the 3D currents were outputted hourly from the model at every grid points of the higher resolution domain, in an offshore line, from the shore, near the locations of the two dipoles, until the shelf break assumed as the 200 m isobath. The shelf break distance from coast in São Jacinto (Dsj) is about twice the one at Sines (Dsi). For this reason and aiming to analyse transports crossing equal areas we caculated the transport for Sines until the 200 m isobath (Hsi) and for São Jacinto until the isobath for which the cross-shelf areas for both regions are equal (the isobath 114 m, Hsj). This choice of the area was arbitrary without any base than the idea that *dop* signals are more influenced by transports near the dipoles. The transports were calculated for the whole simulation period and the resultant time series were considered as the ocean signals (TR hereinafter), ie, the signals that induced the electric *dop* in the onshore dipoles. For this reason these transports are used to calibrate the dipoles.

Thus, as well as for *dop* signals we decompose the TR signals into high frequency and subinercial components. This step was done with the widely used tidal low-pass filter pl33 (Flagg et al. 1976). For São Jacinto the tidal oscillations have a half-amplitude arround 10 $\times 10^4$ m^3 s^{-1}. For Sines, the amplitudes of the tidal component are about half the amplitudes in São Jacinto (Figure 21).

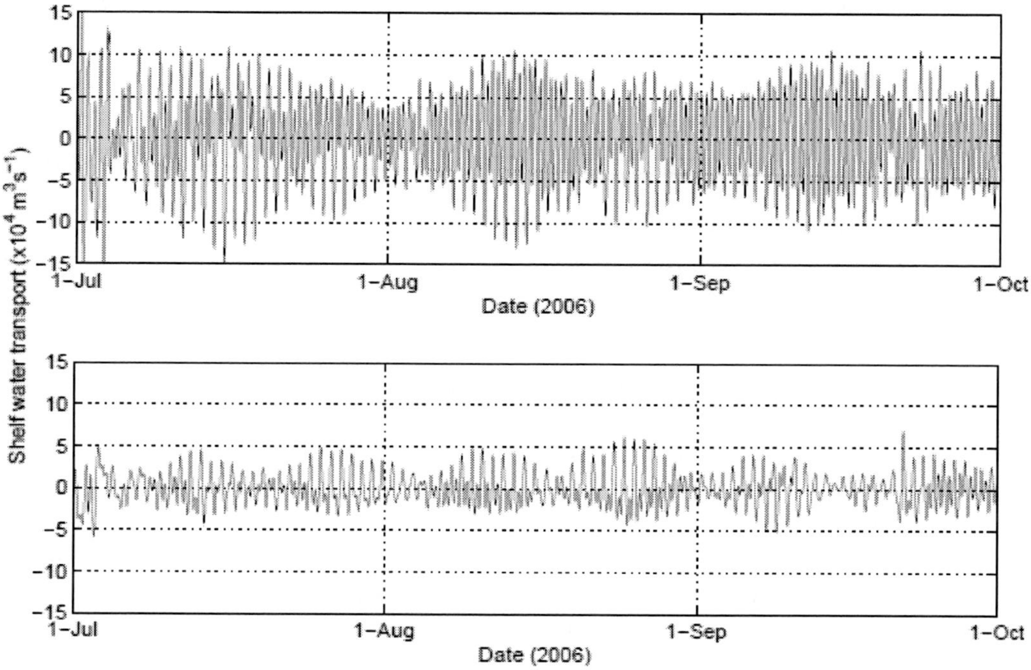

Figure 21. Alongshore shelf water tidal transport component from coast to 200 m isobath adjacent to the dipole of São Jacinto (top) and Sines (bottom).

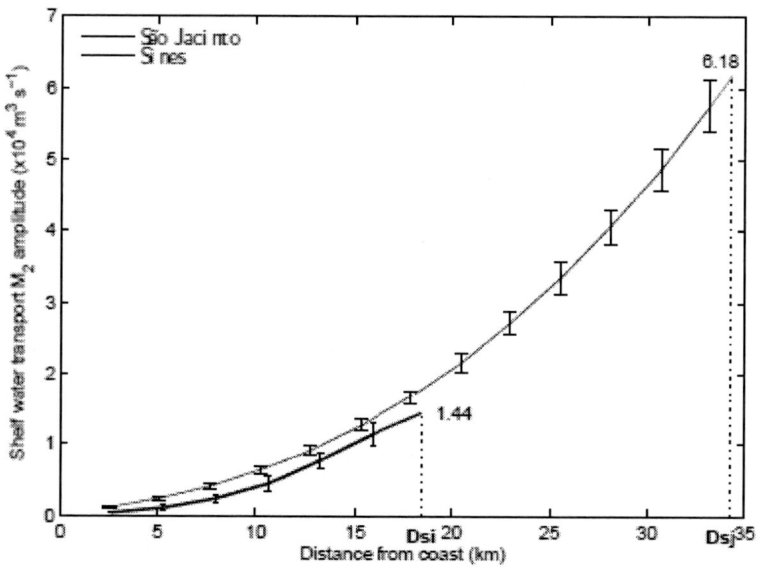

Figure 22. M_2 amplitude of the alongshore shelf sea transport adjacent to the São Jacinto and Sines antennas. The vertical dotted lines indicate the 200 m isobath and the numerical labels at its top are the amplitudes for the distance from coast of the isobath 200 m.

The differences between the two locations, in spite of using the same cross-shelf area, can be explained by the fact that the alongshore current resulting from the adjustment of the M_2

tidal constituent is smaller in the Sines region (Marta-Almeida and Dubert 2006). A tidal decomposition, similar to the described above for *dop* in both sites, was applied to the series of high frequency TR. The amplitudes obtained and its corresponding errors for our tidal constituent of interest, M_2, were $6.18 \pm 0.33 \times 10^4$ m^3 s^{-1} for São Jacinto and $1.44 \pm 0.18 \times 10^4$ m^3 s^{-1} for the Sines transport. The corresponding M_2 Greenwich phases found were 265 ± 3° and 129 ± 6° respectively.

Considering the full series of *dop* signal available and the TR transports we obtain a calibration factor for the main tidal frequency, M_2, of $6.16\times10^4/2.06=3.00\times10^4$ m^3 s^{-1}/mV km^{-1} for São Jacinto and $1.44\times10^4/3.39=4.25\times10^3$ m^3 s^{-1}/mV km^{-1} for Sines (Figure 22).

A Simple Model for MVI Measured on Shore

To predict the on shore electric fields due to the tidal ocean flow, we adopt again, the simple model described by Junge (1988, 2001). The solution allows simulating the electric field, measured on shore due to tidal transport parallel to the coastline. This is a simplification of the ocean-land interaction conditions verified in west Portuguese coast and has the objective to estimate the magnitude of the expected magnitudes of N-S and E-W electric field components originated by M_2 tidal movements.

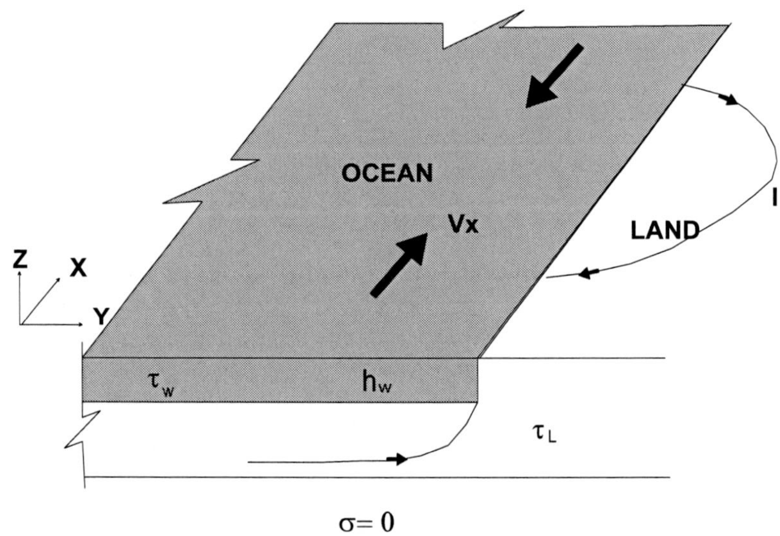

Figure 23. Schematic view of the tidal current in a channel and induced telluric currents at a fixed time (adapted from Junge 2001).

Figure 23 gives a schematic view of the current at the west coast with tidal origin for a fixed time t_0. The scale length of the horizontal water current, characterised by its wavelength λ_x, exceeds the thickness of the uppermost conducting earth layer with an integrated conductivity τ_L. Considering that the current I along the coast varies harmonically with the period T_{M2} (12.4206 h) and the Greenwich phase φ_{M2} (≈20°)

$$I = I_o \sin\left[2\pi(\frac{x_o}{\lambda_x} - \frac{t}{T_{M_2}}) + \varphi_{M_2}\right] \tag{13}$$

where I_o is the mean maximum amplitude of the integrated electric current density and is given by equation (11), the components of the electric field $\vec{E}(P,t)$ at $P(x,y)$ distant of r_P from the current source at x_0 with tidal origin, are given by equation (12) (Junge 2001, Monteiro Santos et al. 2007). In the adopted referential the components of the electric field are

$$E_x(x,y,t) = \frac{I_o}{2\pi\tau_L}\int_{x_A}^{x_B}\frac{x - x_o}{y^2 + (x - x_o)^2}\sin\left[2\pi(\frac{x_o}{\lambda_x} - \frac{t}{T_{M_2}}) + \varphi_{M_2}\right]dx_o$$

(14)

$$E_y(,y,t) = \frac{I_o}{2\pi\tau_L}y\int_{x_A}^{x_B}\frac{1}{y^2 + (x - x_o)^2}\sin\left[2\pi(\frac{x_o}{\lambda_x} - \frac{t}{T_{M_2}}) + \varphi_{M_2}\right]dx_o$$

x_A and x_B are the limits of integration in the water moving direction.

Considering that the wavelength is about 12000 km, v_x=1 m/s, seawater conductivity of 3.33 S/m and B_z = 36200 nT, the mean maximum amplitude of the integrated electric current density is $I_o = 1.2x10^{-2}$ A/m and the water conductance is τ_W = 33.33 S (for a 100 m thick ocean). It must be noted that the earth conductance value (τ_L) is unknown. For this reason we calculated the electric field on shore for different values of τ_L and compared with measured values (Figure 24). Table 2 shows the relationship ($\max E_y / \max E_x$) between the maximum values of the field components measured and calculated in Sines and São Jacinto. There is a very good agreement between observed and calculated values in Sines. At São Jacinto the observed ratio is close to unit, whereas the calculated is 5.72. The reason of this disagreement is the strong influence of the MIV originated by the strong E-W water flow in Ria de Aveiro, which is not considered in this model.

Table 2. Ratio (max E_y)/(maxE_x) in both onshore sites.

	Sines	S. Jacinto
Observed	4.35	1.14
Calculated($\tau_{L=8000\ S}$)	4.92	5.72

At Sines both measured components are compatible with τ_L ranging from 5000 to 6000 S. At São Jacinto only the E-W component is compatible with a value around 8000 S for the earth conductance. The N-S component only can be modelled if a conductance of about 2000 S is chosen. This discrepancy is motivated by the fact that the N-S on shore electric component is strongly affected by the water movement in Ria de Aveiro located very close the on shore dipoles, as demonstrated by Monteiro Santos et al. (2007). As explained above the N-S component in São Jacinto is mainly determined by the water flow in the Ria de

Aveiro. The earth conductance value of 1200 S for presented before is compatible with the result achieved in this paragraph.

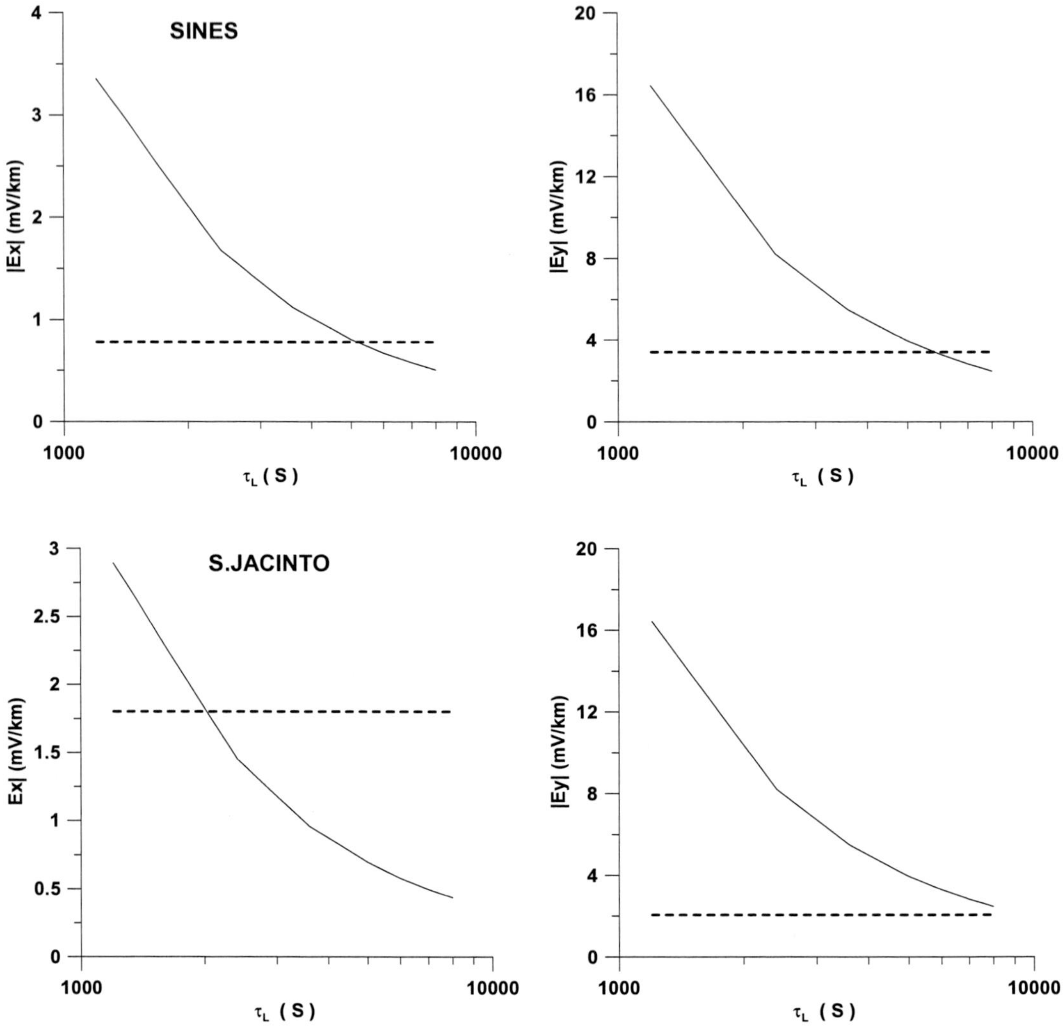

Figure 24. Predicted maximum amplitudes (solid lines) as a functions of log(τ_L) and observed amplitudes (dashed straight lines) of M_2 tide electric field in Sines and São Jacinto. Calculations carried out assuming h_w = 100 m.

These results show that the on shore measurements mostly reflect the effect of the water transport in the direction N-S. The same result, e.g., the major importance of the E-W electric component is obtained considering a 200 m deep ocean. Since the electric field components are proportional to I_o / τ_L the earth conductance will the double of that one considered for a 100 m depth sea.

Conclusion

Electric fields data have been acquired in a submarine cable crossing a channel (the Barra in Aveiro area, northern Portugal) and in two onshore sites installed in the vicinity of the west Portuguese coastline and processed for tidal analysis. The main flows at the Barra channel have tidal origin and represent a key process in the water renewal at Ria de Aveiro lagoon.

It was shown that the dominant tidal component (period) in the submarine cable and onshore antennas data is the semidiurnal lunar period of 12.4602 h (the well known M_2 component). The data analysis reveals that measuring the differences of electric potential across the channel can provide an accurate estimate of the water transport by tidal flows through the Barra channel. A value of 720 $m^3\ s^{-1}\ mV^{-1}$ was estimated for the coefficient relating voltage and water transport at Barra.

It was also demonstrated that one can use the onshore lunar harmonics to estimate the water transport verified through the Barra channel. To achieve this, the onshore N-S antenna installed close to the Barra was calibrated by correlating the water transport tidal lunar and the onshore motionally tidal electric field. A calibration factor of 7350 m^3 s−1 for each mV/km, with motionally induced origin, was estimated for the N–S onshore dipole.

Onshore measurements of telluric field oscillations have been used to estimate the water transport associated to tidal flow in the western Portuguese coast. The onshore E-W dipoles located at São Jacinto and Sines were calibrated by correlating the N-S shelf water transport originated by the main lunar tidal frequency and the corresponding onshore motionally tidal electric field. Calibration factors of 3.00×10^4 and 4.25×10^3 $m^3\ s^{-1}$ for each mV km^{-1}, with motionally induced origin, was estimated for São Jacinto and Sines, respectively.

The possibility of using MIV methods to estimate flow transport on a short scale (about half of km) corresponding to estuarine systems and alongonshore, provides an important advantage over traditional oceanographic methods (classical currentmeter, Acoustic Doppler Currentmeter Profiler, ADCP, etc.). In this kind of environment the use of traditional methodologies is clearly limited in cost and in time, due to batteries and memory limitations, as well as to the very fast growth of algae that will make difficult the proper functioning of the instruments. Furthermore, concerning the water flow is usually determined from discrete measurements while when using ADCP's there are shadow areas close to both the bottom and to the surface, and therefore the obtained results are only approximate values. The MIV methods present low cost for accurate long-term monitoring of integral water transport.

Acknowledgments

This study was developed in the scope of the projects PROTEU (PDCTM/MAR/15275/1999) EMNOCM (POCTI/CTA/44062/2002) and AMDRAPHYD (POCI/AMB/57928/2004) funded by Fundação para a Ciência e Tecnologia (FCT) and by FEDER. One of the authors, R.N., was funded by a grant of the FCT (PRAXIS SFRH/BPD/10256/2002). The authors thank the Administration of the Port of Aveiro, Irmãos Cavaco, Xavisub and Área Militar de S. Jacinto for all aid along the equipment installation and data collection.

References

Chave, A.; Luther, D. Low-frequency, motionally induced electromagnetic fields in the ocean. *J. Geophys. Res*. 1990, 95, 7185–7200.

Dias, J.M., 2001. *Contribution to the Study of the Ria de Aveiro Hydrodynamics, PhD thesis,* Univ. de Aveiro, Portugal.

Dias, J.M.; Lopes, J. F. Implementation and assessment of hydrodynamic, salt and heat transport models: the case of Ria de Aveiro Lagoon). Portugal. *Environ. Model. Softw.* 2006, 21, 1–15.

Dias, J.M.; Lopes, J. F., Dekeyser, I. A numerical model system application to the study of the transport properties in Ria de Aveiro lagoon. *Ocean Dyn*. 2003, 53, 220–231.

Duffus, H. J.; Fowler N. R. On planetary voltages, ocean tides, and electric conductivity below the Pacific. *Can. J. Earth Sci.*1974, 11, 873-892.

Egedal, J. On the lunar-diurnal variation in the earth-currents. *Terr. Magn*. 1937, 42, 179–181.

Erofeeva, S.; Egbert, G. Efficient inverse modeling of barotropic ocean tides. *J. Oceanic Atmosph. Technol.* 2002, 19,183-204.

Filloux, J.H. 1987. Instrumentation and experimental methods for oceanic studies, *New Volumes in Geomagnetism*; 1, Academic, New-York, 143– 246.

Flosadottir, A. H.; Taire, K. *Proceedings of International Workshop on Scientific Use of Submarine Cables 1997*, 10–15. Observations of ocean currents using submarine cables.

Fristedt, T.; Sigray, P., Lundberg, P., Crona, L. Salt-wedge observations in Oresund by direct measurement of the motionally induced voltage. *Cont. Shelf Res*. 2002, 22, 2513–2524.

Fujii, I.; Utada, H. On geoelectric potential variations over a planetary scale. *Mem Kakioka Magn. Obs*. 2000, 29, 1-71.

Harvey, R. R.; Larsen, J. C. and Montaner. R. Electric field recording of tidal currents in the Strait of Magellan. *J. Geophys. Res*. 1977, 82, 3472-3476.

Junge, A. The telluric field in northern Germany induced by tidal motion in the North Sea. *Geophys. J. Int*. 1988, 95, 523–533.

Junge, A. and ISO-3D team Final Report of the ISO-3D project 2001, pp. 101–116: *Oscillations of the onshore telluric field generated by oceanic tides.*

Lanzerotti, L.J., Sayres, C. H., Medford, L.V., Kraus, J. S., and Maclennan, C. Earth potential over 4000 km between Hawaii and California. *J. Geophys. Res. Lett.* 1992, 19, 1177-1180.

Larsen, J.C. Florida Current volume transport from voltage measurements. *Science* 1985, 227, 292–311.

Larsen, J. C. Electric and magnetic fields induced by deep sea tides. *Geophys. J. Int.* 1968, *16*, 40–70.

Larsen, J. C. The electromagnetic field of long and intermediate water waves. *J. Mar. Res.* 1971, *29* (1), 28–45.

Larsen, J. C. Transport and heat flux of the Florida Current at 27 N derived from cross-stream voltages and profiling data: Theory and observations. *Phil. Trans. R. Soc. Lond*. 1992, A, 338, 169–236.

Longuet-Higgens, M. S. The electric and magnetic effects of tidal streams. *Mon. Not. Roy. Astr. Soc. (Geophys. Suppl.)* 1949, 5, 285–307.

Malin, S. R. C. Separation of lunar daily variations into parts of ionospheric and oceanic origin. *Geophys. J. R. Astron. Soc.* 1970, 21, 447–455.

Marques, C. A. F.; Ferreira, J. A., Rocha, A., Castanheira, J. M., Melo-Goncalves, P., Vaz, N. and Dias, J. M. Singular spectrum analysis and forecasting of hydrological time series. *Phys. Chem. Earth* 2006, 31, 1172–1179.

Marta-Almeida, M.; Dubert, J. The structure of tides in the Western Iberian region. *Cont. Shelf Res.* 2006, 26, 385–400.

Marta-Almeida, M.; Dubert, J., Peliz, A., dos Santos, A. and Queiroga, H. A modelling study of the Norway Lobster (Nephrops norvegicus) larval dispersal in southern Portugal: predictions of larval wastage and self-recruitment in the Algarve stock. *Can.* J. *Fish. Aquat. Sci.* 2008, 65, 2253–2268.

Marta-Almeida, M., Monteiro Santos, F.A., Nolasco, R., Dubert, J., Bernardo, I., Soares, A., Peliz, A. and Dias, J., 2010. On the estimation of tidal and subinertial alongshore water transport from onshore telluric measurements. *Physics of the Earth and Planetary Interior* (submitted).

Maus, S.; Kuvshinov, A. Ocean tidal signals in observatory and satellite magnetic measurements. *Geophys. Res. Lett.* 2004, 31, doi: 10.1029/2004GL000634.

McKnight, J. D. Lunar daily variations in New Zealand. *Geophys. J. Int.* 1995, 122, 889–898.

Monteiro Santos, F. A.; Nolasco, R., Soares, A., Dias, J. and Bernardo, I. Estimation of water transport from oscillations of the onshore telluric field generated by tides. *Earth Planet.* Sc. Lett. 2007, 263, 420–428.

Monteiro Santos, F. A.; Soares, A., Trindade, L., Nolasco, R. and Rodrigues, H. Voltage measurements over the CAM-1 submarine cable between Madeira Island and Portugal mainland. *Earth Planets Space* 2002, 54, 393–398.

Nolasco, R.; Soares, A., Dias, J. M., Monteiro Santos, F. A., Palshin, N. A., Represas, P. and Vaz, N. Motional induction voltage measurements in estuarine environments: the Ria de Aveiro Lagoon (Portugal). *Geophys. J. Int.* 2006, 166, 126–134.

Palshin, N. A.; Vanyan, L. L., Kaikkonen P. Onshore amplification of the electric field induced by a coastal sea current. *Physics of the Earth and Planetary Interiors* 1996, 94, 3-4, 269-273.

Palshin, N. A.; Vanyan, L. L., Medzhitov, R. D., Shapiro, G. I., Evdoshenko, M. A., Utada, H., Shimizu, H. and Tanaka, Y. Use of the Nakhodka-Naoetsu submarine cable for studying the temporal variability of the integral water transport in the Sea of Japan. *Oceanology* 2001, 41, 447–453.

Palshin, N. A.; Vanyan, L., Yegorov I., Lebedev, K. Electric field induced by the global ocean circulation. *Physics Solid Earth* 1999, 35, 1028-1035.

Palshin, N. A.; Vanyan, L.L., Poray-Koshits, A.M., Matyushenko, V.A., Kaikkonen, P., Tiikkainen, J. Measurements of motionally induced voltage in the coastal zone of the Throat of the White Sea. *Earth Planets Space* 2002, 54, 433–441.

Pawlowicz, R.; Beardsley, B. and Lentz, S. Classical tidal harmonic analysis including error estimates in MATLAB using T TIDE. *Comput. Geosci.* 2002, 28, 929–937.

Peliz, A.; Dubert, J., Marchesiello, P. and Teles-Machado, A. Surface circulation in the Gulf of Cadiz: Model and mean flow structure. *J. Geophys. Res.* 2007a, 107, C11 015.

Peliz, A., P. Marchesiello, A. M. P. Santos, J. Dubert, A. Teles-Machado, M.Marta-Almeida, and B. Le Cann, 2009: Surface circulation in the Gulf of Cadiz: 2. Inflow-outflow coupling and the Gulf of Cadiz slope current. *J. Geophys. Res.*, 114, C03 011.

Penven, P.; Debreu, L., Marchesiello, P. and McWilliams, J. C. Evaluation and application of the ROMS 1-way embedding procedure to the central california upwelling system. *Ocean. Model.* 2006, 12, 157–187.

Rooney, W.J. Lunar diurnal variation in Earth currents at Huancayo and Tuscon. *J. Geophys. Res*. 1938, 43, 107-118.

Sanford, TMotionally-induced electric and magnetic fields in the sea. . *J. Geophys. Res*. 1971, 76, 3476–3492.

Shchepetkin, A. F.; McWilliams, J. C. A method for computing horizontal pressure-gradient force in an oceanic model with a non-aligned vertical coordinate. *J. Geophys. Res*. 2003, 108, 1–34.

Shchepetkin, A. F.; McWilliams, J. C. The Regional Ocean Modeling System (ROMS): A splitexplicit, free-surface, topography-following coordinates ocean model. *Ocean. Model.* 2005, 9, 347–404.

Skamarock, W. C.; Klemp, J. B., Dudhia, J., Gill, D. O., Barker, D. M., Wang, W. and Powers, J. G. Tech. rep., MMMD-NCAR, Boulder, Colorado, USA. 2005. *A description of the Advanced Research WRF Version 2.*

Szuts, Z., 2008. *The interpretation of motionally induced electric fields in oceans of complex geometry.* PhD Thesis, University of Washigton, 190 p.

Teles-Machado, A.; Peliz, A., Dubert, J., and Sanchez, R. On the onset of the Gulf of Cadiz Coastal Contercurrent. *Geophys. Res. Lett.* 2007, 34.

Tyler, R.; Mysak, L. A., and Oberhuber, J. Electromagnetic fields generated by a 3-D global ocean circulation. *J. Geophys. Res*. 1997, 102, 5531-5551.

Tyler, R.; Maus, S. and Luhr, H. Satellite observations of magnetic fields due to ocean tidal flow. *Science* 2003, 299, 239–241.

Tyler, R.; Oberhuber, J. and Sanford T. The potential for using ocean generated electromagnetic field to remotely sense ocean variability. *Phys. Chem. Earth A* 1999, 24, 429– 432.

In: Sea Level Rise, Coastal Engineering, Shorelines and Tides ISBN: 978-1-61728-655-1
Editor: Linda L. Wright, pp. 235-249 © 2011 Nova Science Publishers, Inc.

Chapter 11

AN EVALUATION OF GLOBAL TIDAL MODELS ON THE INNER CONTINENTAL SHELF OF ARGENTINA, SOUTH AMERICA

Walter Dragani*[*,1,2,4], Enrique D'Onofrio[1,2,3], Monica Fiore[1,2], Walter Grismeyer[1] and Fernando Oreiro[1,3]

[1] Servicio de Hidrografía Naval and ESCM-INUN,
Ciudad Autónoma de Buenos Aires, Argentina
[2] Departamento Ciencias de la Atmósfera y los Océanos, Facultad de Ciencias Exactas y Naturales, Universidad de Buenos Aires, Ciudad Universitaria,
Ciudad Autónoma de Buenos Aires, Argentina
[3] Instituto de Geodesia y Geofísica Aplicadas, Facultad de Ingeniería,
Universidad de Buenos Aires, Ciudad Autónoma de Buenos Aires, Argentina
[4] CONICET, Consejo Nacional de Investigaciones Científicas y Técnicas,
Ciudad Autónoma de Buenos Aires, Argentina

Abstract

Nowadays several global tidal models, most of them freely available for the community, provide tidal sea level heights practically everywhere around the World Ocean, constituting powerful tools for scientists, oceanographers and users of different areas of geosciences. In addition, most of these global models are frequently used to provide tidal heights at the open boundaries of regional high resolution coastal models. This work is an evaluation of the performance of three of the most recognized global tidal models: FES2004 (Lyard et al., 2006), GOT 4.7, an update of GOT00 (Ray, 1999) and TPXO 7.2, an update of TPXO 6 (Egbert et al., 1994; Egbert and Erofeeva, 2002). The tidal harmonic constants from these models are compared to the ones obtained from ten sea level data series gathered at tidal stations located along the Argentine coast (from the Río de la Plata estuary to Tierra del Fuego) and from twenty data series of sea level heights obtained on the crossing of the satellite tracks from the Ocean Topography Experiment (Topex/Poseidon: T/P) on the Argentine continental shelf (in the period 1992–2002). Four semidiurnal constituents (M_2, S_2,

---

* Corresponding author

N_2 and K_2) and four diurnal ones (K_1, O_1, P_1 and Q_1) are studied and the corresponding harmonic vectorial differences of observed sea level and altimeter data series are presented and discussed. The main differences between harmonic amplitudes and phases obtained from observations and from global models are detected in coastal waters of the Río de la Plata (a shallow area characterized by a micro tidal regime). Such differences, in general, significantly decrease towards the south, where the tidal range is greater than 10 m and the tide is practically semidiurnal.

Keywords: Observed tide, Modeled tide, Argentine continental shelf, Harmonic constants.

Introduction

The advantage of having good knowledge about tidal dynamics and ultimately a validated model for tide prediction is considerable not only from the point of view of navigation but also for scientific purposes. Long-term sea level observations are not evenly distributed worldwide. The Argentine Continental Shelf is one of those regions where data are rather scarce. Direct measurements of sea level are sparse; most available measurements are from sites at shore-based ports. There are very few records longer than one year along the Argentine coast. The geographical distribution of all tidal measurements taken along the Argentine coastline, from the Río de la Plata to Cape Horn, is presented in the Fig.1 and their characteristics summarized in Table 1. Sea level records were obtained from two stations (Mar del Plata and Quequén) situated at the coast of Buenos Aires Province and, two others (Buenos Aires and Spar Brasileira, in the Río de la Plata estuary). The stations of Puerto Madryn, Comodoro Rivadavia, Puerto Deseado, Puerto San Julian and Punta Loyola are located along the Patagonian coastline. The tidal station of Caleta La Misión is situated at the coast of Tierra del Fuego (Fig.2). Consequently, the study of tidal dynamics on the inner Argentine Continental Shelf is not possible using these observations only.

A preliminary but complete view of the tidal propagation and amplification could be achieved using products obtained from global tidal numerical models or from altimeters. High resolution global tidal numerical models and the interpretation of satellite altimeter data are frequently used to obtain global features of the circulation and propagation of the main tidal constituents. Such global tidal products give very good results for the deep ocean but, in some cases, they do not have enough resolution to describe the dynamics in shelf areas close to the coast. Problems arise in some specific areas, mainly coastal and shallow waters and step bathymentry regions. This is the case of the coastal waters of the Argentine Continental Shelf.

The Argentine Continental Shelf extends in a broad north–south trend along 2,400 km between the Río de la Plata (35°30' S) and Cape Horn (~57° S) covering an area of 960,000 km^2. The shelf width increases in a north-to-south direction from 170 km at the Río de la Plata mouth to more than 800 km in southern Patagonia. The outer shelf edge (in transition to the slope) has a NNE–SSW direction between the Río de la Plata and 44° S, where it gradually changes to a N–S direction, and then acquires (at 50°S) a striking east direction toward Malvinas (Falkland) Islands, and after surrounding them it turns back again to the west reaching regions near the continent at 51.5° S. The depth of the shelf edge is variable between 115 and 240 m with a general (although irregular) trend of deepening to the south. The continental slope has an area of 700,000 km^2 reaching maximum depths of 3,200– 3,500 m, with a width between 50 and 270 km and an average slope gradient of 2° reaching up to 5°

and even more; the slope shows three major features: a steep upper slope, a gently sloping terrace between 800 and 1,200 m (Ewing Terrace) and a steep lower slope that extends down to the slope foot (Dragani et al., 2008).

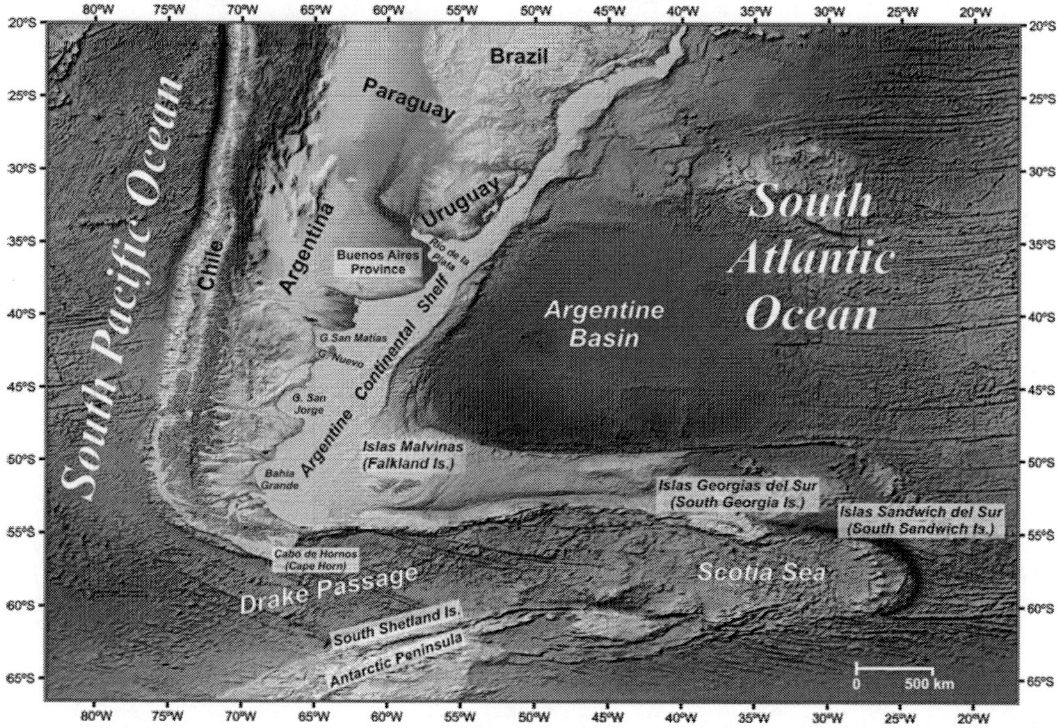

Figure 1. Southwestern Atlantic Ocean and the Argentine Continental Shelf. (Base map taken from: www.ngdc.noaa.gov/mgg/global/global.html).

Table 1. Locations of sea level gauges, duration of the series analyzed to obtain harmonic constants and type of device.

Location	Latitude	Longitude	Lapse	Hours	Type of device
Buenos Aires	34° 32' 42" S	58° 24' 00" W	01-01-82 / 12-31-00	166559	Floater
Spar Brasileira	35° 10' 00" S	56° 37' 00" W	02-01-03 / 01-31-04	8760	Pressure sensor
Mar del Plata	38° 00' 02" S	57° 32' 48" W	01-01-62 / 12-31-80	166559	Floater
Quequén	38° 34' 30" S	58° 42' 48" W	10-01-96 / 12-31-06	89855	Floater
Puerto Madryn	42° 45' 45" S	65° 01' 30" W	01-01-79 / 02-10-82	27273	Floater
Comodoro Rivadavia	45° 51' 38" S	67° 27' 49" W	04-07-80 / 04-30-81	9367	Floater
Puerto Deseado	47° 45' 13" S	65° 54' 53" W	01-01-73 / 12-31-73	8760	Floater
San Julián	49° 14' 50" S	67° 40' 40" W	06-16-88 / 08-03-88	1089	Pressure sensor
Punta Loyola	51° 36' 29" S	69° 00' 52" W	01-01-72 / 12-31-72	8760	Floater
Caleta La Misión	53° 42' 19" S	67° 49' 45" W	05-02-73 / 06-01-73	721	Tide pole

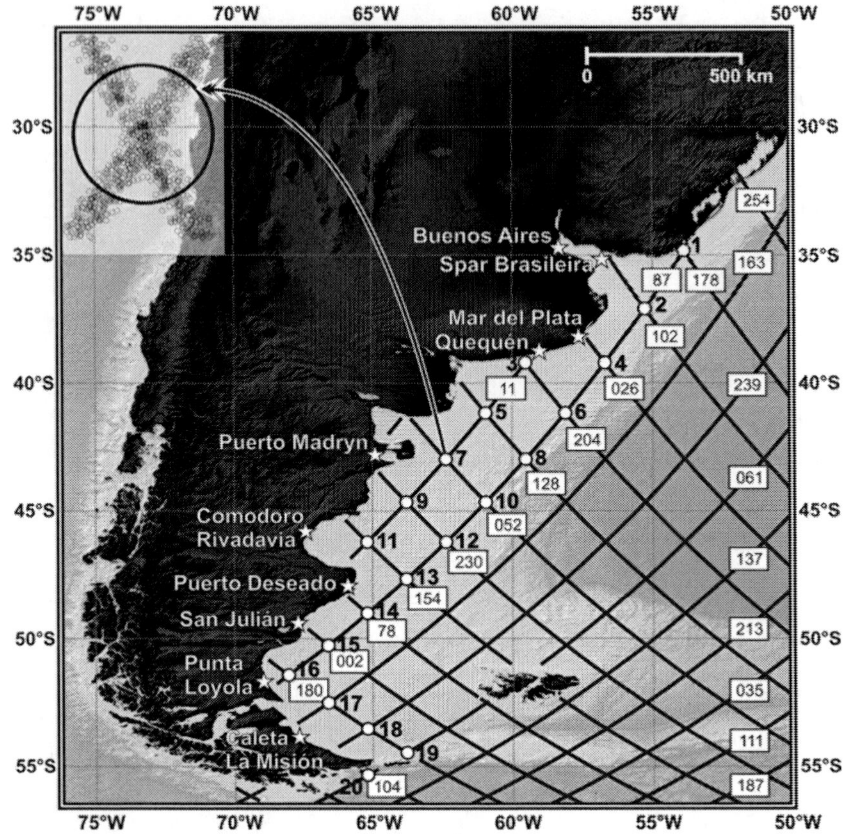

Figure 2. Map of the study area showing locations of the tidal gauge stations and the crossing tracks (points ranging from 1 to 20) used for the comparison between observed and modeled harmonic constants. The geographical position of the locations can be found in Table 1. Satellite tracks and the corresponding numeration are also included. A detail of a crossing of the satellite tracks can be appreciated in the upper left. (Base map taken from: www.ngdc.noaa.gov/mgg/global/global.html).

The amplification and propagation of the tide in the Argentine Continental Shelf has been studied by analysis of measurements (Balay, 1956) and by means of some validated regional numerical models (Glorioso and Simpson, 1994, Glorioso and Flather, 1995, 1997, Glorioso, 2000, Simionato et al., 2004 and Palma et al., 2004). Tidal waves propagate northward with amplitudes growing towards the coast along the Southeastern Continental Shelf of South America. The highest amplitudes occur at around 50° S. Amphidromic points have been reported, from south to north, southward Malvinas/Falkland Islands (at around 52° S), southeastward San Jorge Gulf (at approximately 48° S) and between San Matías Gulf and Bahía Blanca (near 40° S). Glorioso and Flather (1995) argued that one more amphidromic point at the Río de la Plata could be located near Punta del Este but, O'Connor (1991) presented theoretical arguments supporting that amphidrome should not occur.

Nowadays several validated global tidal models provide tidal sea level heights practically everywhere around the World Ocean, constituting powerful tools for scientists, oceanographers and users of different areas of Geosciences. These models present a very high performance at deep or open water but, close to the coast, in shallow waters, they show significant differences. In this work, the performance of three of the most recognized global tidal models, FES2004 (Lyard et al., 2006), GOT 4.7, an update of GOT00 (Ray, 1999) and TPXO 7.2, an update of TPXO6 (Egbert et al., 1994; Egbert and Erofeeva, 2002) are evaluated by comparing them to the tidal harmonic constants obtained from ten sea level data series gathered at tidal stations located along the Argentine coast (from the Río de la Plata estuary to Tierra del Fuego) and from twenty data series of sea level heights obtained on the crossing of the satellite tracks from the Ocean Topography Experiment (Topex/Poseidon: T/P) at the Argentine Continental Shelf in the period 1992-2002.

Data

Sea level data described in this section were used to compute the harmonic constants presented in this chapter. Tidal records were measured at standard tidal stations at seven locations, where sea levels were gathered by basic tide gauges with a floater and counterweight inside a vertical tube (UNESCO, 1985). Harmonic constants at Caleta La Misión were calculated from hourly sea level data taken from tide pole measurements. Pressure sensors, installed in Spar Brasileira and Puerto San Julián, were programmed with an interval sampling of 30 minutes. The geographical distribution of locations presented in Table 1 is shown in Fig. 2.

Launched in 1992, TOPEX/Poseidon was a joint satellite mission between NASA, the U.S. Space Agency, and CNES, the French Space Agency, to map ocean surface topography. Sea level data obtained from the Ocean Topography Experiment were also explored in this work (http://www.jpl.nasa.gov/missions/missiondetails.cfm?mission=Topex).

Available altimetry sea level data (September, 1992 – August, 2002) were gathered on circles (2.5 km radius) centered at the crossing of the satellite tracks, (one cycle every 9.915642 days, and, approximately, 550 observations per decade). The position of each crossing of tracks where sea levels were obtained (indicated from 1 to 20) and the numbers of each satellite track are included in Fig. 2. Crossings of satellite track areas (instead of single tracks) were chosen in order to have a better sampling (two values every cycle) and, consequently, a greater amount of sea level data. A detail of a crossing satellite tracks showing the circle of 2.5 km radius and the locations of the individual satellite observations are shown in the upper left Fig. 2.

Sea level data gathered at tidal stations and from altimeters were processed according to standard techniques based on the least square method (see, for example, Foreman, 1977) in order to obtain the tidal harmonic constants. Four semidiurnal constituents (M_2, S_2, N_2 and K_2) and four diurnal ones (K_1, O_1, P_1 and Q_1) were computed and compared in this work. Fang et al. (2004) showed that among these eight constituents the longest Topex-Poseidon alias synodic period is 9.317 years (between P_1 and K_2). Therefore the data length in the present study (ten years) is sufficient for resolving all the aforementioned constituents.

Global Tidal Models

There are several global tidal models which are freely available for download over the internet and have been extensively validated against altimetric, and coastal and pelagic tide gauge data (Shum et al., 1997; Zahran et al., 2006). The models typically perform well, and are very similar to one another (within a few cm) in the deep ocean, but modeled sea level hights can be very different in coastal and shallow shelf regions (maximum differences of over 1 m).(http://amcg.ese.ic.ac.uk/index.php?title=Local:Global_Tidal_Models).

The performance of three of the most recognized global tidal models, FES2004 (Lyard et al., 2006), GOT 4.7, an update of GOT00 (Ray, 1999) and TPXO 7.2, an update of TPXO 6 (Egbert et al., 1994; Egbert and Erofeeva, 2002) is evaluated by comparing the tidal harmonic constants to the ones computed from ten long term sea level data series gathered at tidal stations and from twenty data series of sea level heights obtained on the crossing of the satellite tracks from the Ocean Topography Experiment (Topex/Poseidon: T/P) at the Argentine Continental Shelf in the period 1992-2002.

The Goddard Ocean Tide (GOT) model (Ray, 1999) is based on the analysis of altimetry data to extract various tidal signals. This model followed the TOPEX/POSEIDON (T/P) mission which led to the development of many global tidal models. The inverse model of Egbert et al. (1994), TPXO, and the more recent French tidal group models (FES02, FES2004) are dynamical models with assimilation from observed tidal data (altimeter, and coastal and pelagic tide gauges) to improve their solution (Lefevre et al., 2002; Lyard et al., 2006). In this work, GOT 4.7, TPXO 7.2 and FES2004 are named as GOT, TPX and FES, respectively.

Harmonic constants of each global tidal model are regularly distributed on a spatial grid on the World Ocean. Amplitudes and phases corresponding to the eight constituents studied were spatially interpolated using the inverse distance method in order to obtain the values in the positions corresponding to the tidal stations and the crossing of each satellite track. In general, the nearest four nodes of the grid were used, but, in some cases, very close to the coast, only two or three nodes were used because some of the nearest nodes were located on land.

Results

Relative errors (%) between observed (*Ho*) and modeled amplitudes (*Hm*), computed as 100(*Ho*-*Hm*)/*Ho*, and temporal differences (h) between observed (*Go*) and modeled (*Gm*) Greenwich phases, given by $(Go\text{-}Gm)/\sigma_C$ (σ_C: theoretical speed of the wave or celerity given in °/h, see, for example, Schureman, 1988) for GOT, FES and TPX models at the locations shown in Fig. 2 are presented in Tables 2 to 5. For reasons of space, only the tables corresponding to the main semidiurnal constituents (M_2 and S_2) and the main diurnal constituents (K_1 and O_1) are presented in this chapter. Cells (Tables 1 to 5) containing relative errors greater than ±10% for amplitudes and temporal differences greater than ±0.5 h (±1 h) for semidiurnal (diurnal) harmonic constituents are highlighted in gray and cells containing maximum differences between modeled and observed harmonic constants for each model are highlighted in black.

Table 2 presents a comparison of amplitudes and phases corresponding to the M_2 constituent (the most important constituent, especially on the Patagonian Continental Shelf). In general, the three analyzed models are in good agreement with the observations. It can be seen that modeled amplitudes are slightly overestimated in the crossing of satellite tracks (Point 1 to 20) but they are a little underestimated in the tidal stations located on the coast. The best agreement between observed and modeled amplitudes corresponds to the GOT model for M_2 constituent. Three models present significant differences in the upper Río de la Plata estuary (Buenos Aires).

Table 2. Observed M_2 harmonic amplitudes and phases. Relative errors (%) and differences (h) between observed and modeled amplitudes and phases for GOT, FES and TPX models are shown. Cells containing relative errors greater than ±10% and temporal differences greater than ±0.5 h are highlighted in gray; cells containing the maximum differences are highlighted in black.

Station	Observed		Rel. error GOT 4.7 H (%)	Temp. diff. GOT 4.7 G (h)	Rel. error FES 2004 H (%)	Temp. diff. FES 2004 G (h)	Rel. error TPXO7.2 H (%)	Temp. diff. TPXO7.2 G (h)
	H_O (cm)	G_O (°)						
Buenos Aires	26.8	303.40	11.5	1.70	-44.0	3.53	32.8	3.44
Spar Brasileira	24.8	114.50	0.0	0.23	61.4	0.09	53.4	1.51
Mar del Plata	37.2	302.20	0.4	0.06	7.7	0.12	-10.2	0.08
Quequén	46.4	279.70	-0.1	0.08	-3.5	0.14	-7.1	0.14
Puerto Madryn	184.5	320.62	1.6	-0.06	-20.1	-0.32	-89.8	-1.05
Comodoro Rivadavia	193.9	227.95	0.9	0.26	2.3	0.27	4.0	0.22
Puerto Deseado	178.8	138.85	13.5	0.16	4.9	0.16	-3.9	0.73
San Julián	281.7	75.40	4.8	0.14	8.6	0.02	5.2	0.17
Punta Loyola	373.4	32.25	11.1	0.28	5.8	0.15	9.6	0.22
Caleta La Misión	286.8	341.65	2.9	-0.09	1.1	-0.03	-4.2	-0.11
Point 1	9.4	297.21	-10.9	-0.11	-2.8	0.26	-1.3	-0.16
Point 2	16.7	308.46	-5.0	-0.06	-0.6	0.00	-4.6	-0.27
Point 3	48.9	258.95	-7.3	-0.08	-3.4	0.03	-2.9	0.08
Point 4	25.8	287.93	-5.8	-0.14	-1.0	-0.04	-6.7	-0.19
Point 5	12.3	104.69	-6.5	-0.14	-27.2	0.12	-1.9	-0.71
Point 6	21.0	278.11	-7.3	-0.23	-2.6	-0.19	-16.6	-0.34
Point 7	68.7	345.04	-6.4	-0.40	4.2	-0.48	-7.7	-0.22
Point 8	20.8	299.70	-8.5	-0.20	6.9	-0.19	-14.5	-0.22
Point 9	92.6	269.75	-3.6	-0.12	5.0	-0.04	-4.9	-0.14
Point 10	35.9	289.05	-4.5	-0.10	18.9	-0.05	-1.7	-0.10
Point 11	116.6	213.32	-3.7	-0.14	-3.6	-0.20	-5.1	-0.21
Point 12	34.5	240.36	-6.6	-0.20	-8.0	-0.02	-9.4	-0.20
Point 13	47.8	135.35	-8.7	-0.04	-6.8	-0.40	-3.0	-0.17
Point 14	141.3	75.86	-6.4	0.03	1.5	-0.18	-2.8	-0.06
Point 15	226.2	48.58	-4.9	-0.08	-2.3	-0.10	-3.2	-0.03
Point 16	294.3	23.79	-5.9	-0.18	-7.9	-0.20	-4.8	-0.17
Point 17	214.2	1.08	-3.9	-0.08	-7.7	-0.19	-2.0	-0.11
Point 18	136.1	332.88	-4.9	-0.04	-7.1	-0.19	-3.3	-0.13
Point 19	72.9	293.65	-7.1	-0.42	-4.2	-0.16	0.0	-0.38
Point 20	46.6	253.89	-4.9	-0.25	-5.0	-0.14	-14.2	-0.18

The agreement between observed and modeled amplitudes corresponding to the S_2 constituent (Table 3) is also reasonably good for GOT model. Maximum relative error (-160.9%) can be appreciated at Buenos Aires. Two coastal tidal stations and three crossing satellite tracks present temporal differences greater than 0.5 h. The highest difference at the coast is detected at Buenos Aires (1.35 h) and, in the inner continental shelf, at Point 1 (13.5 h). Amplitudes modeled with FES present five coastal locations and fifteen crossing satellite tracks where relative errors in amplitudes are higher than 10% and two coastal locations and three crossing satellite tracks where the temporal differences are higher than 0.5 h. Maximum differences are detected at the upper Río de la Plata (Buenos Aires, -84.1% and 3.40 h). Amplitudes modeled with TPX present three coastal locations and six crossing satellite tracks with differences greater than 10%. TPX model presented conspicuous relative errors at Buenos Aires and Spar Brasileira (approximately 45%). Amplitude errors are also significant at Punta Loyola and Points 2, 5, 7, 8, 11 and 12 (with relative errors ranging from 10 to 25%). Observed and modeled phases are in very good agreement except at Buenos Aires (-5.98 h), Spar Barsileira (4.68 h), Puerto Madryn (-1.77 h) and Point 5 (1.36 h).

Table 3. Observed S_2 harmonic amplitudes and phases. Relative errors (%) and differences (h) between observed and modeled amplitudes and phases for GOT, FES and TPX models are shown. Cells containing relative errors greater than ±10% and temporal differences greater than ±0.5 h are highlighted in gray; cells containing the maximum differences are highlighted in black.

Station	Observed		Rel. error GOT 4.7 H (%)	Temp. diff. GOT 4.7 G (h)	Rel. error FES 2004 H (%)	Temp. diff. FES 2004 G (h)	Rel. error TPXO7.2 H (%)	Temp. diff. TPXO7.2 G (h)
	H_O (cm)	G_O (°)						
Buenos Aires	4.2	40.60	-160.9	1.35	-84.1	3.40	45.0	-5.98
Spar Brasileira	3.9	202.00	3.4	-1.17	56.2	0.39	45.1	4.68
Mar del Plata	6.6	8.10	-21.8	-0.25	-12.6	-0.37	9.4	0.38
Quequén	7.6	7.90	-4.6	-0.32	-51.4	0.27	-1.6	0.37
Puerto Madryn	42.9	33.69	-3.3	-0.09	-26.6	-0.11	-8.1	-1.77
Comodoro Rivadavia	35.8	315.60	-5.1	0.28	-3.7	0.31	-1.1	0.44
Puerto Deseado	32.9	198.10	13.4	0.39	2.6	0.58	-9.0	0.88
San Julián	67.7	108.56	3.0	-0.34	2.9	-0.40	8.1	-0.41
Punta Loyola	100.7	87.90	16.3	0.40	4.0	0.37	14.8	0.37
Caleta La Misión	81.5	31.40	16.8	-0.10	6.1	-0.01	7.0	-0.24
Point 1	1.5	54.83	-8.3	1.40	-39.0	-1.15	5.8	0.08
Point 2	2.8	353.16	-38.4	-0.31	5.5	-1.36	-21.5	-0.26
Point 3	9.2	359.00	18.1	-0.29	-38.7	0.36	9.6	0.05
Point 4	5.1	357.37	3.3	0.57	-13.8	-0.02	-3.7	0.31
Point 5	2.6	35.89	6.6	-0.41	-77.9	0.92	24.3	1.36
Point 6	5.5	346.50	-4.7	0.12	-20.3	0.13	-2.0	-0.06
Point 7	16.7	47.26	1.7	-0.45	-11.8	0.41	-16.0	-0.29
Point 8	5.0	339.83	-8.8	-0.22	-35.7	-0.30	-11.8	-0.57
Point 9	23.3	351.32	-0.9	-0.08	-10.9	0.10	-5.3	-0.03
Point 10	10.1	350.43	7.8	-0.09	-11.8	0.04	3.0	-0.12
Point 11	22.5	302.87	-9.4	-0.11	-9.1	-0.22	-12.0	-0.11
Point 12	10.2	335.23	-2.0	0.11	-11.8	0.14	-14.0	0.21
Point 13	3.7	176.24	-45.6	-1.14	13.5	-0.43	-2.7	-0.76
Point 14	30.7	122.22	-10.8	0.05	-10.3	0.07	-6.0	-0.03
Point 15	57.4	98.55	-2.2	-0.09	-9.6	0.01	-2.3	-0.03
Point 16	76.6	74.88	-3.1	-0.19	-14.7	-0.12	-3.0	-0.18
Point 17	55.7	53.39	-1.8	-0.06	-13.9	-0.06	-0.1	-0.09
Point 18	35.5	22.64	0.1	-0.03	-11.8	-0.12	-0.3	-0.14
Point 19	18.4	342.03	-3.4	-0.37	-8.0	-0.12	-1.1	-0.36
Point 20	10.2	299.00	-1.4	-0.10	-5.3	-0.25	-8.3	-0.24

Table 4. Observed K_1 harmonic amplitudes and phases. Relative errors (%) and differences (h) between observed and modeled amplitudes and phases for GOT, FES and TPX models are shown. Cells containing relative errors greater than ±10% and temporal differences greater than ±1.0 h are highlighted in gray; cells containing the maximum differences are highlighted in black.

Station	Observed		Rel. error GOT 4.7 H (%)	Temp. diff. GOT 4.7 G (h)	Rel. error FES 2004 H (%)	Temp. diff. FES 2004 G (h)	Rel. error TPXO7.2 H (%)	Temp. diff. TPXO7.2 G (h)
	H_O (cm)	G_O (°)						
Buenos Aires	8.2	85.00	**-23.1**	**3.63**	**38.0**	**10.80**	**52.9**	**3.78**
Spar Brasileira	4.9	345.40	-8.4	0.86	**32.7**	**5.40**	**-26.9**	0.38
Mar del Plata	14.8	172.80	3.7	0.41	7.8	0.02	**23.9**	0.93
Quequén	15.5	125.30	**13.9**	0.23	2.7	-0.46	**14.3**	-0.45
Puerto Madryn	19.1	309.53	0.2	-0.16	-3.7	0.08	**-27.8**	-0.69
Comodoro Rivadavia	19.6	253.02	-0.7	0.25	-4.2	0.51	**10.3**	0.25
Puerto Deseado	23.5	206.32	8.3	0.15	-5.1	0.17	4.6	0.71
San Julián	24.8	173.84	**10.5**	0.00	-4.7	-0.19	5.4	0.15
Punta Loyola	26.3	146.12	6.2	0.25	-1.1	0.25	3.2	0.30
Caleta La Misión	22.4	122.82	-4.1	0.55	-4.9	0.62	-3.6	0.44
Point 1	6.2	129.25	**-10.8**	0.14	**11.0**	**-1.39**	**25.4**	**-1.03**
Point 2	6.5	154.73	7.5	-0.45	10.0	-0.75	**10.2**	-0.47
Point 3	16.0	110.23	**13.3**	0.32	**17.4**	-0.28	**14.0**	0.01
Point 4	7.7	146.85	-3.7	0.22	4.0	-0.25	0.4	-0.14
Point 5	10.7	60.46	5.3	0.20	**21.6**	0.14	1.3	-0.01
Point 6	6.0	129.44	7.0	0.08	3.8	-0.52	4.5	0.17
Point 7	9.0	342.06	**-13.8**	-0.09	**17.7**	0.80	-3.9	0.32
Point 8	5.8	136.22	**17.0**	0.54	**11.3**	-0.36	**24.2**	0.46
Point 9	10.2	277.77	0.2	-0.12	-8.3	0.30	2.9	-0.22
Point 10	2.5	154.95	**-13.1**	0.37	**-49.2**	**-1.16**	-9.4	0.72
Point 11	13.7	240.05	**-10.5**	-0.46	**-16.4**	-0.19	-2.1	-0.47
Point 12	5.2	202.96	**21.1**	0.42	7.3	-0.51	**24.9**	-0.47
Point 13	12.0	201.91	**12.3**	0.00	-2.1	-0.40	7.3	0.06
Point 14	18.6	179.91	**12.1**	0.06	-0.3	-0.37	**10.8**	-0.03
Point 15	20.8	162.24	-3.1	-0.10	**-14.4**	-0.24	0.0	-0.01
Point 16	25.6	147.28	5.3	0.00	0.9	0.06	5.6	0.13
Point 17	21.3	133.76	2.7	0.17	-0.9	-0.02	5.2	0.12
Point 18	19.3	116.39	2.9	0.04	-0.7	-0.17	5.4	-0.18
Point 19	18.6	101.88	**10.8**	-0.35	5.3	-0.09	5.4	-0.29
Point 20	17.0	102.13	3.5	0.06	-1.6	0.13	3.4	0.14

The agreement between observed and modeled amplitudes corresponding to the K_1 constituent (Table 4) is good for GOT model. Larger relative errors can be seen at three coastal locations (Buenos Aires, Quequén and San Julian) and at ten crossing satellite tracks. Regarding the phases, temporal differences are, in general, lower than 0.5 h, except at Buenos Aires, Spar Brasileira, Caleta La Misión and Point 8. FES model presents maximum relative errors in amplitudes at Buenos Aires and Spar Brasileira and significant errors at ten crossing satellite tracks in the continental shelf (maximum difference is detected at Point 10, -49.2%). Observed and modeled phases (FES) are in very good agreement except at Buenos Aires and Spar Brasileira, where the temporal differences are 10.80 h and 5.40 h, respectively. TPX model presents large relative errors practically in all tidal stations located along the Argentine coast, except the ones located southward of Puerto Deseado. Relative errors on the continental shelf are quite low, except at Point 1, 2, 3, 8, 12 and 14. Observed and modeled phases (TPX) are also in very good agreement except at Buenos Aires and Point 1, where the temporal differences are 3.78 h and -1.03 h respectively.

Table 5. Observed O_1 harmonic amplitudes and phases. Relative errors (%) and differences (h) between observed and modeled amplitudes and phases for GOT, FES and TPX models are shown. Cells containing relative errors greater than ±10% and temporal differences greater than ±1.0 h are highlighted in gray; cells containing the maximum differences are highlighted in black.

Station	Observed		Rel. error GOT 4.7 H (%)	Temp. diff. GOT 4.7 G (h)	Rel. error FES 2004 H (%)	Temp. diff. FES 2004 G (h)	Rel. error TPXO7.2 H (%)	Temp. diff. TPXO7.2 G (h)
	H_O (cm)	G_O (°)						
Buenos Aires	13.4	258.90	60.6	0.75	-15.6	5.47	50.2	-0.88
Spar Brasileira	12.3	171.60	8.5	0.32	10.9	0.11	24.2	-1.30
Mar del Plata	15.7	66.00	-18.6	-0.81	-5.0	-0.74	3.2	-0.75
Quequén	14.7	57.80	-5.2	0.30	-0.8	0.26	3.8	-0.03
Puerto Madryn	7.5	222.78	-15.9	0.47	-22.6	0.98	-79.3	-1.35
Comodoro Rivadavia	12.2	154.33	-0.6	0.17	5.1	0.52	1.6	0.22
Puerto Deseado	18.7	126.43	10.4	0.08	7.5	0.37	5.9	0.46
San Julián	20.5	104.70	-1.3	0.49	2.5	0.27	3.6	0.20
Punta Loyola	21.1	82.73	-2.6	0.57	-1.4	0.44	-4.5	0.24
Caleta La Misión	20.2	51.43	-1.6	-0.09	0.6	-0.41	0.6	-0.47
Point 1	4.6	87.69	-55.2	0.29	-64.8	0.59	-33.0	-1.47
Point 2	11.0	99.72	-1.4	0.69	-0.1	0.83	-2.9	0.52
Point 3	13.9	37.69	-2.5	-0.31	6.8	-0.64	11.7	-0.66
Point 4	12.5	78.52	0.5	0.15	6.5	0.14	3.8	0.03
Point 5	7.7	15.76	1.6	-0.14	3.1	-0.37	14.5	0.40
Point 6	10.6	76.95	-1.7	0.37	-0.3	0.36	0.9	0.36
Point 7	2.2	295.63	-8.0	-0.88	84.6	2.79	12.2	0.02
Point 8	10.6	72.05	-2.3	0.05	-0.6	-0.02	5.6	-0.21
Point 9	4.6	159.65	-13.1	-0.48	-23.4	0.35	-7.2	-0.65
Point 10	8.3	76.79	-9.4	0.09	-7.6	-0.29	-9.1	-0.07
Point 11	10.2	149.22	5.5	0.16	3.7	0.63	2.7	0.30
Point 12	8.1	88.23	4.1	0.00	7.1	-0.20	4.3	-0.24
Point 13	10.8	105.86	0.3	-0.25	1.8	-0.26	-2.0	-0.21
Point 14	15.9	95.69	5.5	-0.59	3.5	-0.58	8.1	-0.57
Point 15	22.2	94.84	16.2	0.14	13.6	0.18	17.2	0.13
Point 16	22.0	79.92	4.6	0.14	6.7	0.04	4.4	-0.16
Point 17	19.5	74.83	4.0	0.53	1.6	0.32	4.9	0.34
Point 18	19.8	63.65	5.9	0.25	6.1	0.17	8.4	0.18
Point 19	17.2	53.08	1.3	-0.13	3.4	0.06	-0.3	0.07
Point 20	16.6	52.44	3.0	0.05	2.8	0.04	5.2	0.03

A comparison between observed and modeled amplitudes corresponding to the O_1 constituent is presented in Table 5. For GOT model, the highest relative errors can be appreciated at Buenos Aires (60.6%) and Point 1 (-55.2%). Significant errors can be also seen at Mar del Plata, Puerto Madryn, Puerto Deseado and at the Points 9 and 15. With regard to the phases, temporal differences are, in general, lower than 0.5 h, except at Buenos Aires, Mar del Plata, Punta Loyola, Point 7 and Point 14. FES model presented maximum relative errors at Buenos Aires, Spar Brasileira, Puerto Madryn and Points 1, 7, 9 and 15. Observed and model phases are in very good agreement except in Buenos Aires (5.47 h) and Point 7 (2.79 h). TPX model presents maximum relative errors at Buenos Aires, Spar Brasileira,

Puerto Madryn and Points 1, 3, 5, 7 and 15. Observed and modeled phases are in very good agreement except at Spar Brasileira, Puerto Madryn and Point 1.

Table 6. Number of cases where relative errors (amplitudes) and temporal differences (phases) are greater than ±10% and ±0.5 h (±1.0 h) for the N_2 and K_2 (P_1 and Q_1) constituents.

MODEL	N_2		K_2		P_1		Q_1	
	Hm rel error > 10%	Temp dif >0.5 h	Hm rel error > 10%	Temp dif >0.5 h	Hm rel error > 10%	Temp dif >1 h	Hm rel error > 10%	Temp dif >1 h
GOT	11	3	18	7	17	7	22	16
FES	11	4	20	11	21	15	22	13
TPX	14	4	10	11	19	11	26	17

Table 6 presents the number of cases (from a total of 30) where relative errors (amplitudes) and temporal differences (phases) are greater than ±10% and ±0.5 h (±1.0 h) for the N_2 and K_2 (P_1 and Q_1) constituents. It can be seen that the N_2 constituent is the best modeled. On the other hand, the K_2 and P_1 constituents present a similar performance and the Q_1 present the highest amount of differences. Results obtained in this study reveal that N_2, K_2, P_1 and Q_1 constituents are not recommended to be used in the Río de la Plata estuary.

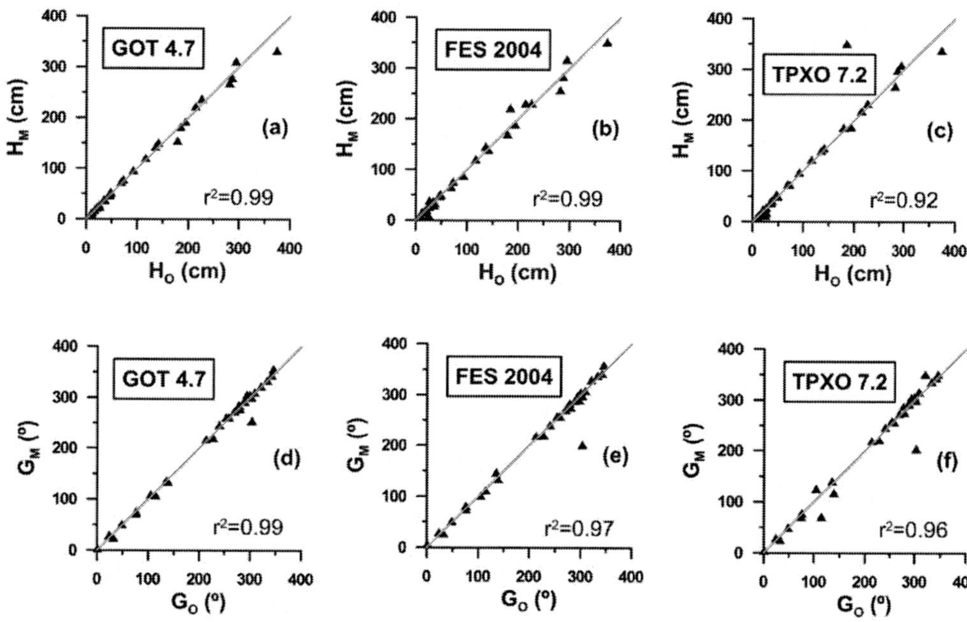

Figure 3. Comparison between observed and modeled amplitudes (*Ho* and *Hm*) and phases (*Go* and *Gm*) for the M_2 constituent. Upper panel: Scatter plots of modeled vs. observed amplitudes for GOT (left), FES (middle) and TPX (right) models; lower panel: Idem for phases; the full line indicates the perfect fit.

In order to simplify the interpretation of the results, Figs. 3 and 4 display the scatter plots of the modeled and observed values of amplitude and phase for the M_2 and O_1 tidal constituents (for reasons of space, only two constituents are shown in this work). The solid line, representing the ideal fit, and the linear determination coefficient (r^2) are included in each case. From this comparison it can be seen that GOT and FES present the best agreements between observed and modeled harmonic constants; GOT presents values of r^2 equal to 0.99 for amplitudes and phases and FES 0.99 and 0.97, respectively. On the other hand, TPX presents a relatively lower agreement for the amplitudes (0.92) and phases (0.96). Neither significant overestimation nor sub estimation is appreciated in the amplitudes and phase modeled with GOT, FES and TPX models. With regards to the O_1 constituent, the amplitudes obtained from GOT and TPX models present more dispersion (r^2 equal to 0.88 and 0.87, respectively) than the obtained from FES model (r^2 = 0.96). It can be noted that the amplitudes of the O_1 constituent (and of any diurnal constituent) are one magnitude of order lower than the amplitude of the M_2 constituent, especially on the Patagonian continental shelf.

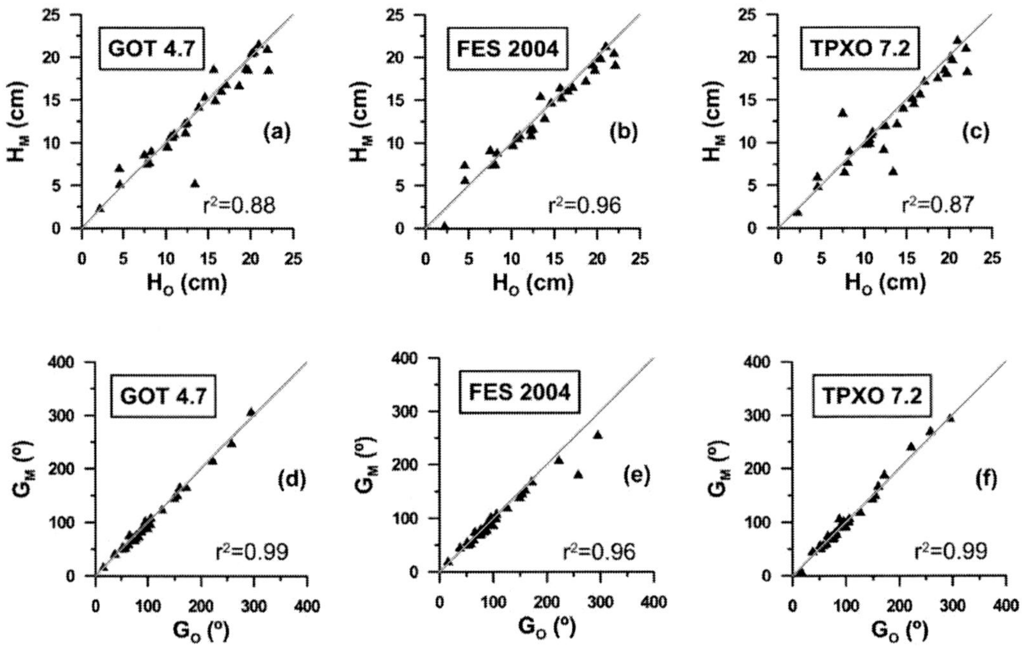

Figure 4. Comparison between observed and modeled amplitudes (Ho and Hm) and phases (Go and Gm) for the O_1 constituent. Upper panel: Scatter plots of modeled vs. observed amplitudes for GOT (left), FES (middle) and TPX (right) models; lower panel: Idem for phases; the full line indicates the perfect fit.

Table 7 presents the root mean square errors (*RMSE*) for the vectorial differences (cm), worked out as $\{[Ho\ \cos(Go) - Hm\ \cos(Gm)]^2 + [Ho\ \sin(Go) - Hm\ \sin(Gm)]^2\}^{1/2}$, corresponding to the eight analyzed constituents, considering all locations (30) listed in Table 1. *RMSE* gives a regional measure of the combined errors of amplitudes and phases. GOT and FES models present the lowest errors (17.6 and 19.7 cm, respectively) and the error associated to TPX model is more than double the error of GOT and FES models (44.0 cm) for the M_2 constituent. For the S_2 constituent GOT and FES models also present very similar vectorial differences (Erms are 6.6 and 6.7 cm, respectively) but the value associated to TPX model is larger than the one associated to GOT and FES models (10.1 cm). For the K_1 and O_1 constituents, GOT, FES and TPX models present very similar vectorial differences, *RMSE* are

2.2, 3.1, 2.6 cm and 2.3, 3.7 and 2.5 cm, respectively. Vectorial differences for the Q_1, P_1 and K_2 constituents are, in general, lower than 2-3 cm (Table 7) and, for the N_2 constituent, they are a little larger: 5-7 cm for GOT and FES models and 11.1 cm for TPX model.

Table 7. *RMSE* of the modeled vectorial differences with respect to the observed vectorial difference. Harmonic constants corresponding to the locations presented in Table 1 were used. *ERMS* corresponding to the eight analyzed harmonic constituents for GOT, FES and TPX models are shown.

Constituent	RMSE GOT4.7 (cm)	RMSE FES2004 (cm)	RMSE TPXO7.2 (cm)
Q_1	1.4	1.5	1.5
O_1	2.3	3.7	2.5
P_1	1.7	2.1	1.9
K_1	2.2	3.1	2.6
N_2	5.7	6.5	11.1
M_2	17.6	19.7	44.0
S_2	6.6	6.7	10.1
K_2	2.0	2.8	3.7

Discussion and Conclusion

Even though the performance of GOT, FES and TPX models is reasonably good in the Argentine continental shelf, the results obtained in this study reveal that there are some specific regions where they can not be reliably applied. Amplitudes and phases obtained with each of the three analyzed models give significant differences in the Río de la Plata estuary, especially when they are compared with the harmonic constants corresponding to Buenos Aires. The agreement is a little better at Spar Brasileira.

At the coast of Buenos Aires Province (Mar del Plata and Quequén), where the diurnal and semidiurnal amplitudes are of the same order of magnitude, the three global models work reasonably well showing some amplitude and phase differences for each constituent and, consequently, it is hard to determine which of the three models present the best performance. On the other hand, the tide is practically regulated by the M_2 constituent along the Patagonian Coast. Consequently, from the analysis of Table 2, it can be easily concluded that the three global models present a very good performance along the Patagonian coast. Nevertheless, it should be highlighted that the GOT model is the single model that satisfactorily works at Puerto Madryn (located inside of Nuevo Gulf). GOT model presents an excellent agreement at the points located off-shore (Points 6, 8, 10 and 12) especially in the semidiurnal constituents.

Finally, even though the three analyzed models could be used for the prediction of sea levels at the inner continental shelf of Argentine, from the results raised in this work, GOT 4.7 model is the most recommended, except in the upper Río de la Plata (Buenos Aires). On the other hand, it is necessary to highlight that all the analyzed models could be used satisfactorily as boundary conditions to force regional or local high-resolution models, which nowadays are mainly implemented to simulate not only sea levels but also currents at coves, small gulfs, bays and in estuaries or inlets located along the Argentine continental shelf.

Acknowledgments

This work is a contribution to the PICT200532606, UBACyT I014 and CONICET PIP112-200801-02599 projects.

References

Balay, M. A., 1956. Determination of mean sea level of Argentine Sea. Influences of oscillations of the sea not caused by tides. *International Hydrographic Review (Monaco),* 33(2), 31–65.

Dragani, W., D'Onofrio, E., Grismeyer, W., Fiore M., Violante, R. and Rovere, E., 2008. Vulnerability of the Atlantic Patagonian coast to tsunamis generated by submarine earthquakes located in the Scotia Arc region. Some numerical experiments. *Natural Hazards*, 49, 437-458. DOI: 10.1007/s11069-008-9289-4

Egbert, G., A. Bennett, and M. Foreman, 1994. TOPEX/POSEIDON tides estimated using a global inverse model. *Journal of Geophysical Research*, 99(C12):24821–24852

Egbert G.D. and S.Y. Erofeeva, 2002. Efficient inverse modeling of barotropic ocean tides, *Journal of Atmospheric and Oceanic Technology*, 19(2), 183-204

Fang, G., Y. Wang, Z. Wei, B. H. Choi, X. Wang, and J. Wang, 2004. Empirical cotidal charts of the Bohai, Yellow, and East China Seas from 10 years of TOPEX/Poseidon altimetry, *J. Geophys. Res.*, 109, C11006, doi:10.1029/2004JC002484.

Foreman, M.G.G., 1977. *Manual for tidal heights analysis and prediction*. Pac. Mar. Sci. Rep. 77-10, Inst. of Ocean Sci., Patricia Bay, Sidney, B. C., Canada, 97 pp.

Glorioso P.D. and Simpson, J.H., 1994. Numerical modeling of the M_2 tide on the northern Patagonian shelf. *Continental Shelf Research*, 14, 267–278.

Glorioso P.D. and Flather, R.A., 1995. A barotropic model of the currents off SE South America. *Journal of Geophysical Research*, 100, 13427–13440.

Glorioso P.D. and Flather, R.A., 1997. The Patagonian Shelf tides. *Progress in Oceanography*, 40, 263–283.

Glorioso, P.D., 2000. Patagonian Shelf 3-D tide and surge model. *Journal of Marine Systems,* 24, 141–151.

Lefevre, F., F.H. Lyard, C. Le Provost, and E.J.O Schrama, 2002. FES99: a global tide finite element solution assimilating tide gauge and altimetric information. *Journal of Atmospheric and Oceanic Technology*, 19:1345–1356.

Lyard F., F. Lefevre, T. Letellier, and O. Francis, 2006. Modeling the global ocean tides: modern insights from FES2004. *Ocean Dynamics*, 56:394–415.

O'Connor, W. P., 1991. A numerical model of tides and storm surges in the Río de la Plata estuary. *Continental Shelf Research,* 11, 1491–1508.

Palma E.D., R.P. Matano y A.R. Piola, 2004. A numerical study of the Southwestern Atlantic Shelf circulation: Barotropic response to tidal and wind forcing, *J. Geophys. Res.*, 109, C08014, doi: 10.1029/2004JC002315.

Ray. R.D. , 1999. *A global ocean tide model from TOPEX/POSEIDON altimetry: GOT99. Technical Report NASA Tech. Mem. 209478*, Goddard Space Flight Centre, Greenbelt, MD, USA.

Schureman, P., 1988. *Manual of harmonic analysis and prediction of tides*. Washington, U.S. Government Printing Office, Special Publ. No. 98, 317 p.

Simionato, C., Dragani, W. C., Nuñez, M. N. y Engel, M., 2004. A set of 3-d nested models for tidal propagation from the Argentinean Continental Shelf to Río de la Plata Estuary, *Journal of Coastal Research*, (20) 3, 893-912

Shum, C. K., P. L. Woodworth, O. B. Andersen, G. D. Egbert, O. Francis, C. King, S. M. Klosko, C. Le Provost, X. Li, J-M. Molines, M. E. Parke, R. D. Ray, M.G. Schlax, D. Stammer, C. C. Tierney, P. Vincent, and C.I. Wunsch, 1997. Accuracy assessment of recent ocean tide models. *Journal of Geophysical Research*, 102 (C11):25,173–25,194.

UNESCO, 1985. Manual on sea level measurement and interpretation. *Intergovernmental Oceanographic Commission*, 83 p.

Zahran, K. H., G. Jentzsch, and G. Seeber, 2006. Accuracy assessment of ocean tide loading computations for precise geodetic observations. *Journal of Geodynamics,* 42:159–174.

In: Sea Level Rise, Coastal Engineering, Shorelines and Tides ISBN: 978-1-61728-655-1
Editor: Linda L. Wright, pp. 251-261 © 2011 Nova Science Publishers, Inc.

Chapter 12

A SIMULATION MODEL FOR CORAL REEF FORMATION: REEF TOPOGRAPHIES AND GROWTH PATTERNS RESPONDING TO RELATIVE SEA-LEVEL HISTORIES

Takashi Nakamura[1,*] and Toru Nakamori[2]
[1] Department of Mechanical and Environmental Informatics, Graduate School of Information Science and Engineering, Tokyo Institute of Technology, O-okayama 2-12-1, Meguro-ku, Tokyo 152-8552 Japan
[2] Institute of Geology and Paleontology, Graduate School of Science, Tohoku University, 6-3, Aoba, Aramaki, Aoba-ku, Sendai 980-8578, Japan

Abstract

Coral reef topographies and reef growth patterns are influenced from relative sea-level histories. Several types of reef growth patterns responding to the relative sea-level histories, e.g. balanced aggrading/onlapping, seaward prograding, back stepping, etc., have been identified in previous studies. Recently, Nakamura and Nakamori (*Coral Reefs* 2007, 26, 741–755) developed a geochemical model for coral reef formation based on diffusion-limited and light-enhanced calcification, and the model reconstructed well the reef topography and Holocene reef-growth history. In this paper, we modified the model, and simulated it on four scenarios of relative sea-level histories. The simulation result on the first scenario, which is similar with global sea level history between 8,000 years B.P. and present, is very similar with balanced aggrading/onlapping type of Holocene fringing reefs. The simulations on the scenarios of stable sea level and gradual sea level falling are well-reconstructed seaward prograding type reefs. The result of the simulation on the scenario of faster sea-level rising is similar with back-stepping type reefs. Therefore, the reef topographies and growth pattern responding to relative sea level histories simulated by the model were in general well-reconstructed concerning Holocene reefs observed in nature.

* E-mail address: takashishishi@gmail.com, TEL.: +81-3-5734-2949, FAX: +81-3-5734-2650. (Corresponding author)

Introduction

Coral reef topographies and reef growth patterns are influenced from relative sea-level histories. Several types of reef growth patterns are defined by Montaggioni (2005). In the fringing reefs, there are mainly three types of reef growth patterns. In the first type defined as "balanced aggradation/onlapping type", deposition within lagoons operated at rates close to those of adjacent reef rim allowing the different parts of the system to reach sea level at the same type. There was a fine balance between reef rim growth through aggregation and back-reef accumulation through onlapping. Pointe-au-Sable reef, Mauritius island (Montaggioni and Faure 1997), Kabira reef, Ishigaki island (Yamano et al 2001), Nishimezaki reef, Kume island (Takahashi et al. 1988), Hanauma reef, Oahu, Hawaiian island (Easton and Olson 1976) and Punta islotes reef, Costa Rica (Cortés et al. 1994) are examples of this model. Most of the reefs are characterized by both a convex "reef crest" and a concave "moat" (shallow lagoon) on the reef flat. The second type defined as "seaward prograding type" relate to reef systems that developed seaward by lateral accretion of the fore-reef. They started to grow in settings where vertical accommodation space was limited or missing. Such Holocene reefs developed mostly close to the shore when the sea level was around its present-day position. Fringing reef at Phuket (Tudhope and Scoffin 1994), Fantome island (Johnson and Risk 1987), and Hikauhi, Molokai island (Engels et al. 2004) are typical of seaward prograding systems. In the third type defined as "back stepping type", reef initiation occurred below modern sea level at depths where the rates of reef aggradation during the early Holocene times were not efficient enough to fill available accommodation space. The entire reef body therefore caught up to sea level through successive accretion centres stepping backward; it finally comprises a series of superimposed retrograding units. The Hale O Lono reef, Molokai island (Engels et al. 2004) is an example of this architecture.

Coral calcification strongly depends on the light intensity because of the symbiotic relationship between corals and photosynthetic zooxanthellae algae (Gattuso et al. 1999). This phenomenon is called "light-enhanced calcification" (Goreau and Goreau 1959; Barnes and Chalker 1990), and has been reported on the basis of observations carried out at different spatial scales; for example, a coral colony (Goreau 1959), a coral community (Suzuki et al. 1995), and a complete reef (Barnes and Devereux 1984). Bosscher and Schlager (1992) proposed a reef growth model based on the light-enhanced calcification. The model successfully simulates the growth and demise of framework-dominated Holocene reefs. Webster et al. (2007) also reconstructed the submerged reef development around Hawaii during the last two glacial cycles (250 kyr) by the model simulation. Overall, the light-enhanced growth model well-reconstructed long term and large-scale topography. So the model is useful for such large-scale problems, but the middle scale topography, such as reef crest and moat, could not be reconstructed only by the model.

Calcification of corals depends on not only light (or photosynthesis) but also CO_2 system parameters, such as CO_3^{2-} concentration or aragonite saturation state (Gattuso et al. 1998; Marubini and Thake 1999; Ohde and van Woesik 1999; Langdon et al. 2000; Leclercq et al. 2000, 2002; Ohde and Hossain 2004; Langdon and Atkinson 2005; Hossain and Ohde 2006). On the other hand, distributions of the light and the inorganic carbon species are also affected by reef topography and reef metabolisms. Efflux of carbon species by calcification is balanced with an influx from the ocean by advection and diffusion of carbonate species.

These geochemical settings suggest that reef crest might be formed by a morphogenesis-like mechanism originated from reaction-diffusion system. Reef growth mechanism is considered as the mix of two phenomena; i.e., light-enhanced calcification and reaction-diffusion of carbon species. Nakamura and Nakamori (2007) proposed a light-enhanced and diffusion-limited growth model for reef formation based on these mechanisms. This model well reconstructs the actual reef topography, which includes moat and crest structure. In this study, we modified this model and demonstrated in some geological setting and sea level histories.

Light-Enhanced and Diffusion-Limited Growth Model

Calcification Model

Coral calcification depends strongly on the photosynthesis (Goreau 1959; Barnes and Devereux 1984; Suzuki et al. 1995; Gattuso et al. 1999). In addition, coral calcification also depends on the aragonite saturation state (Gattuso et al. 1998; Marubini and Thake 1999; Ohde and van Woesik 1999; Langdon et al. 2000; Leclercq et al. 2000, 2002; Ohde and Hossain 2004; Langdon and Atkinson 2005; Hossain and Ohde 2006). The aragonite saturation state is almost proportional to concentration of CO_3^{2-} ($[CO_3^{2-}]$). Thus, it is expected that calcification rate is expressed by the function of photosynthesis rate and $[CO_3^{2-}]$, and the rate has to depend on the coral biomass. Therefore, the simple model is here proposed:

$$G = (a \cdot P \cdot [\mathrm{CO}_3^{\ 2-}] + b) \cdot M \ , \tag{1}$$

where G is coral calcification rate per unit area, P is photosynthesis rate per unit area, M is coral biomass per unit area, a and b are coefficients. Coral photosynthesis clearly depends on light intensity (or photon flux density) (Chalker 1981; Barnes and Devereux 1984), and the photosynthesis rate is also limited by HCO_3^-, because coral symbiotic unit as well as isolated zooxanthellae is HCO_3^--users (Burris et al. 1983; Al-Moghrabi et al. 1996; Goiran et al. 1996; Marubini et al. 2008). Therefore, coral P_n should depend on both photon flux density (F_P) and concentration of HCO_3^- ($[HCO_3^-]$). For simplicity's sake, P model were described by the linear function of $F_P \cdot [HCO_3^-]$. Therefore, the P in equation (1) can be replaced as following:

$$G = \{a \cdot (c \cdot F_P \cdot [\mathrm{HCO}_3^{\ -}] + d) \cdot [\mathrm{CO}_3^{\ 2-}] + b\} \cdot M \ , \tag{2}$$

where c and d are coefficients. Using map view coverage of corals (Cov; %), we expressed the function of $M = \mathrm{Cov}^{1.5}$, because coral coverage is two-dimensional projection but coral colons have spread three dimensionally. The dataset of G, F_P, $[HCO_3^-]$, $[CO_3^{2-}]$ and Cov by *in situ* incubation experiments is already provided by Nakamura and Nakamori (2007, 2009), and the unknown parameters are estimated by the nonlinear least squares fitting (listed in Table 1). The model is described well (Figure 1) with high coefficient of determination (r^2 = 0.88).

Table 1. Parameters used for computer simulations

Symbol	Parameter	Value	Unit
a	Constant in equation (2)	1.8×10^{-3}	kg-soln μmol^{-1}
b	Constant in equation (2)	0.23	mol m^{-2} yr^{-1}
c	Constant in equation (2)	2.4×10^{-8}	kg-soln μmol^{-1}
d	Constant in equation (2)	−0.27	mol m^{-2} yr^{-1}
A_{T0}	Initial total alkalinity*	2,200	μmol kg^{-1}
C_{T0}	Initial total dissolved inorganic carbon dioxide*	1,900	μmol kg^{-1}
F_{P0}	Photon flux density on sea surface*	2.5×10^4	mol m^{-2} yr^{-1}
α	Albedo of sea surface*	0.07	
λ	Extinction coefficient of seawater*	0.2	
D	Diffusion coefficient in equation (4)	0.1	m^2 yr^{-1}
K_x	Horizontal eddy diffusion coefficient	1.0×10^8	m^2 yr^{-1}
K_v	Vertical eddy diffusion coefficient	1.0×10^4	m^2 yr^{-1}

* These parameters are detailed in Nakamura and Nakamori (2007).

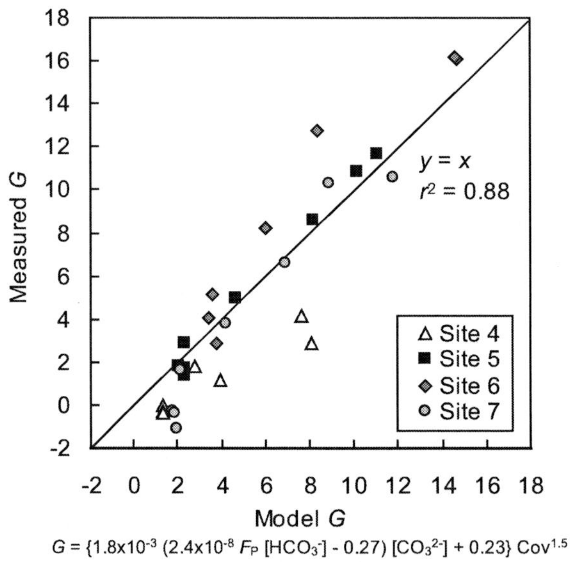

Figure 1. Comparison of measured calcification rate (G) with model G as the function of photon flux density (F_P), concentration of HCO_3^- ($[HCO_3^-]$), concentration of CO_3^{2-} ($[CO_3^{2-}]$) and coral coverage (Cov).

The F_P received by corals (in equation (2)) can be calculated from water depth (h) by Beer-Lambert low, and the $[HCO_3^-]$ and $[CO_3^{2-}]$ in equation (2) can be computed from total dissolved inorganic carbon dioxide (C_T), if some assumptions are made (detailed in Nakamura and Nakamori 2007). Modeling of Cov in equation (2) is difficult, because Cov is affected by damage from various sources, e.g., by bottom sedimentation, turbidity, and nutrients. But for sake of simplicity, the function of Cov is adopted, a constant value of 50 %:

Furthermore, corals tend to not inhabit an environment that is shallower than the low-tide level. So, the following condition was applied:

$$\mathrm{Cov} = \begin{cases} 0\,\%, & \text{if } h < 0.6\,\mathrm{m} \\ 50\,\%, & \text{otherwise} \end{cases}, \tag{3}$$

In conclusion, G as given by equation (2) can be written as a simple function of two variables, i.e. h and C_T, and the G values are disposed to decrease with increased h and decreased C_T.

Reef Profile

Depth of coral reef (h) is considered to influence various phenomena in a reef environment; for example, reef growth, sea-level change, and mass movement by erosion or deposition. When two-dimensional profile is supported, the changes in h can be expressed by the following equation:

$$\frac{\partial h}{\partial t} = -V_R + V_S + D\frac{\partial h}{\partial x^2}, \tag{4}$$

where x is a coordinate on the horizontal axis, V_R is the growth rate of the reef, V_S is the rate of relative sea-level rise, and D is a constant. The first term of right hand side represents a decrease in h as a result of upward reef growth. The second term represents an increase in h as a result of sea-level rise. The last term has the effect of surface smoothing, and represents transportation of carbonate mass by erosion and secondary. Scheidegger (1970) proposed this term by drawing an analogy between geomorphology and thermodynamics. The V_R of equation (4) can be derived from the variable G by using the following equation:

$$V_R = \frac{M}{\rho_R} G, \tag{5}$$

where M is molecular mass of $CaCO_3$ and ρ_R is the density of the bulk reef rock. Buddemeier and Smith (1988) suggested that the average value of ρ_R within reef systems is 1.4 g cm^{-3}.

Distribution of C_T

The distribution of C_T in seawater in a reef environment may exhibit a complicated pattern, caused mainly by tidal effects. Total fluxes due to ebb and flow are the same. It implies that the averaged advection should be zero. If considered over long periods (years), and at large scale (meters to kilometers), the tidal effects can be represented by a simple eddy, and approximated by eddy diffusion. The C_T distribution can then be determined using the following two-dimensional diffusion equation:

$$\frac{\partial C_{\mathrm{T}}}{\partial t} = K_x \frac{\partial C_{\mathrm{T}}}{\partial x^2} + K_y \frac{\partial C_{\mathrm{T}}}{\partial y^2}, \tag{6}$$

where x indicate horizontal position (right angle to the shore) and y indicate vertical positions, and K_x and K_y are eddy diffusion coefficients of x and y axis, respectively. The reef growth is much later than C_{T} diffusion, therefore time derivation term at the left hand side in equation (6) is negligible, and as a result, C_{T} satisfies the Laplace equation:

$$K_x \frac{\partial C_{\mathrm{T}}}{\partial x^2} + K_y \frac{\partial C_{\mathrm{T}}}{\partial y^2} = 0, \tag{7}$$

Calcification of reef communities alters C_{T}, such that when 1 mol of $CaCO_3$ is produced, the C_{T} decreases by 1 mol. So a calcification rate per unit area is just comparable to C_{T} flux through seawater-reef boundary. The boundary condition between seawater and coral reef ($y = h$) is given by Fick's first law of diffusion:

$$G = -\rho_{\mathrm{W}} K_y \frac{\partial C_{\mathrm{T}}}{\partial y}\bigg|_{y=h}, \tag{8}$$

where ρ_{W} is seawater density, and the G indicates the C_{T} flux as a result of calcification.

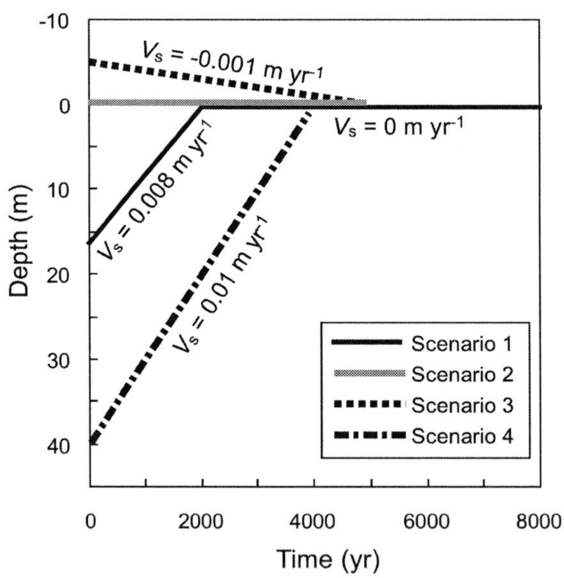

Figure 2. Scenarios of relative sea-level history used for simulations. Scenario 1: the rate of relative sea-level rise (V_S) is constructed by two phases; rising constantly at 0.008 m yr^{-1} for first 2,000 years, then stable (V_s = 0 m yr^{-1}) between 2,000 and 8,000 years. Scenario 2: Sea level is stable (V_s = 0 m yr^{-1}) for 5,000 years. Scenario 3: Sea level is falling constantly at 0.001 m yr^{-1} for 5,000 years. Scenario 4: Sea level is rising constantly at 0.01 m yr^{-1} for 4,000 years.

Simulations of Coral Reef Formation

Computer simulations were carried out by applying the finite difference method both on variables indicating space and time to the geochemical model for coral reef formation described above. Depth profiles given by equation (4) were calculated every 3 m along the *x*-axis ($0 \leq x \leq 1{,}800$ m) on two-dimensional field. For the sake of simplicity, the initial profile of h is represented by an inclined flat plane. The calculations of C_T distribution from equation (6) were carried out in two-dimensional space (1,800 × 45 m). Grid dimensions of 3 m horizontal (x) and 0.3 m vertical (y) on two-dimensional field was chosen to provide adequate resolution of reef topography. The reef–seawater boundary was mapped onto the lattice, and was updated for every output of topography in the simulation. The initial condition for C_T of seawater was set at a homogeneous concentration of 1,900 μmol kg^{-1}. Atmosphere–seawater interfaces was set zero flux boundary. The vertical component of the flux at the reef–seawater boundary was determined from equation (8). The Dirichlet condition ($C_T = 1{,}900$ μmol kg^{-1}) was employed at the boundary 1,800 m distant from the shore. The interval of every simulation step was set to 2.0 year. The values of other parameters used in the simulation are provided in Table 1.

Results and Discussion

The simulations were executed on four scenarios of sea-level history detailed in Figure 2. First exam is the evolution of the reef on Scenario 1. This scenario is similar with global sea-level history between 8,000 years B.P. and present (Bard et al. 1996). Snapshots of the topography and C_T distribution between $t = 0$ and $t = 8{,}000$ years are shown in Figure 3. The following process is evident in the series of simulations. The horizontal surface of the carbonate platform is constructed as sea level rises, and the C_T gradient from offshore to shore becomes steeper over the first 1,000 years. As a result, the near-shore growth rate of the reef falls, and crest-like convex structures develop at around 1,000 to 2,000 years. The seaward crest inhibits carbon supply to the near-shore part and the C_T gradient becomes progressively steeper. At the same time, the growth rate of the near-shore part decreases because of a lower supply of light and carbon. As a result, both the seaward crest and the moat are clearly defined. After 3,000 years, when sea level is stable, crest expands to the moat, and the moat also becoming shallow by accumulation. Then, fore-reef is progressing to offshore through 3,000 to 8,000 years. This evolution pattern is very similar to balanced aggrading/onlapping type (Montaggioni 2005) of Holocene fringing reefs. In this simulation, the moat is fully filled with sediment, but the moat of many actual reefs in this type is still present. Defective coverage modeling may cause this difference. Coral coverage is adopted constant value (Cov = 50 %) in this simulation, but actual coverage is affected by damage from various sources, e.g., by bottom sedimentation, turbidity, and nutrients. In fact, back-reef moat and deeper part of fore-reef are lower coverage because of sand and gravel sedimentation or some other factors. To construct the more realistic model, understanding of coverage dynamics and modifying the coverage model are important.

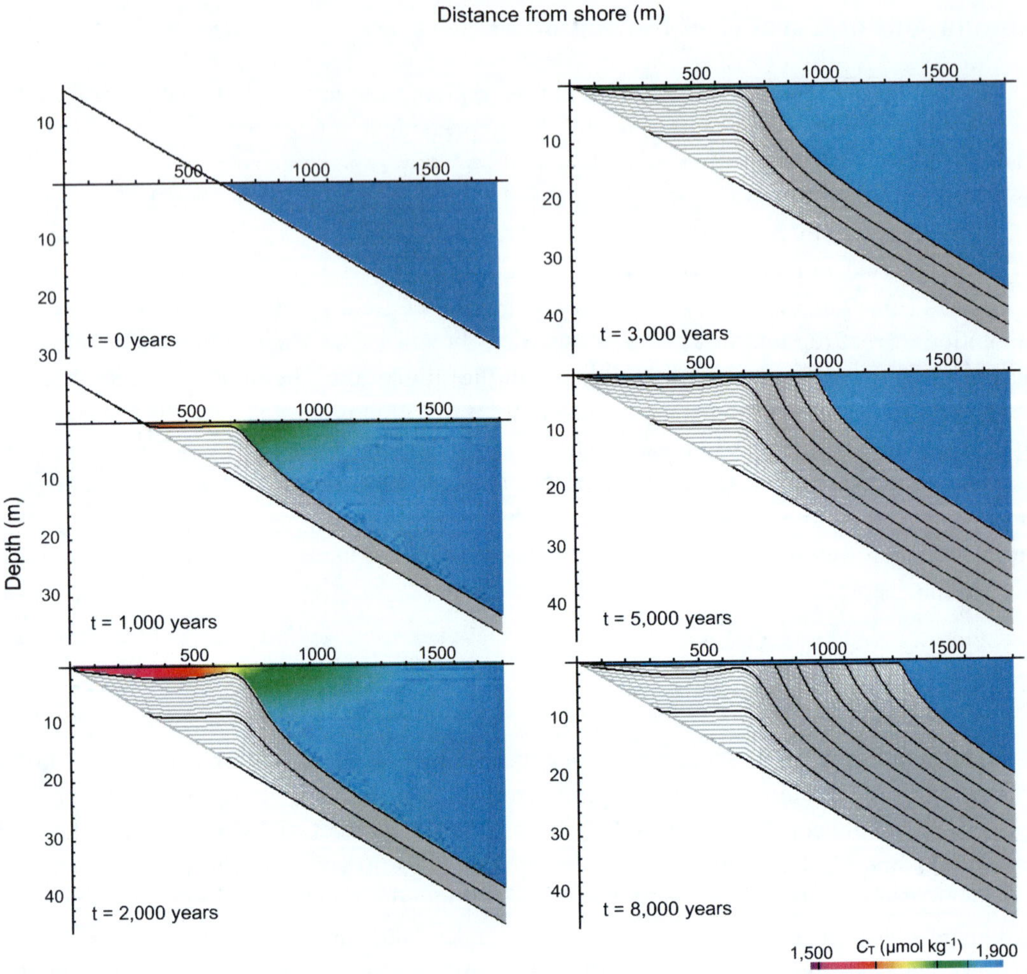

Figure 3. Evolutions of coral reef topography and concentration distributions of total inorganic carbon dioxide (C_T) from 0 to 8,000 years obtained from simulations on sea level Scenario 1.

We also examined the model simulations on other sea-level scenarios. The results of the simulations on Scenario 2 (sea level is stable ($V_s = 0$ m yr^{-1}) for 5,000 years) and Scenario 3 (sea level is falling constantly at 0.001 m yr^{-1} for 5,000 years) (Figure 2) are shown in Figure 4a and b, respectively. In these scenarios, reef cannot growth upper because of limitation by the sea level, and these growth patterns look like that the fore-reef is progressing to offshore. These evolution patterns are similar to seaward prograding type (Montaggioni 2005) of Holocene reefs. The result of the simulation on Scenario 4 (sea level is rising faster at 0.01 m yr^{-1} for 4,000 years) (Figure 2) is shown in Figure 4c. In this case, reef crest cannot keep-up with sea level rising, and only shoreward shallower part is keeping up. Therefore accretion front of the reef looks like stepping back because shoreline is stepping back with sea level rising. This evolution pattern is similar to back-stepping type (Montaggioni 2005) of Holocene reefs.

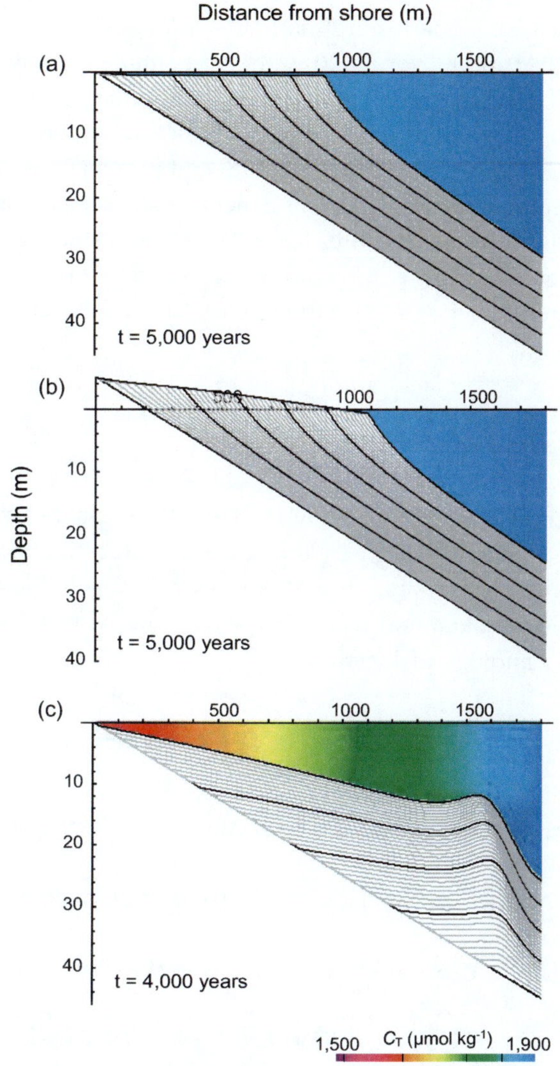

Figure 4. Evolutions of coral reef topography and concentration distributions of total inorganic carbon dioxide (C_T). Period of the simulation is 5,000 years and sea-level scenario of the simulation is scenario 2 (a), the period is 5,000 years and the scenario is Scenario 3 (b), and the period is 4,000 years and the scenario is Scenario 4 (c).

Conclusion

We developed light-enhanced and diffusion-limited growth model for coral reef formation based on calcification model; depending on light intensity and concentrations of carbon species, diffusion of carbon species, and reef growth due to calcification. This model was verified on four scenarios of relative sea-level history used for simulations. The first scenario, which are similar with global sea level history between 8,000 years B.P. and present, is that the relative sea-level is constructed by two phases; rising constantly at 0.008 m yr^{-1} for first 2,000 years, then stable between 2,000 and 8,000 years. The simulation result of this scenario is very similar with balanced aggrading/onlapping type of Holocene fringing reefs. The second and third scenarios are that the sea level is stable for 5,000 years, and is falling constantly at 0.001 m yr^{-1} for 5,000 years. The simulation patterns of these scenarios are well-reconstructed seaward prograding type reefs. Fourth scenario is that the sea level is rising constantly at 0.01 m yr^{-1} for 4,000 years. This simulation pattern is similar with back-stepping type reefs. Therefore, the reef topographies and growth pattern responding to relative sea level histories simulated by the model in general well-reconstructed the Holocene reefs observed in nature. The model, therefore, has potential applications within a range of reef studies in many different contexts, e.g. reconstruction of the Holocene histories of many reef systems worldwide, and prediction of reef topography under the conditions of future sea level rise.

Acknowledgments

We gratefully acknowledge K. Otsuki, M. Uehara, K. Sugihara and T. Watanabe for their advice. This work was supported partly by the Sasakawa Scientific Research Grant from The Japan Science Society (to Nakamura, 19-619) and by a Grant-in-Aid for Scientific Research on Innovative Areas (to Nakamura, 21121501) from the Ministry of Education, Culture, Sports, Science and Technology of Japan.

References

Al-Moghrabi, S; Goiran, C; Allemand, D; Speziale, N; Jaubert, J. *J. Exp. Mar. Biol. Ecol.*, 1996, 199, 227-248.

Bard, E; Hamelin, B; Arnold, M; Montaggioni, L; Cabioch, G; Faure, G; Rougerie, F. *Nature*, 1996, 382, 241-244.

Barnes, DJ; Chalker, BE. In *Coral reefs: Ecosystems of the World*; Dubinsky, Z; Ed; Elsevier: Amsterdam, 1990, 109-131.

Barnes, DJ; Devereux, MJ. *J. Exp. Mar. Biol. Ecol.*, 1984, 79, 213-231.

Bosscher, H; Schlager, W. *Sedimentology*, 1992, 39, 503-512.

Buddemeier, RW; Smith, SV. *Coral Reefs*, 1988, 7, 51-56.

Burris, JE; Porter, JW; Laing, WA. *Mar. Biol.*, 1983, 75, 113-116.

Chalker, BE. *Mar. Biol.*, 1981, 63, 135-141.

Cortés, J; Macintyre, IG; Glynn, PW. *Coral Reefs*, 1994, 13, 65-73.

Easton, WH; Olson, EA. *Geol. Soc. Am. Bull*, 1976, 87, 711-719.

Engels, MS; Fletcher, CH; Field, ME; Storlazzi, CD; Grossman, EE; Rooney, JJB; Conger, CL. *J. sediment. Res.*, 2004, 74, 255-269.

Gattuso, JP; Allemand, D; Frankignoulle, M. *Am. Zool*, 1999, 39, 160-183.

Gattuso, JP; Frankignoulle, M; Bourge, I; Romaine, S; Buddemeier, RW. *Global. Planet. Change*, 1998, 18, 37-46.

Goiran, C; Al-Moghrabi, S; Allemand, D; Jaubert, J. *J. Exp. Mar. Biol. Ecol.*, 1996, 199, 207-225.

Goreau, TF. *Biol. Bull*, 1959, 116, 59-75.

Goreau, TF; Goreau, NI. *Biol. Bull*, 1959, 117, 239-250.

Hossain, MMM; Ohde, S. *Proc. 10th Int. Coral Reef Symp*, 2006, 597-606.
Johnson, DP; Risk, MJ. *Sedimentology*, 1987, 31, 515-529.
Langdon, C. *Proc. 9th Int. Coral Reef Symp*, 2000, 2, 1091-1098.
Langdon, C; Atkinson, MJ. *J. Geophys. Res*., 2005, 110, doi: 10.1029/2004JC002576.
Leclercq, N; Gattuso, JP; Jaubert, J. *Global. Change. Biol*., 2000, 6, 329-334.
Leclercq, N; Gattuso, JP; Jaubert, J. *Limnol. Oceanogr*, 2002, 47, 558-564.
Marubini, F; Ferrier–Pagès, C; Furla, P; Allemand, D. *Coral Reefs*., 2008, 27, 491-499.
Marubini, F; Thake, B. *Limnol. Oceanogr*, 1999, 44, 716-720.
Montaggioni, LF. *Earth-Sci. Rev*., 2005, 71, 1-75.
Montaggioni, LF; Faure, G. *Sedimentology*, 1997, 44, 1053-1070.
Nakamura, T; Nakamori, T. *Coral Reefs*, 2007, 26, 741-755.
Nakamura, T; Nakamori, T. *Coral Reefs*, 2009, 28, 229-250.
Ohde, S; Hossain, MMM. *Geochem. J*, 2004, 38, 613-621.
Ohde, S; van, Woesik, R. *Bull. Mar. Sci*., 1999, 65, 559-576.
Scheidegger, AE. *Theoretical geomorphology* (2nd ed.); Springer: New York, 1970, 435.
Suzuki, A; Nakamori, T; Kayanne, H. *Sediment. Geol*, 1995, 99, 259-280.
Takahashi, T; Koba, M; Kan, H. *Proc. 6th Int. Coral Reef Symp*. 1988, 3, 194-196.
Tudhope, AW; Scoffin, TP. *J. Sediment. Res. A*, 1994, 64, 752-764.
Webster, JM; Wallace, LM; Clague, DA; Braga, JC. *Geochemistry. Geophysics. Geosystems*. 2007, 8, doi: 10.1029/2006GC001415.
Yamano, H; Kayanne, H; Yonekura, N. *J. Sediment. Res*., 2001, 71, 295-304.

In: Sea Level Rise, Coastal Engineering, Shoreslines and Tides ISBN: 978-1-61728-655-1
Editor: Linda L. Wright, pp. 263-268 © 2011 Nova Science Publishers, Inc.

Chapter 13

SEA LEVEL TREND CHANGES IN THE IONIAN SEA IN RELATION TO THE EASTERN MEDITERRANEAN TRANSIENT

***J. Soto-Navarro*[1]*, J. Del Río Vera*[1]*, F. Criado-Aldeanueva*[1]*, J. García-Lafuente*[1]*, C. Naranjo-Rosa*[1]*, C. Calero-Quesada*[1] *and A. Sanchez-Roman*[1]**

[1]Physical Oceanography Group, University of Malaga, Spain.

Abstract

Altimetry measurements over the Ionian region have been used to analyse the negative sea level trend over the Ionian basin in the last decades. The apparent decreasing trend should be better understood as an abrupt sea level drop in 1998 probably linked to changes in the surface circulation in the Ionian basin induced by the Eastern Mediterranean Transient, which changed from anticyclonic to cyclonic about March 1998. From then onwards, a rising rate of 7.9 ± 0.9 mm/year is observed over the basin.

Keywords: sea level trends; altimetry; Eastern Mediterranean Transient; Ionian Sea

1. Introduction

The Mediterranean Sea, a semi-enclosed basin that extends over 3000 km in longitude and over 1500 km in latitude, is separated into the eastern and western basins by the Strait of Sicily (Figure 1). The average sea level trends estimated for the last years in the entire basin range from 7±1.5 mm/year reported by Cazenave et al. (2002) from 6 years (1993–1998) of T/P and ERS-1 merged altimetry data to 2.2 mm/year and 2.1±0.6 mm/year obtained by Fenoglio-Marc (2002) and Criado-Aldeanueva et al. (2008) from 8 years (09/1992–08/2000) and 13 years (1993–2005) of altimetry data, respectively. Trends considerably vary with location: sea level rise of thermosteric origin is particularly important in the Levantine basin south of Crete with values up to 10±1 mm/year; in contrast, the Ionian basin has been identified as a falling sea level region. Cazenave et al. (2002) reported a decreasing trend of 15–20 mm/year over the area for the period 1993–1998. Using longer data series, Fenoglio-Marc (2002) and Criado-Aldeanueva et al. (2008) found more moderate values: −11.9 mm/year for 1992–2000 and −10±0.8 mm/year for 1993–2005, respectively. Due to the relatively short time series, regional trends depend both on the length of the series and on the boundaries of the area selected to compute the regional mean sea level. The Ionian negative trend is more probably related to mass redistribution than to a thermosteric forcing, which suggests an origin related to changes of the circulation pattern in this basin during the last decade, probably linked to the Eastern Mediterranean Climate Transient (EMT, Pinardi et al., 1997; Malanotte-Rizzoli et al., 1997, 1999; Manca, 2000). In this paper, some features of this decreasing trend are revisited using altimetry records over the Ionian region.

2. Data

The total sea level has been determined from altimetry data from diverse satellite/missions (T/P, ERS-1/2, GFO, ENVISAT and JASON 1) collected through the merged AVISO (Archiving Validation and Interpretation of Satellite Oceanographic data) products, freely available on www.aviso.oceanobs.com. The data consist of sea level anomalies referred to a 7-year average (1993–1999) and combine information from different missions, which significantly improves the estimation of mesoscale signals (Le Traon and Dibarboure, 1999; Le Traon et al., 2001). The AVISO regional solution for the Mediterranean Sea for the period 1992–2007, with 1/8×1/8° spatial resolution and weekly time resolution has been used. To focus on seasonal variations, a low-pass Butterworth filter (cut-off frequency 60 days−1) has been applied. All standard geophysical and environmental corrections including ionospheric, dry and wet tropospheric corrections, solid Earth and ocean tides, ocean tide loading, pole tide, electromagnetic bias, instrumental corrections, orbit error reduction and inverse barometer have been applied by AVISO processing algorithms (AVISO, 1996).

3. Results and Discussion

Figure 1A displays the total spatially averaged sea level anomaly over the Ionian Sea (solid rectangle) from altimetry data. More than 700 altimetry grid points have been extracted so the results are thought to be representative of the Ionian basin. To diagnose trends, the signal has been least squares fitted to the following function:

$$y(t) = a_0 + a_1 t + A_1 \cos(\omega_1 t - \varphi_1) + A_2 \cos(\omega_2 t - \varphi_2) \tag{1}$$

that includes annual (ω1) and semi-annual (ω2) frequencies. The semi-annual amplitude turned out to be an order of magnitude smaller than the annual amplitude and results do not vary significantly when considering only the annual oscillation. A total trend of −4.6±0.7 mm/year (95% confidence interval) is computed for the entire period superposed to the seasonal oscillation. This value is lower than reported in the literature (Cazenave et al., 2002; Fenoglio-Marc, 2002; Criado-Aldeanueva et al., 2008) because sea level has a positive trend in the last years (Figure 1A), thus reducing the overall trend. A remarkable feature is the presence of two well-differentiated periods before and after 1998. If the fitting process is performed separately, no significant trend and an annual amplitude of 9.1 cm is observed before 1998, whereas from 1998 onwards a positive trend of 7.9±0.9 mm/year is observed superposed to an annual amplitude of 7.1 cm. In 1998, an abrupt sea level change occurred with a drop of 10–15 cm between maxima that appears to be responsible for the negative trend reported by different authors (Cazenave et al., 2002; Fenoglio-Marc, 2002; Criado-Aldeanueva et al., 2008) using time series spanning over this jump. Actually, the decreasing trend over the Ionian basin in the last decades is somewhat deceitful since from 1998 onwards, sea level is rising in the Ionian basin as well. This picture would explain the anomalously high negative trend of Cazenave et al. (2002), since the abrupt fall was at the end of the time series usedin their analysis.

The question then arises as which could have been the physical forcing mechanism behind this sudden sea level drop. We propose that it was related to the EMT. A shift in the formation site of deep and bottom waters from the Adriatic to the Aegean Sea (Roether et al., 1996; Malanotte-Rizzoli et al., 1999; Theocharis et al., 1999) have altered both the deep and upper conveyor belts of the Eastern Mediterranean. Different hypothesis such as internal redistribution of salt (Klein et al., 1999), changes in the local atmospheric forcing combined with long term salinity change (Theocharis et al., 1999) or changes in circulation patterns (Malanotte-Rizzoli et al., 1999) have been proposed as possible causes of this unique event. At the same time, noticeable changes in circulation patterns over the Ionian basin have been reported in the last decades: a switch of the surface circulation in the Ionian basin from cyclonic to anticyclonic took place around 1986–87 (Pinardi et al., 1997; Malanotte-Rizzoli et al., 1997, 1999) and forced variations in the path of the Atlantic Ionian Stream (AIS) along the Ionian basin. This circulation pattern was well established until about March 1998, when it returned to cyclonic (Manca, 2000). The coincidence in time of the sea level drop and the switch of the circulation pattern over the Ionian basin suggests a cause–effect relationship between these two events. A rough estimation of the sea level drop associated to such change in the circulation pattern can be achieved by means of the geostrophic balance:

$$\Delta h = \frac{\Delta x}{g} \cdot fv \tag{2}$$

where v is a reference velocity of the flow (~0.3 m s^{-1} is representative for the mean velocity of the AIS in the Ionian Sea (Robinson et al., 1999)), Δx is a typical across stream distance over which the velocity vanishes (~5 104 m for the core of the AIS), f the Coriolis parameter (~9 10^{-5} s^{-1} for a latitude of 38 °N) and g the gravity acceleration. With these values, Eq. (2) gives Δh~0.13 m, which must be doubled to account for the change from anticyclonic to cyclonic circulation. Therefore, the computed drop is about 0.3 m, a value in good agreement (within the same order of magnitude) with the observed drop. This simple estimation supports the physical mechanism invoked.

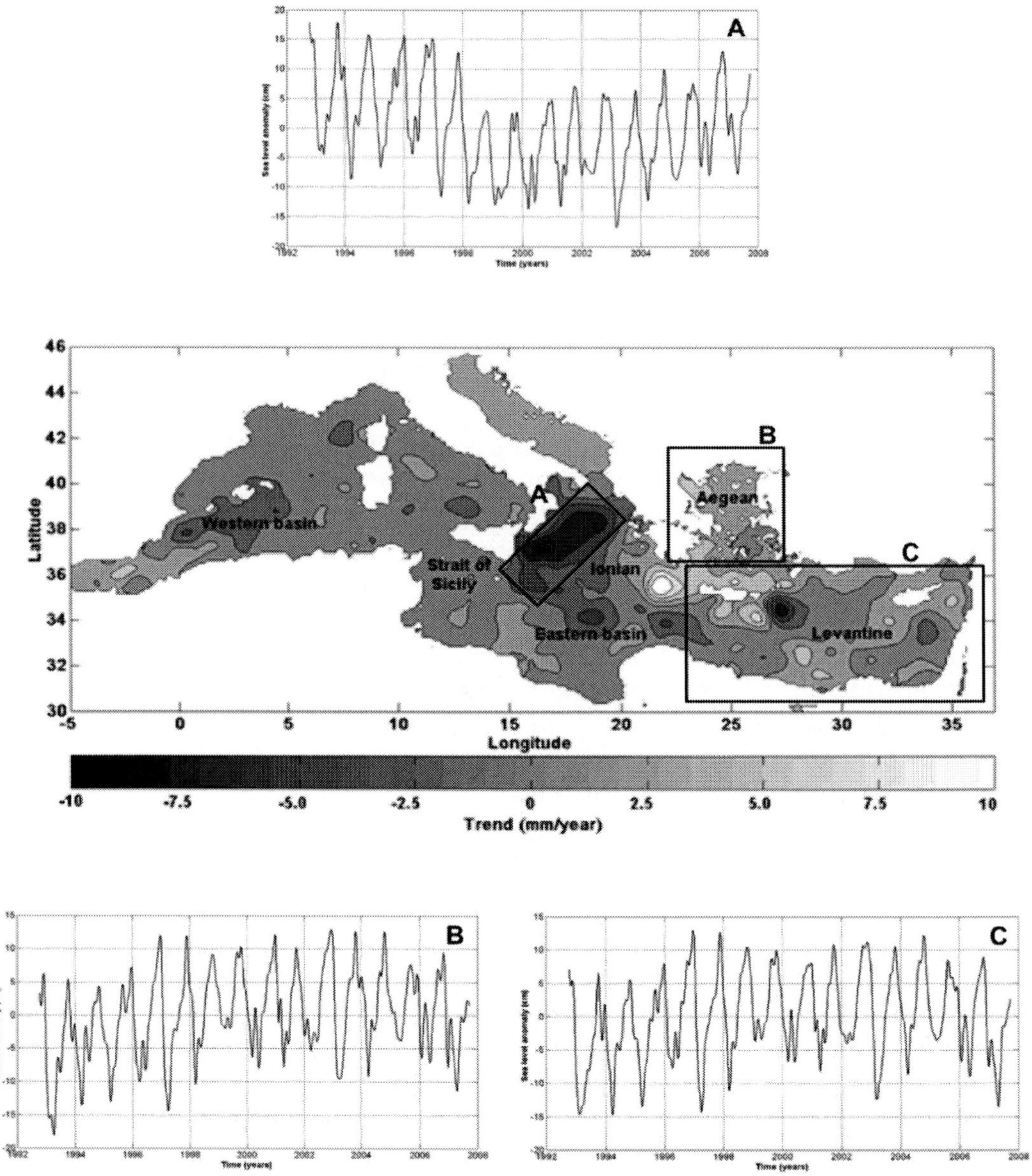

Figure 1. Spatial distribution of sea level trend in the Mediterranean Sea from 1992 to 2007 altimetry data. The rectangles limit the areas selected for the low-passed spatially averaged altimetry time series shown in panels A (Ionian basin), B (Aegean basin) and C (Levantine basin).

The EMT did not only affect the Ionian Sea but the whole Eastern basin. Larnicol et al. (2002) suggested that the changes observed in the sea level during the period between 1995 and 1999 in the Levantine basin and between 1997 and 1999 in the Ionian Sea are related to variations in the deep and intermediate water masses distribution in the whole Eastern basin. Vigo et al. (2005) show a sort of time–space coupled oscillation between the Ionian Sea and the Levantine basin in which a sea level rise in the Levantine basin is connected to a sea level drop in the Ionian Sea before 1998, while after 1999 the behavior seems to be the opposite. With longer time series available, we can now confirm and specify these findings. Figure 1B–

C shows the low-passed sea level anomaly over the Aegean (panel B) and Levantine (panel C) basins. Different behaviours before and after 1998 are again revealed in both regions. Positive trends of 12±2.5 mm/year and 14±2.5 mm/year are obtained for the period 1993–1998 in the Aegean and Levantine basins that change into negative trends of −2.0±0.8 mm/year and −2.9±0.8 mm/year, respectively for the period 1998–2008, which strongly contrast with the Ionian sea level rise. Trends before 1998 are in good agreement with previous results of Vigo et al. (2005) but those for the period 1998–2008 are lower than reported by these authors from time series until 11/2003. An explanation for this could be that the rebound effect after the 1998 event extends during the 3–4 subsequent years (see, for instance the reduction of amplitude in the seasonal oscillations in Figure 1B–C), then recovering the usual behaviour. Although the change in trend has also been evidenced in the eastern basins, its magnitude was not as significant as in the Ionian basin (Figure 1A), this suggesting that this basin was more sensitive to the 1998 event, probably due to the changes of the path of the AIS flowing through the Ionian Sea.

4. Conclusions

In this work altimetry data from 1992 to 2007 have been analysed to investigate an abrupt sea level drop in the Ionian basin by the late 90s, concretely a ~10 cm drop found in 1998. The data analysis shows that the historically reported sea level decreasing trend over the Ionian basin in the last decades stems from this punctual event rather than a continuous negative trend. Actually, the sharp drop was followed by a sea level rise from 1998 onwards. This unique event was provably linked to the EMT that induced noticeable changes in the surface circulation in the Ionian basin, which changed from anticyclonic to cyclonic about 1998.

Acknowledgment

This work has been carried out in the frame of the Junta de Andalucía Spanish funded P07-RNM-02938 project.

References

AVISO,(1996). *AVISOUserHandbook: Merged TOPEX/POSEIDON Products. 201.* Available at www.aviso.oceanobs. com/ donnees/tools/hdbk._gdrm.pdf.

Cazenave, A., Bonnefond, P., Mercier, F., Dominh, K. & Toumazou, V. (2002). Sea level variations in the Mediterranean Sea and Black Sea from satellite altimetry and tide gauges. *Global and Planetary Change,* 34, 59-86.

Criado-Aldeanueva, F., Del Río, J. & García-Lafuente, J. (2008). Steric and mass induced Mediterranean Sea level trends from 14 years of altimetry data. *Global and Planetary Change,* 60, 563-575.

Del Rio, J. (2007). *Satellite and In Situ Monitoring of Ocean Magnitudes: Application to the Mediterranean Sea.* PhD University of Malaga, 234.

Fenoglio-Marc, L. (2002). Long-term sea level change in the Mediterranean Sea from multi-satellite altimetry and tide gauges. *Physics and Chemistry of the Earth,* 27, 1419-1431.

Klein, B., Roether,W., Manca, B., Bregant, D., Beitzel, V., Kovacevic, V. & Luchetta, A. (1999). The large deep water transient in the Eastern Mediterranean. *Deep-Sea Research I,* 46, 371-414.

Larnicol, G., Ayoub, N. & Le Traon, P. Y. (2002). Major changes in Mediterranean Sea level variability from 7 years of TOPEX/Poseidon and ERS-1/2 data. *Journal of Marine Systems,* 33-34, 63-89.

Le Traon, P. Y. & Dibarboure, G. (1999). Mesoscale mapping capabilities of multi-satellite altimeter missions. *Journal of Atmospheric and Oceanic Technology,* 16, 1208-1223.

Le Traon, P. Y., Dibarboure, G. & Ducet, N. (2001). Use of a high resolution model to analyse the mapping capabilities of multi-altimeter missions. *Journal of Atmospheric and Oceanic Technology,* 18, 1277-1288.

Malanotte-Rizzoli, P.,Manca, B. B., D'Alcala, M. R., Theocharis, A., Bergamasco, A., Bregant, D., Budillon, G., Civitarese, G., Georgopoulos, D., Michelato, A., Sansone, E., Scarazzato, P. & Souvermezoglou, E. (1997). A synthesis of the Ionian hydrography, circulation and water mass pathways during POEM-phase I. *Progress in Oceanography,* 39, 153-204.

Malanotte-Rizzoli, P., Manca, B. B., D'Alcala, M. R., Theocharis, A., Brenner, S., Budillon, G. & Ozsoy, E. (1999). The Eastern Mediterranean in the 80s and in the 90s: the big transition in the intermediate and deep circulations. *Dynamics of Atmospheres and Oceans,* 29(2-4), 365-395.

Manca, B. M. (2000). Recent changes in dynamics of the EasternMediterranean affecting the water characteristics of adjacent basins. *CIESMWorkshop Series,* 2000, 27-31.

Pinardi, N., Korres, G., Lascaratos, A., Stanev, E. & Roussenov, V. (1997). Evidence for interannual variability of the Mediterranean Sea upper ocean circulation. *Geophysical Research Letters,* 24(4), 245-247.

Robinson, A. R., Sellschopp, J., Warn-Varnas, A., Leslie, W. G., Lozano, C. J., Haley Jr., P. J., Anderson, L. A. & Lermusiaux, P. F. J. (1999). The Atlantic Ionian Stream. *Journal of Marine Systems,* 20, 129-156.

Roether, W., Manca, B. B., Klein, B., Bregant, D., Georgopoulos, D., Beitzel, V., Kovacevic, V. & Luchetta, A. (1996). Recent changes in eastern Mediterranean Deep Waters. *Science,* 271, 333-335.

Theocharis, A., Nittis, K., Kontoyiannis, H., Papageorgiou, E. & Balopoulos, E. (1999). Climatic changes in the Aegean Sea influence the Eastern Mediterranean thermohaline circulation (1986-1997). *Geophysical Research Letters,* 26(11), 1617-1620.

Vigo, I., García, D. & Chao, B. F. (2005). Change of sea level trend in the Mediterranean and Black seas. *Journal of Marine Research,* 63, 1085-1100.

In: Sea Level Rise, Coastal Engineering, Shorelines and Tides ISBN: 978-1-61728-655-1
Editor: Linda L. Wright, pp. 269-276 © 2011 Nova Science Publishers, Inc.

Chapter 14

CHARACTERISTIC AND MOMENT GENERATING FUNCTIONS OF GENERALISED EXTREME VALUE DISTRIBUTION (GEV)

G. Muraleedharan*[*], *C. Guedes Soares and Claudia Lucas
Centre for Marine Technology and Engineering (CENTEC),
Technical University of Lisbon, Instituto Superior Tecnico, Lisboa, Portugal

Abstract

Recently generalised extreme value distribution is widely used for extreme event modeling and ocean engineering applications. Hence the characteristic function (CF) of generalized extreme value distribution is derived and the moment generating function (MGF) is deduced from it. It generates all the moments of the distribution and satisfies the tests to verify a function to be a characteristic function. Expressions for mean, variance, skewness and kurtosis are obtained from MGF. The GEV is suggested as the sampling distribution of the newly defined significant wave height distribution by the method of characteristic function. The CF of Frechet distribution, a special case of GEV is also derived. The MGF deduced from it generates all the moments of the distribution. The characteristic function has many useful and important properties which gives it a central role in statistical theory.

Keywords: Generalized extreme value distribution, Frechet distribution, Characteristic function, Moment generating function

1. Introduction

The generalized extreme value distribution is widely used in ocean engineering as well as in extreme event modeling. But the characteristic function of the distribution was not available. The characteristic function has great theoretical importance. It also yields many valuable results in the theory of sampling (Kendall and Stuart 1964).

---

[*] E-mail address: muralg@mar.ist.utl.pt. (Corresponding author)

The characteristic function is also known to be the Inverse Fourier Transform of the probability density function with Fourier Transform parameters (a,b) equal to (1,1). Then the probability density function is the Forward Fourier Transform of the characteristic function.

A characteristic function exists for all probability distributions. However this is not the case for moment generating function. It is possible to determine the characteristics of a random variable from its characteristic function without actually knowing its distribution function. Also, two probability distributions never share the same characteristic function. In other words, a characteristic function uniquely determines its probability distribution. The Levy continuity theorem in probability theory is the basis for one approach to prove the central limit theorem and it is the one the central theorems concerning characteristic functions. It is also used to prove the weak law of large numbers using convergence of characteristic functions. The characteristic function has many useful and important properties which gives it a central role in statistical theory. Here an attempt is made to derive the characteristic function (CF) of the generalized extreme value distribution and the moment generating function (MGF) is deduced from it. Expressions for mean, variance, skewness and kurtosis are also obtained from MGF.

2. Mathematical Derivation of the Characteristic Function of Generalised Extreme Value Distribution

The characteristic function of the generalized extreme value probability density function is given by

$$\varphi(t) = \mathrm{E}(\exp(itX)) \tag{1}$$

E-usual expectation operator
X- GEV random variable
i=$\sqrt{-1}$
t-any arbitrary real constant

The probability density function of generalized extreme value distribution is given by

$$f(z) = \frac{1}{\sigma}(1+\xi z)^{\frac{-1}{\xi}-1} \exp-(1+\xi z)^{\frac{-1}{\xi}} \tag{2}$$

Where

$$z = \frac{x-\mu}{\sigma}, \mathrm{x} > \mu - \frac{\sigma}{\xi} \text{for } \xi > 0,$$

$\mu(-\infty,\infty)$, $\sigma(0,\infty)$, $\xi(0,\infty)$ are respectively the location, scale and shape parameters of the distribution. Depending upon the value of the shape parameter (ξ) of the distribution it is known as Type-I (Gumbel, $\xi = 0$), Type-II (Frechet, $\xi > 0$) and Type-III (Weibull, $\xi < 0$). The generalized extreme value distribution unites the type-I, type-II and type-III extreme value distributions into a single family, to allow a continuous range of possible shapes. It is also known as Fisher-Tippet distribution. Its cumulative distribution function is given by

$$F(z) = \exp-(1+\xi z)^{\frac{-1}{\xi}} \tag{3}$$

The characteristic function is mathematically derived as

$$\varphi(t) = \mathrm{E}(\exp(itX)) = \int_{\infty}^{\mu-\frac{\sigma}{\xi}} \exp(itx) f(x) dx$$

$$= \int_{\infty}^{\frac{-1}{\xi}} \exp[it(\sigma z+\mu)](1+\xi z)^{\frac{-1}{\xi}-1} \exp-(1+\xi z)^{\frac{-1}{\xi}} dz$$

$$= \exp(it\mu)\left[\underbrace{\int_{\infty}^{\frac{-1}{\xi}} \cos(t\sigma z)(1+\xi z)^{\frac{-1}{\xi}-1} \exp-(1+\xi z)^{\frac{-1}{\xi}} dz}_{\mathbf{I}} + \underbrace{\int_{\infty}^{\frac{-1}{\xi}} i\sin(t\sigma z)(1+\xi z)^{\frac{-1}{\xi}-1} \exp-(1+\xi z)^{\frac{-1}{\xi}} dz}_{\mathbf{II}}\right]$$

Integrating the **I**st expression by parts, ⇒

$$\exp(it\mu)\left[\cos\left(\frac{t\sigma}{\xi}\right) + \int_{\infty}^{\frac{-1}{\xi}} (t\sigma)\sin(t\sigma z)\exp-(1+\xi z)^{\frac{-1}{\xi}} dz\right] \tag{A}$$

Now integrating the **II** expression by parts ⇒

$$\exp(it\mu)\left[-i\sin\left(\frac{t\sigma}{\xi}\right) - \int_{\infty}^{\frac{-1}{\xi}} (it\sigma)\cos(t\sigma z)\exp-(1+\xi z)^{\frac{-1}{\xi}} dz\right] \tag{B}$$

(A).+ (B) ⇒

$$\exp(it\mu)\left[\exp-\left(\frac{it\sigma}{\xi}\right) - it\sigma\left(\sum_{r=0}^{\infty} \frac{(it\sigma)^r}{r!} \int_{\infty}^{\frac{-1}{\xi}} z^r \exp-(1+\xi z)^{\frac{-1}{\xi}} dz\right)\right]$$

$$\varphi(t) = \exp(it\mu)\left[\exp-\left(\frac{it\sigma}{\xi}\right) - \sum_{r=0}^{\infty} \frac{(-it\sigma)^{r+1}}{r!\,\xi^{r}} \sum_{n=0}^{r} (-1)^n \, {}^{r}C_n \frac{\Gamma(1-(n+1)\xi)}{(n+1)\xi}\right]; r = 0,1,2,. \tag{4}$$

Expression (4) is the characteristic function of the generalized extreme value distribution. Its expansion will be of the form

$$\left(1+\frac{it\mu}{1!}-\frac{t^2\mu^2}{2!}-......\right)\left\{\left(1-\frac{it\sigma}{1!\xi}-\frac{t^2\sigma^2}{2!\xi^2}+....\right)-\left[\begin{array}{l}\frac{-it\sigma}{0!}\left(\frac{\Gamma(1-\xi)}{\xi}\right)-\frac{t^2\sigma^2}{1!\xi}\left(\frac{\Gamma(1-\xi)}{\xi}-\frac{\Gamma(1-2\xi)}{2\xi}\right)+\\ \frac{it^3\sigma^3}{2!\xi^2}\left(\frac{\Gamma(1-\xi)}{\xi}-\frac{\Gamma(1-2\xi)}{\xi}+\frac{\Gamma(1-3\xi)}{3\xi}\right)+\frac{t^4\sigma^4}{3!\xi^3}(...)\end{array}\right]\right\}$$

Where

$${}^rC_n=\frac{r!}{n!(r\text{-}n)!}$$

It also satisfies the tests to verify whether a function is a characteristic function (Kendall and Stuart, 1964; Ochi, 1998)) such as

1) That φ (t) must be continuous in t
2) That φ (t) is defined in every finite t interval
3) That φ (0) =1
4) That φ (t) and φ $(-t)$ shall be conjugate quantities.

3. Discussion

Characteristic functions are also known as moment generating functions (MGF). Ie., by arranging $it=\theta$,

$$M(\theta)=\exp(\theta\mu)\left[\exp-\left(\frac{\theta\sigma}{\xi}\right)-\sum_{r=0}^{\infty}\frac{(-\theta\sigma)^{r+1}}{r!\xi^r}\sum_{n=0}^{r}(-1)^n\,{}^rC_n\frac{\Gamma(1-(n+1)\xi)}{(n+1)\xi}\right] \quad (5)$$

which is the moment generating function of the generalized extreme value distribution.

M (θ) can generate all the raw moments of the distribution. Ie

$$M_X^{(n)}(0)=\mu_n'=E(X^n)=\left[\left(\frac{d^n(M(\theta))}{d\theta^n}\right)\right]_{\theta=0}$$

μ_n' -n^{th} arbitrary moment or n^{th} raw moment.

Accordingly

$$Mean=\mu_1'=\mu-\frac{\sigma}{\xi}+\frac{\sigma}{\xi}g_1$$

$$Var(x) = \mu_2 = \mu_2^{'} - \left(\mu_1^{'}\right)^2 = \frac{\sigma^2}{\xi^2}\left(g_2 - g_1^2\right)$$

$$Skewness = \frac{\mu_3}{\left(\mu_2\right)^{\frac{3}{2}}} = \frac{g_3 - 3g_1g_2 + 2g_1^3}{\left(g_2 - g_1^2\right)^{\frac{3}{2}}}$$

$$Kurtosis = \frac{\mu_4}{\left(\mu_2\right)^2} = \frac{g_4 - 4g_1g_3 + 6g_2g_1^2 - 3g_1^4}{\left(g_2 - g_1^2\right)^2}$$

μ_2, μ_3 *and* μ_4 are the central moments.

$$\mu_2^{'} = \frac{\sigma^2}{\xi^2}\left(1 - 2g_1 + g_2\right) - \frac{2\mu\sigma}{\xi}\left(1 - g_1\right) + \mu^2$$

$$\mu_3^{'} = -\frac{\sigma^3}{\xi^3}\left(1 - 3g_1 + 3g_2 - g_3\right) + 3\frac{\mu\sigma^2}{\xi^2}\left(1 - 2g_1 + g_2\right) - \frac{3\mu^2\sigma}{\xi}\left(1 - g_1\right) + \mu^3$$

$$\mu_4^{'} = \frac{\sigma^4}{\xi^4}\left(1 - 4g_1 + 6g_2 - 4g_3 + g_4\right) - \frac{4\mu\sigma^3}{\xi^3}\left(1 - 3g_1 + 3g_2 - g_3\right) +$$

$$\frac{6\mu^2\sigma^2}{\xi^2}\left(1 - 2g_1 + g_2\right) - \frac{4\mu^3\sigma}{\xi}\left(1 - g_1\right) + \mu^4$$

$$\mu_3 = \mu_3^{'} - 3\mu_2^{'}\mu_1^{'} + 2\left(\mu_1^{'}\right)^3$$

$$\mu_4 = \mu_4^{'} - 4\mu_3^{'}\mu_1^{'} + 6\mu_2^{'}\left(\mu_1^{'}\right)^2 - 3\left(\mu_1^{'}\right)^4$$

$$g_k = \Gamma\left(1 - k\xi\right)$$

The kurtosis for a standard normal distribution is 3. For this reason the following definition of kurtosis is also used.

$$kurtosis = \frac{\mu_4}{\mu_2^2} - 3$$

With this definition positive kurtosis indicates a peaked distribution and negative kurtosis indicates a flat distribution.

By Taylor's theorem for complex function, the characteristic function of any random variable, *X*, with finite mean $\bar{x}$, can be written as

$$\varphi(t) = 1 + it\bar{x} + 0(t);\ t \to 0$$

This is also true for the characteristic function $\varphi(t)$ of generalized extreme value distribution as

$$\bar{x} = \mu - \frac{\sigma}{\xi} + \frac{\sigma}{\xi} g_1$$

Hence $\varphi(t)$ is the characteristic function of generalized extreme value distribution.

Another condition for a function to be a characteristic function is that if $\varphi(t)$ be a characteristic function of a non-negative random variable X with finite non-zero first moment, then

$$\psi(t) = \frac{\varphi'(t)}{\varphi'(0)}$$

(Kubik, 1975) is also a characteristic function.

This condition is also satisfied by the characteristic function (4) derived for the generalized extreme value distribution.

If the characteristic function of the parent distribution is known, then it is possible to obtain the sampling distribution of the statistic by the method of characteristic function discussed below. It is to be noted that the CF of the sum of n independent variables is the product of their CF. This simple property enables us to find out the sampling distribution of the newly defined significant wave height (Muraleedharan et.al., 2007) which is an average of the highest one-third of a constant number of wave heights in a wave record. This definition resembles the statistical average of wave heights and is more meaningful.

The characteristic function of sum of n independent generalized extreme value wave height variables is given by

$$[\varphi(t)]^n$$

Then characteristic function of $\frac{1}{n}$ of the sum is given by

$$\left[\varphi\left(\frac{t}{n}\right)\right]^n$$

Accordingly

$$\left[\varphi\left(\frac{t}{n}\right)\right]^n = \left[1+\frac{it}{n}\left(\mu-\frac{\sigma}{\xi}+\frac{\sigma}{\xi}g_1\right)\right]^n$$

Neglecting the higher order terms for $t \to 0$ and for large n, we get

$$\left[\varphi\left(\frac{t}{n}\right)\right]^n = 1+it\left(\mu-\frac{\sigma}{\xi}+\frac{\sigma}{\xi}g_1\right)$$

This is by Taylor's Theorem, the characteristic function of the GEV distribution. The GEV distribution is thus can also be suggested for simulating newly defined significant wave height distribution.

The probability density function of one parameter Frechet distribution is given by

$$f(x)=\xi x^{-\xi-1}\exp\left[-(x)^{-\xi}\right];\ \xi>0,\ 0<x<\infty$$

$\xi - \textit{shape parameter}$

And cumulative distribution is

$$F(x)=\exp\left[-(x)^{-\xi}\right]$$

The characteristic function can also be obtained by the same procedure used to derive the CF of GEV as

$$\phi(t)=1+\sum_{r=0}^{\infty}\frac{(it)^{r+1}}{(r+1)!}\Gamma\left(1-\left(\frac{r+1}{\xi}\right)\right);\ r=0,1,2,\ldots\ldots. \qquad (6)$$

Accordingly, its moment generating function is deduced from its characteristic function to be

$$N(\theta)=1+\sum_{r=0}^{\infty}\frac{(\theta)^{r+1}}{(r+1)!}\Gamma\left(1-\left(\frac{r+1}{\xi}\right)\right) \qquad (7)$$

Hence

Mean = $\mu_1' = g_1'$

Variance = $g_2' - (g_1')^2$

Skewness = $\dfrac{g_3' - 3g_2'g_1' + 2(g_1')^3}{\left(g_2' - (g_1')^2\right)^{\frac{3}{2}}}$

$$\text{Kurtosis} = \frac{g_4' - 4g_3'g_1' + 6g_2'\left(g_1'\right)^2 - 3\left(g_1'\right)^4}{\left(g_2' - \left(g_1'\right)^2\right)^2}$$

$$\mu_2' = g_2'$$

$$\mu_3' = g_3'$$

$$\mu_4' = g_4'$$

$$g_k' = \Gamma\left(1 - \frac{k}{\xi}\right)$$

The characteristic function of the Frechet distribution also satisfies all the tests for a function to be a characteristic function.

4. Conclusion

The generalized extreme value distribution is widely used in extreme event modeling. The characteristic function derived satisfies all the conditions for a function to be a characteristic function. It is able to generate all the moments of GEV from its MGF. The GEV is suggested as a sampling distribution for the newly defined significant wave height distribution by the method of characteristic function. The characteristic function of the one parameter Frechet distribution is also derived. Expressions for skewness and kurtosis etc are obtained from its MGF.

Acknowledgment

Muraleedharan is thankful to the HIPOCAS Project, Portugal for financial assistance.

References

Kendall, GM; Stuart, A. *The Advanced Theory of Statistics; Distribution Theory*; Charles Griffin and Co. Ltd: London, UK, 1964, vol. 1, 341-342.

Kubik, L. *J. Periodica Mathematica Hungarica*., 1975, 6(2), 185-190.

Muraleedharan, G; Rao, AD; Kurup, PG; Unnikrishnan Nair, N; Mourani, S. *Coast Engng.*, 2007, 54(8), 630-638.

Ochi, M. Ocean Waves; *Ocean Technology Series-6*; Cambridge University Press: Cambridge, UK, 1998, 230.

In: Sea Level Rise, Coastal Engineering, Shorelines and Tides ISBN: 978-1-61728-655-1
Editor: Linda L. Wright, pp. 277-293 © 2011 Nova Science Publishers, Inc.

Chapter 15

MODEL-BASED CLUSTERING OF EXTREME SEA LEVEL HEIGHTS

Manuel G. Scotto[1,*]***, Susana M. Barbosa***[2] ***and Andrés M. Alonso***[3]
[1]Departamento de Matemática
Universidade de Aveiro, Portugal
[2]IDL - Laboratorio Associado
Universidade de Lisboa, Portugal
[3]Departamento de Estadística
Universidad Carlos III de Madrid, Spain

Abstract

A topic of current interest in the analysis of sea-level states is to investigate the occurrence of future rare events which is essential for the prediction of flooding risks, coastal management and in the design of coastal defences and offshore structures. Nowadays, it is widely believed that the frequency of such rare events is increasing as a result of climatic and other changes, although they are hard to predict and their effects are, yet, poorly understood. Recent developments in multivariate statistical techniques for discrimination, clustering and dimension reduction for time series, have the potential to aid on the construction of new tools and models for forecasting the occurrence and impact of such future rare events. In studies of regional sea-level variability, tidal measurements are often analyzed individually for characterizing sea-level variability at each location. Marginal analysis, however, is in itself insufficient to come with an accurate description of regional sea-level variability. An alternative approach is to consider simultaneously the whole data set of sea-level records from a given region, and characterize regional variability in terms of locations exhibiting similar behavior through clustering techniques. Cluster analysis is a useful approach for characterizing regional variability of locations exhibiting similar behavior in terms of, for example, short-term or long-term predictions of extreme values. In this work, time series clustering is applied to the analysis of long tide gauge records from the Baltic Sea. In order to describe the regional variability of Baltic sea-level, tide gauge measurements are clustered on the basis of their corresponding predictive distributions for 25-, 50- and 100-years return values. This is relevant for the design of marine systems and

*E-mail address: mscotto@ua.pt

coastal structures, which requires a good knowledge of the most severe sea-level conditions that they need to withstand during their lifetime, and also for describing and understanding the variability of extreme sea heights in a climate change context.

Keywords: extreme value theory; hydrology, hydrography, oceanography

AMS Subject Classification: 60G70; 86A05.

1. Introduction

Sea-level is a fundamental indicator of the state of the Earth's system. Oceanographic and atmospheric changes influence sea-level variability on a wide range of scales, from local to global spatial scales and from minutes to millennia temporal scales. The height of the sea surface provides therefore an integrated record of changes of the Earth's climate. Sea-level is a particularly relevant parameter in a climate change context, since it is expected to rise in a warming world due to thermal expansion of the ocean's water and melting of continental glaciers and ice sheets. Sea-level rise can impact low-lying, highly populated areas of the Earth's coasts, with considerable social and economic impacts. However, it is the impact of extreme sea-levels, more than global mean sea-level rise, that is associated with acute societal risks. For example, changes in extreme high waters can impact considerably coastal locations and populations due to flooding and consequent damaging of coastal infrastructure and loss of crops. The analysis of extreme sea-levels is therefore essential for the prediction of flooding risks and the development of adaptation strategies.

This work focuses on the analysis of the regional variability of extreme sea-levels by combining a Bayesian extreme value analysis with a time series clustering procedure. Extreme sea-levels are expected to be non-stationary due to long-term changes in the intensity and frequency of storminess (Bijl *et al.*, 1999; Meehl *et al.*, 2000; Woodworth and Blackman, 2004). The extreme value theory is particularly appropriate for the analysis of extremes in a non-stationary context, since trends and cycles can be incorporated as parameters of the extreme value distribution, allowing for trends and cycles in the upper tail which are not necessarily of the same form as in the center of the distribution (Smith, 1989; Katz *et al.*, 2002; 2005). Furthermore, methods based on extreme value theory avoid the decomposition of sea-level records into tidal and non-tidal components (Tsimplis and Blackman, 1997; Unnikrishnan *et al.*, 2004; Mendez *et al.*, 2007). While the application of extreme value theory is often based on likelihood approaches, Bayesian methods have the advantage of a substantially more flexible inference. The application of Bayesian extreme value analysis to wave and tide gauge records (e.g. Egozcue *et al.*, 2005; Scotto and Guedes Soares, 2007; Scotto *et al.*, 2010) allows to retrieve the complete posterior density of the parameters and the return values, and therefore to quantify the degree of estimation uncertainty.

The present study addresses the variability of extreme sea-levels in the Baltic Sea region. The Baltic Sea is the world's water area with the longest observational records of the height of the sea surface (Ekman, 1988; 1999). Since it is a shallow, semi-enclosed sea, connected to the North Sea through shallow and narrow straits, the water exchange is constrained, and sea-level extremes are mainly driven by regional atmospheric events occurring in scales of a few days (Stigebrandt, 1984). The spatial distribution of extreme sea-levels in the Baltic area is of both geophysical and practical interest. Thus, instead of considering each tide gauge record individually, summarizing the information for the region of interest in terms of the spatial distribution of individual features (e.g. Barbosa, 2008), the present work takes the whole dataset of tide gauge records and considers cluster analysis for assessing the spatial features of sea-level extremes.

The rest os the chapter is organized as follows: Sections 2. and 3. briefly introduce basic concepts related with extreme value theory and Bayesian methodology for extreme value models. The time series clustering approach is described in Section 4. In Section 5., an application of this approach for clustering daily tide gauge records of relative sea-level measurements collected in the Baltic Sea area, is presented. Finally, some concluding remarks are given in Section 6.

2. Extreme Value Theory

The numerous approaches by which extreme values may be statistically modeled come in two forms: methods for maxima over fixed intervals, and methods for exceedances over high thresholds. We outline the fundamental aspects of each method. Further details can be found in Reiss and Thomas (2007), Castillo *et al.* (2004) and Coles (2001).

The limit theory for the maximum of a sample of n independent random variables, say $(X_1,\dots,X_n)$, having a common distribution F, is based on a location-scale normalization of the maximum, $M_n(X) := \max(X_1,\dots,X_n)$, so that its distribution is non-degenerate as $n \to \infty$. Provided a non-degenerate limit can be archived, the limit distribution of the maximum must be the Gumbel, the Fréchet or the negative Weibull distributions. The generalized extreme value distribution (GEV) is a unifying model that encompasses these three types of extreme value distributions, having the form

$$G(x) := \exp\left\{-\left[1+\xi\left(\frac{x-\mu}{\sigma}\right)\right]^{-\frac{1}{\xi}}\right\}, \tag{1}$$

defined on $\{x : 1+\xi(x-\mu)/\sigma > 0\}$ where $-\infty < \mu < \infty$, $\sigma > 0$ and $-\infty < \xi < \infty$. It has three parameters, μ, σ and ξ, denoting location, scale and shape parameters, respectively. The shape parameter ξ, also called the tail index, determines the three extreme value types. Specifically, when ξ takes negative values, positives values or when $\xi = 0$, interpreted by taking the limit of equation (1) as $\xi \to 0$, the GEV distribution is the negative Weibull, the Fréchet or the Gumbel distribution, respectively. Although the unification of the original three families of extreme value distribution into a single family greatly simplifies statistical implementation, the convention often remains of choosing between these three subfamilies and, in particular, of deciding whether the Gumbel family provides and adequate simplified description of data relative to the full GEV family. The simplification to the Gumbel family is also often justified by the argument that there are many standard distributions for which the Gumbel distribution is the appropriate limit for the maximum. This class includes distribution functions with very different tails, ranging from moderately heavy (such as the lognormal distribution) to light (such as the Normal distribution). Moreover, the implications of a fitted extreme value model are usually made with reference to extreme quantiles. By inversion of the GEV distribution function, the quantile x_p for a specified exceedance probability p is

$$x_p := \mu - \frac{\sigma}{\xi}\left\{1-(-\log(1-p))^{-\xi}\right\},$$

where $G(x_p) = 1-p$. Roughly speaking, x_p is the return level that is associated with the return period $1/p$ for small p, in units of, say, years, if the GEV corresponds to the annual maximum. The estimate of x_p is

$$\hat{x}_p := \hat{\mu} - \frac{\hat{\sigma}}{\hat{\xi}}\left\{1-(-\log(1-p))^{-\hat{\xi}}\right\},$$

being the standard errors based on the observed information matrix.

A limitation with (1) as an inferential model is its restriction to the sample maxima when if fact the series $(X_1,\ldots,X_n)$ may contain many other data which are informative about the tail of F. Consequently, several other models of a similar genesis have been proposed to incorporate a greater amount of the data into the inference. Conceptually easiest is to consider the joint asymptotic distribution of the r-largest order statistics. The origins of this method can be traced back to Weissman (1978). Through his theoretical developments, Smith (1986) proposed a method for extending the classical analysis for the case when the r largest values are available for each year. Smith only considered these ideas for the limiting joint Gumbel distribution, but Tawn (1988) dealt with extensions of the limiting joint generalized extreme value distribution to the r-largest order statistics. The idea behind the r-largest order statistics method is the calculation of the limiting behavior of the variable

$$M_{n,k}(X) := k\text{th largest of}\,(X_1,\ldots,X_n),$$

for a fixed value of k, s $n \to \infty$. If (1) holds, then, for a fixed value of k the asymptotic distribution of $M_{n,k}$, properly normalized, converges to $G_k(x)$ with

$$G_k(x) := \exp\left\{-\left[1+\xi\left(\frac{x-\mu}{\sigma}\right)\right]^{-\frac{1}{\xi}}\right\}\sum_{s=0}^{k-1}\frac{\left[1+\xi\left(\frac{x-\mu}{\sigma}\right)\right]^{-\frac{s}{\xi}}}{s!},$$

defined on $\{x : 1+\xi(x-\mu)/\sigma > 0\}$. Note that $G_1(x) \equiv G(x)$. Hence, the joint probability density function of $M_n^{(r)} := (M_{n,1}\ldots,M_{n,r})$, properly normalized, is in the form

$$f(x_1,\ldots,x_r) = \exp\left\{-\left[1+\xi\left(\frac{x_r-\mu}{\sigma}\right)\right]^{-\frac{1}{\xi}}\right\} \times \prod_{k=1}^{r}\frac{1}{\sigma}\left[1+\xi\left(\frac{x_k-\mu}{\sigma}\right)\right]^{-\frac{1}{\xi}-1}, \qquad (2)$$

where $-\infty \leq \mu \leq \infty, \sigma > 0$ and $-\infty \leq \xi \leq \infty, x_r \leq x_{r-1} \leq \cdots \leq x_1$, and $\{x_r : 1+\xi(x_r-\mu)/\sigma > 0\}$ for $k = 1,\ldots,r$.

Standard statistical methodology from parametric estimation is available if the data consist of a sample of independent and identical distributed random variables. In this case, assuming that the r largest values for each of N years are known, equation (2) provides an approximate form of the joint density of the observations within a calendar year, in terms of parameters μ, σ, and ξ. Assuming that data from separated years are independent, then the product of these joint densities is the appropriate joint density for the set of whole observations. This may be treated as a likelihood function and hence the unknown parameters can be estimated by maximum likelihood methodology. For the asymptotic arguments to hold, the number of order statistics, r, used in each year must be chosen carefully, since small values of it will produce likelihood estimators with high variance whereas large values of r are likely to violate the asymptotic support for the model, leading to bias. In practice it is usual to select r as large as possible subject to adequate model diagnostics. Moreover, in assessing the accuracy of the maximum likelihood estimators, use is made of the observed information matrix, which is the Hessian of the negative log likelihood; its inverse provides a good approximation to the variance-covariance matrix of the estimates.

In order to extract the r-largest independent values it is necessary to use the concept of the *standard storm length*, first introduced by Tawn (1988). The standard storm length τ, is estimated through the analysis of the autocorrelation function of the process. As an estimator of τ, say $\hat{\tau}$, Tawn (1988) proposed the lag for which the autocorrelation at only a few lags greater that $\hat{\tau}$ exceed the confidence bound. Thus, for a given standard storm length, the extraction of the r-largest independent values is as follows:

1. pick out the largest daily value from within the year of interest;
2. discard values with a lag of $\hat{\tau}/2$ and less from either side of the value chosen in step 1;
3. repeat until the r-largest are extracted.

An extension of the classical approaches is to use threshold methods. These methods are based on fitting a stochastic model to either the exceedances or the peaks over a threshold. Threshold methods are based on the fact that if block maxima have approximating distribution GEV, then threshold excesses have a corresponding approximate distribution within the Generalized Pareto (GP) family, which has the representation

$$P(X > u+x|X > u) \approx \left(1+\xi^* \frac{x}{\sigma^*}^{\frac{1}{\xi^*}}\right), \tag{3}$$

for large thresholds u. Taking the limit $\xi^* \to 0$ gives the exponential distribution as a special case. EVT provides an asymptotic justification for the GP distribution to be an appropriate model for the distribution of excesses over a suitably chosen high threshold. This theory states that under appropriate normalization of the conditional excess that ensures non-degeneracy of the limit distribution, and under conditions on the smoothness of the distribution of the original variable, the only family of non-degenerate limit distributions is a GP distribution. Thus whatever distribution significant wave heights follow, provided a large enough threshold is used, significant wave heights exceeding that threshold should be well approximated by a GP distribution. The GP distribution offers additional flexibility in the range of tail behavior it can be described. So even when the limit theory justification is not considered appropriate, the flexibility of the upper tail of the model provides an alternative basis for use of this distribution. There is a mutual consistency between the GEV and GP distribution representations: if (3) applies with parameters ξ^* and σ^* and a *year* consists on the N independent observations $(Y_1,\dots,Y_N)$ so that the limiting distribution of $M_N(Y) := \max(Y_1,\dots,Y_N)$ is GEV(μ,σ,ξ) with parameters related by $\xi^* \equiv \xi$ and $\sigma^* \equiv \sigma + \xi(u-\mu)$. A third parameter in the GP distribution model specification is the exceedance rate of u, $\lambda := P(X > u)$, which is related to the GEV parameters by

$$\lambda = 1 - \exp\left\{ -\frac{1}{N}\left[1+\xi\left(\frac{u-\mu}{\sigma}\right)\right]^{-\frac{1}{\xi}} \right\}.$$

It follows that the mean crossing rate of any level $x(> u)$ is $\lambda\{1+\xi^*(x-u)/\sigma\}^{\frac{1}{\xi^*}}$. Setting this equal to $1/m$ we obtain the m-year return level, q_m, as

$$q_m := u + \frac{\sigma^*}{\xi^*}\left\{(\lambda m)^{\xi^*} - 1\right\}.$$

3. Bayesian Setting for Extreme Value Theory

Bayes' theorem converts a prior distribution for a parameter vector $\boldsymbol{\theta}$, $\pi(\boldsymbol{\theta})$, on the availability of historical data $\boldsymbol{x} := (x_1,\dots,x_n)$, into a posterior distribution $\pi(\boldsymbol{\theta}|\boldsymbol{x})$. This takes the form

$$\pi(\boldsymbol{\theta}|\boldsymbol{x}) := \frac{L(\boldsymbol{x}|\boldsymbol{\theta})\pi(\boldsymbol{\theta})}{\int_{\boldsymbol{\theta}} L(\boldsymbol{x}|\boldsymbol{\theta})}\pi(\boldsymbol{\theta})d\boldsymbol{\theta} \propto L(\boldsymbol{x}|\theta)\pi(\boldsymbol{\theta}),$$

where $L(\boldsymbol{x}|\boldsymbol{\theta})$ is the likelihood for the historical data. Note that the posterior distribution includes the additional information provided by the data $\boldsymbol{x}$. The prior distribution is intended to represent beliefs about parameter values, prior to the availability of data. In some contexts it may be appropriate to use the prior distribution to apply genuine knowledge about the process under study. In general, however, there is no such knowledge, so essentially arbitrary distributions with a large variance are adopted to reflect this prior ignorance. Since the GEV model admits no conjugate prior distributions, throughout this paper we adopt the approach used by Coles and Powell (1996); see also Stephenson and Tawn (2004) for further details. They suggest the trivariate Normal prior distribution on $\boldsymbol{\theta}^* := (\mu, \log\sigma, \xi)'$, which contains nine hyperparameters, and takes the form

$$\pi(\boldsymbol{\theta}) \propto \frac{1}{\sigma}\exp\left\{ -\frac{1}{2}(\boldsymbol{\theta}^* - \nu)^T \Sigma^{-1}(\boldsymbol{\theta}^* - \nu) \right\},$$

where ν and Σ represent the mean vector and the covariance matrix respectively. The reason for working with $\log\sigma$ is that it is an easier parametrization with which to respect the positivity on σ. The trivariate Normal distribution it is the only who enables the specification of independent parameters. This distribution is often used for a *naive* analysis where there is no external information with which to formulate a dependence structure.

Another important aspect when dealing with the posterior distribution, is the computation of the proportionality constant which can be problematic, specially for high-dimensional θ. Analytic computations on the distribution would be avoided if we could simulate form the posterior distribution directly and apply Monte Carlo integration techniques, but even this is unwieldy. An alternative procedure is to employ Markov Chain

Monte Carlo (MCMC) methods. The key idea is to construct a Markov chain, which has to be straightforward to simulate from, and whose equilibrium distribution is proportional to a required function, in our case the posterior distribution $\pi(\boldsymbol{\theta}|\boldsymbol{x})$. In this context $\pi(\boldsymbol{\theta}|\boldsymbol{x})$ is known as the *target distribution* of the Markov Chain. The innovations of these methods comes from the fact that the simulated values are drawn sequentially with distribution depending on the last value drawing; hence the simulated values form a Markov chain. It is important to stress that the key to the method's success is not the Markov property but rather that the approximate distributions are improved at each step in the simulation, in the sense of converging to the target distribution. This implies that if we run the chain for a long time, the chain can be used as a basis for summarizing features of the target distribution. To implement this strategy, we simply need algorithms for constructing chains with specified equilibrium distributions. The most general MCMC method is the Metropolis-Hastings algorithm which is a general term for a family of Markov chain simulation methods that are useful for drawing samples from Bayesian posterior distributions. The algorithm proceeds as follows: given a current value $\boldsymbol{\theta}^{(i)}$, a proposition value, say $\boldsymbol{\theta}^{\dagger}$, is simulated according to an arbitrary transition density, say $q(\boldsymbol{\theta}^{\dagger}|\boldsymbol{\theta}^{(i)})$. Then $\boldsymbol{\theta}^{(i+1)}$ is adopted according to the following rule:

$$\boldsymbol{\theta}^{(i+1)} = \begin{cases} \boldsymbol{\theta}^{\dagger} & \text{with probability } \Delta(\boldsymbol{\theta}^{(i)}, \boldsymbol{\theta}^{\dagger}) \\ \boldsymbol{\theta}^{(i)} & \text{with probability } 1 - \Delta(\boldsymbol{\theta}^{(i)}, \boldsymbol{\theta}^{\dagger}) \end{cases}$$

with

$$\Delta(\boldsymbol{\theta}^{(i)}, \boldsymbol{\theta}^{\dagger}) := \min\left\{1, \frac{\pi(\boldsymbol{\theta}^{\dagger})q(\boldsymbol{\theta}^{(i)}|\boldsymbol{\theta}^{\dagger})}{\pi(\boldsymbol{\theta}^{(i)})q(\boldsymbol{\theta}^{\dagger}|\boldsymbol{\theta}^{(i)})}\right\}.$$

Under mild conditions on q, the sequence $\{\boldsymbol{\theta}^{(i)}\}$ is bound to have equilibrium distribution proportional to the target distribution. The point of this algorithm is that the transition density q can be chosen almost arbitrarily, so that choices can be made which facilitates straightforward simulations. Common choices include random walks densities or *independence* models, in which $q(\boldsymbol{\theta}^{\dagger}|\boldsymbol{\theta}^{(i)})$ has fixed form independent of $\boldsymbol{\theta}^{(i)}$. Throughout this work, standard random walks with Gaussian steps were considered. A small amount of trial and error was needed to adapt the variances of the random walk terms to obtain a sequence that seemed to have reasonable mixing properties. Moreover, the rejection criterion for $\boldsymbol{\theta}^{(i+1)}$ then ensures that the required target distribution is obtained. A drawback is, however, that some choices for q can lead to very poor performances, so considerable experimentation is needed to obtain a scheme which is easy to implement and which also provides accurate inferences with relatively short simulation series. For further details see Smith and Roberts (1993).

An important issue when implementing MCMC methods is the assessment of the convergence of the algorithm. Although there is a reassuring theoretical literature concerning the convergence of MCMC methods (see, for example, Mergensen *et al.*, 1999), results do not easily translate into clear guidelines for the practitioners. Two of the most popular test of diagnostic were proposed by Geweke (1992) and by Raftery and Lewis (1992). The diagnostic of Geweke is particularly simple: for each parameter the means of the first and last parts of the chain are tested for equality. By default, the first 10% and the last 50% are used. The difference between the two means is divided by its estimated standard errors. The estimation of the standard errors attempts to take into account the autocorrelations. If the chain has reach equilibrium the distribution of such statistics is approximately standard Normal. The diagnostic of Raftery and Lewis attempts to assess how long the chain should be in order to achieve a given aim. Specifically, it gives the number of samples that are needed to estimate a quantile (on each margin) within a certain accuracy with at least probability 0.95.

As with all MCMC output, a major benefit of this approach is that prior and posterior distributions of functions of the parameters are easily evaluated as empirical estimates of the quantities of interest. Quantities of interest to us include prior and posterior distributions of quantiles and predictive distributions.

4. Time Series Cluster Methodology

A time series clustering procedure based on long-term predictions of extreme values of Baltic sea-level heights, is applied in the present study to describe regional sea-level variability in the Baltic. We outline here the essential of the method.

Our starting point is a panel of T time series $\mathbf{X} := (\mathbf{X}^{(1)},\ldots,\mathbf{X}^{(T)})$, observed for, say $n_1,\ldots,n_T$, time units respectively, that is $\mathbf{X}^{(i)} := (X_{1,i},\ldots,X_{n_i,i})$ for $i = 1,\ldots,T$. The implementation of the method for clustering time series is carried out in three stages: first, the algorithm starts with the estimation of the posterior predictive distributions for each time series, for 25-, 50-, and 100-years return values by means of the approach described in Section 3.. In order to accommodate features often exhibited by temperature records such as trends or seasonality, the location and scale parameters are allowed to vary in time (in this case years). Specifically, the location parameter is modeled as $\mu(t) := \beta_0 + \beta_1 t$. Moreover, the scale parameter is expressed as $\sigma(t) := \exp(\gamma_0 + \gamma_1 t)$, where the exponential function is used to ensure that the positivity of σ is maintained for all values of t. On the other hand we took a near-flat Normal multivariate distribution for $\boldsymbol{\theta} := (\beta_0,\beta_1,\gamma_0,\gamma_1,\xi)$ as a prior distribution, which reflect the absence of external information. A Markov chain is generated (by means of the Metropolis-Hastings algorithm) of length $n = 10000$, $(\boldsymbol{\theta}_0,\ldots,\boldsymbol{\theta}_n)$, including a *burn-in* period of 5000 observations with target distribution $\pi(\boldsymbol{\theta}|\boldsymbol{x})$ being the initial values $\boldsymbol{\theta}_0$ the maximum log-likelihood estimates obtained from the distribution of the r-largest order statistic model. Furthermore, only every fifth iteration is stored in order to obtain an independent and identically distributed sample. From the Markov Chain sequence a sample from the posterior predictive distribution of the return value $x_p^{(i)}$, associated with the ith time series, say $\boldsymbol{x}_p^{(i)} := (x_{p,1}^{(i)}, x_{p,2}^{(i)},\ldots,x_{p,B}^{(i)})$, can be generated as follows: let $G^{-1}(\cdot|\boldsymbol{\theta}_j)$ be the inverse of the extreme value distribution in (1) then $x_{p,j}^{(i)} = G^{-1}(1-p|\boldsymbol{\theta}_j)$, for $j = 1,\ldots,B$. Thus, an estimate of $F_{x_p^{(i)}}(z|\boldsymbol{x})$ is given by

$$\widehat{F}_{x_p^{(i)}}(z|\boldsymbol{x}) := \frac{1}{B}\sum_{j=1}^{B} I_z(x_{p,j}^{(i)}),\ z > 0,$$

where $I_z(\cdot)$ represents the indicator function. Next, we compute the *dissimilarity matrix*, $D := (D_{ij})$, $j,i = 1,\ldots,T$. In doing so, an adequate metric between univariate distribution functions is required. The choice of this metric should reflect the final goal of the clustering procedure in the sense that the distances capture the discrepancies between predictive distributions of return values. In this case, the weighted L_2-Wasserstein distance between posterior predictive distributions is adopted. This means that the distance between two time series, say, $\mathbf{X}^{(i)}$ and $\mathbf{X}^{(j)}$ is defined as

$$D_{ij} := \int_0^1 \left(F^{-1}_{x_p^{(i)}}(y|\boldsymbol{x}) - F^{-1}_{x_p^{(j)}}(y|\boldsymbol{x}) \right)^2 y(1-y)\,dy,$$

where $F_{x_p^{(i)}}(\cdot|\boldsymbol{x})$ and $F_{x_p^{(j)}}(\cdot|\boldsymbol{x})$ denote the posterior predictive distribution functions of the return values of the ith and jth time series $x_p^{(i)}$ and $x_p^{(j)}$ respectively, with $p = 1/m$, corresponding to a return period of m years. The distances D_{ij} are estimated through the expression

$$\widehat{D}^2_{ij} := \sum_{\ell=1}^{h} \left(\widehat{F}^{-1}_{x_p^{(i)}}(s_\ell|\boldsymbol{x}) - \widehat{F}^{-1}_{x_p^{(j)}}(s_\ell|\boldsymbol{x}) \right)^2 s_\ell(1-s_\ell), \tag{4}$$

where $s_\ell := \ell/h$, i.e., $(s_1, s_2, \ldots, s_h)$ is a regular grid in the interval $(0,1)$. In the present analysis, we have considered $h = 100$. Notice that the expression in (4) is a weighted sum of the squared difference of the percentiles for the distribution of the ith and jth series. Finally a dendrogram based on the application of classical cluster techniques to the dissimilarity matrix is built and that gives us the different clusters formed by the predictive distributions for 25-, 50- and 100-years return values. In particular, agglomerative hierarchical methods with nearest distance (single linkage), furthest distance (complete linkage) and unweighted average distance (average linkage) are used as grouping criteria. In the single linkage case, each member of a cluster has another member at a distance lower than the vertical distance at which they are joined. In the complete linkage case all cluster members are at a distance lower than the distance at which they are joined. Finally in the average linkage clustering the average distance from one member to the remaining members is lower than the distance at which they are joined. Further details can be found in Everitt *et al.* (2001) and Jolliffe and Philipp (2009). In the next section, the results obtained by applying the average linkage procedure are presented. Similar conclusions are obtained using the other two methods. The corresponding diagrams are available upon request to the authors.

5. Exploring the Baltic Data Set

Sea-level monitoring has a long tradition in the Baltic Sea area, in part motivated by the study of the postglacial rebound of Fennoscandia (e.g. Ekman, 1996). Thus a number of very long and high-quality records of relative sea-level measurements are available from tide gauges in the Baltic sea region. A total of 8 stations (see Table 1 and Figure 1) with daily records extending at least up to 2005 (clustering based on forecast densities requires time series data ending at the same time) are considered. The data from Gedser and Hornbæk are obtained from the Danish Meteorological Institute, DMI (Hansen, 2007). Tide gauge data from Goteborg and Stockholm are obtained from the University of Hawai Sea Level Center, UHSLC. Finally the daily records from Furuögrund, Kungsholmsfort, Ölands Norra Ude and Ratan stations are provided by the Swedish Meteorological and Hydrological Institute, SMHI. The mean and a linear trend have been removed from the data in order to eliminate postglacial rebound signals from the tide gauge records. The analyzed time series are shown in Figure 2.

Table 1. Analyzed tide gauge records.

Tide gauge	Station code	Period	Longitude	Latitude
Furuögrund	FUR	Jan 1900 - Dec 2007	21.23	64.92
Gedser	GED	Jan 1900 - Dec 2005	11.93	54.57
Goteborg	GOT	Jan 1967 - Dec 2006	57.68	11.80
Hornbæk	HOR	Jan 1900 - Dec 2005	12.47	56.10
Kungsholmsfort	KUN	Jan 1900 - Dec 2007	15.58	56.10
Ölands Norra Udde	OLA	Jan 1900 - Dec 2007	17.10	57.37
Ratan	RAT	Jan 1900 - Dec 2007	20.92	64.00
Stockholm	STO	Jan 1889 - Dec 2007	18.08	59.32

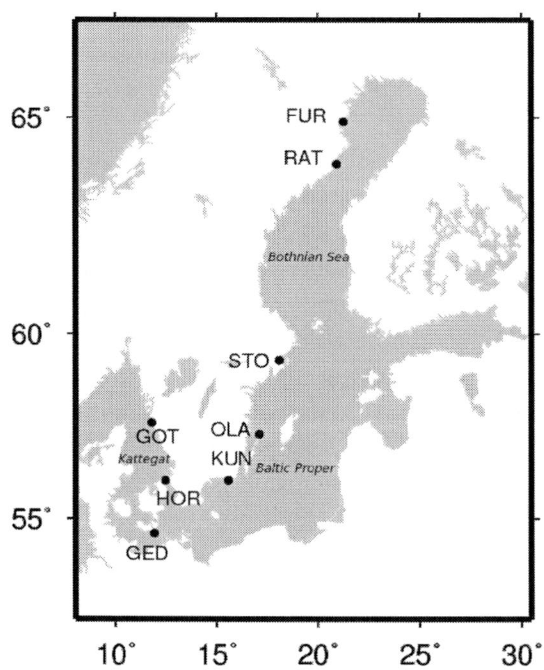

Figure 1. Baltic Sea area and tide gauge locations. The station names are as in Table 1.

The method described in Sections 2., 3. and 4. is applied to compute 25-, 50- and 100-years return values (i.e. for 2030, 2055 and 2105). Table 2 below summarizes the results of the Bayesian analysis, including the r-largest order statistic and the estimates (i.e. mean of the marginal posterior distributions) of the location, μ, scale ,σ, and shape, ξ, parameters of the GEV distribution, with the location and scale parameters assumed to evolve in time as $\mu(t) := \beta_0 + \beta_1 t$ and $\sigma(t) := \exp(\gamma_0 + \gamma_1 t)$, respectively. The choice of r was carried out through the use of the likelihood ratio test which is based on the set of statistics $S_{(i,j)} := 2\{\ell(r_i) - \ell(r_j)\} \sim \chi_1^2$, where $\ell(r_i)$ is the log-likelihood function of r_i. A test of the validity of model based on r_j relative to r_i at the α level of significance is to

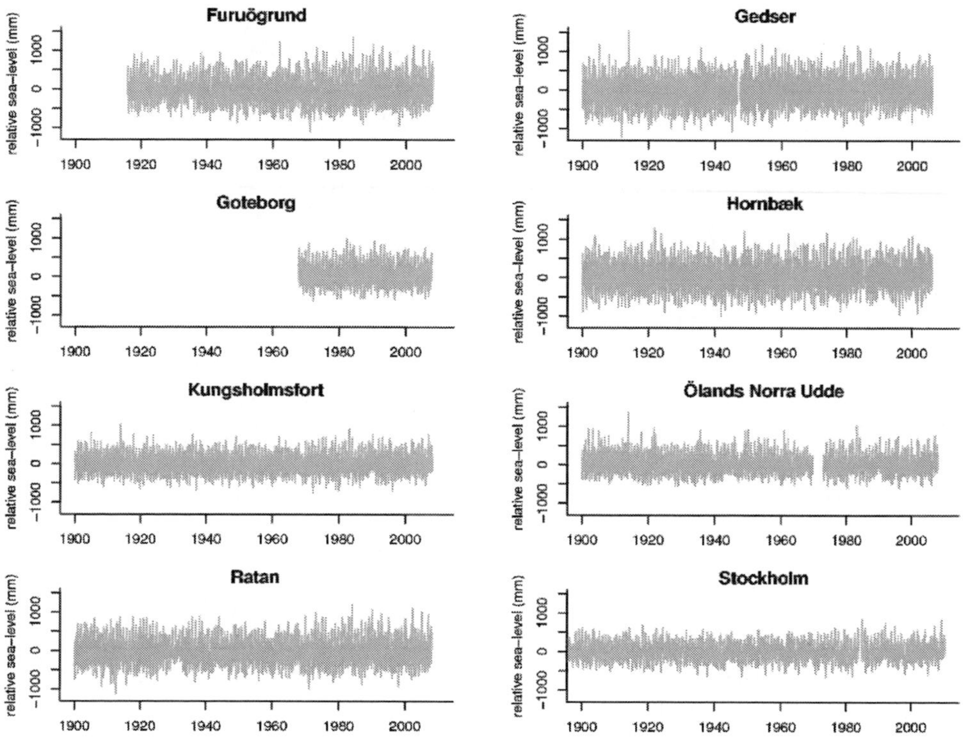

Figure 2. Time series of relative sea-level measurements.

reject r_j if $S_{(i,j)} > c_\alpha$ where c_α is the $(1-\alpha)$ quantile of the χ_1^2 distribution. Moreover the validity of the models was checked trough the application of graphical methods, in particular the probability plot, the *Q-Q* plot, the return level plot, and the comparison of the probability density function of the fitted model with a histogram of the data. For all cases considered the accuracy of the fit is satisfactory.

A closer look at Table 2 reveals that for all locations, the posterior mean of ξ is negative. Furthermore the posterior probability of non-negative values for ξ in Furuögrund, Hornbæk, Kungsholmsfort, Ratan, and Stockholm (FUR, HOR, KUN, RAT, STO) is negligible clearly indicating that a bounded upper tail distribution is a reasonable choice to fit the data sets corresponding to these locations. This is in contrast with the posterior mean of ξ for Gedser, Goteborg, and Ölands Norra Udde (GED, GOT, OLA), for which the posterior probability of non-negative values for ξ is non-negligible so that distributions with light tails (bounded or unbounded) are also consistent with the data. The location parameter for Goteborg (GOT) exhibits a statistically significant trend of 14 cm per year, and a positive trend of nearly 13 cm is also found at Ratan, indicating a considerable increase in extreme sea-levels at these two tide gauges. For the remaining stations the slopes of the location parameter are much smaller or even not significant. Table 3 presents predictive return values estimates for 25-, 50- and 100-years, including posterior standard deviations. One of the advantages of the Bayesian framework is that it allows to retrieve

Table 2. Parameter estimates and posterior standard deviations (in parentheses).

		Parameter estimates				
Location	r	$\hat{\beta}_0$	$\hat{\beta}_1$	$\hat{\gamma}_0$	$\hat{\gamma}_1$	$\hat{\xi}$
FUR	5	744.07(11.87)	24.14(41.20)	4.89(0.04)	-0.02(0.13)	-0.15(0.02)
GED	7	702.53(10.50)	44.95(37.21)	2.13(0.72)	3.29(2.63)	-0.06(0.02)
GOT	7	556.70(8.49)	141.06(29.47)	4.46(0.04)	0.35(0.23)	-0.04(0.03)
HOR	6	727.03(9.63)	28.28(31.95)	2.10(0.63)	2.48(2.73)	-0.14(0.02)
KUN	7	572.80(7.43)	31.50(23.60)	1.97(0.58)	1.24(2.22)	-0.11(0.02)
OLA	7	629.00(9.56)	13.58(31.43)	4.77(1.02)	0.08(0.10)	-0.10(0.03)
RAT	5	701.83(9.43)	129.37(29.29)	4.75(1.38)	3.55(3.09)	-0.22(0.02)
STO	6	525.52(6.23)	59.97(19.33)	4.41(0.97)	2.22(1.93)	-0.20(0.02)

the complete posterior density of the return values, instead of a single return value. Figure 3 displays the predictive distributions for 25-, 50- and 100-years, and illustrates the relevance of considering the complete density of returns, for example Goteborg (GOT) and Hornabæk (HOR) display similar 25-years return values but the corresponding densities are very different, with a much higher dispersion in the case of Goteborg. Note also that while for 25-years and 50-years returns Goteborg and Hornabæk exhibit very similar values, for 100-years returns the sea-level height at Goteborg is much larger (by about 30 cm) that at Hornabæk.

For a spatial interpretation of the estimated return levels, time series clustering is applied as described in Section 4. The results of the clustering procedure are represented by a tree diagram known as dendrogram which displays the arrangement of the clusters produced by hierarchical agglomerative clustering. Figure 4 shows the dendrograms based on average linkage obtained for 25-, 50- and 100-years horizons. The horizontal axis represents the distance at which two clusters are joined using the unweighted average distance as a grouping criteria. For 25-years returns, the group corresponding to the smallest distance between stations includes the stations at both extremities of the Baltic Sea, both in the Bothnian area (Furuögrun, FUR and Ratan, RAT) and in the western Baltic and Kattegat (Gedser, GED, Goteborg, GOT and Hornabæk, HOR). This group is further joined with Ölands Norra Udde (OLA) and finally with the group of the lowest return values, that contains Kungsholmsfort (KUN) and Stockholm (STO) stations. For 100-years returns a similar structure is found, but Goteborg (GOT) joins Gedser and Ratan in the cluster of higher return values, while Ölands Norra Udde is included in the intermediate group with Hornabæk and Furuögrund.

6. Conclusions

This work presents a Bayesian analysis of extreme sea-level heights in the Baltic area. First a Bayesian extreme value analysis is carried out for deriving the complete distributions

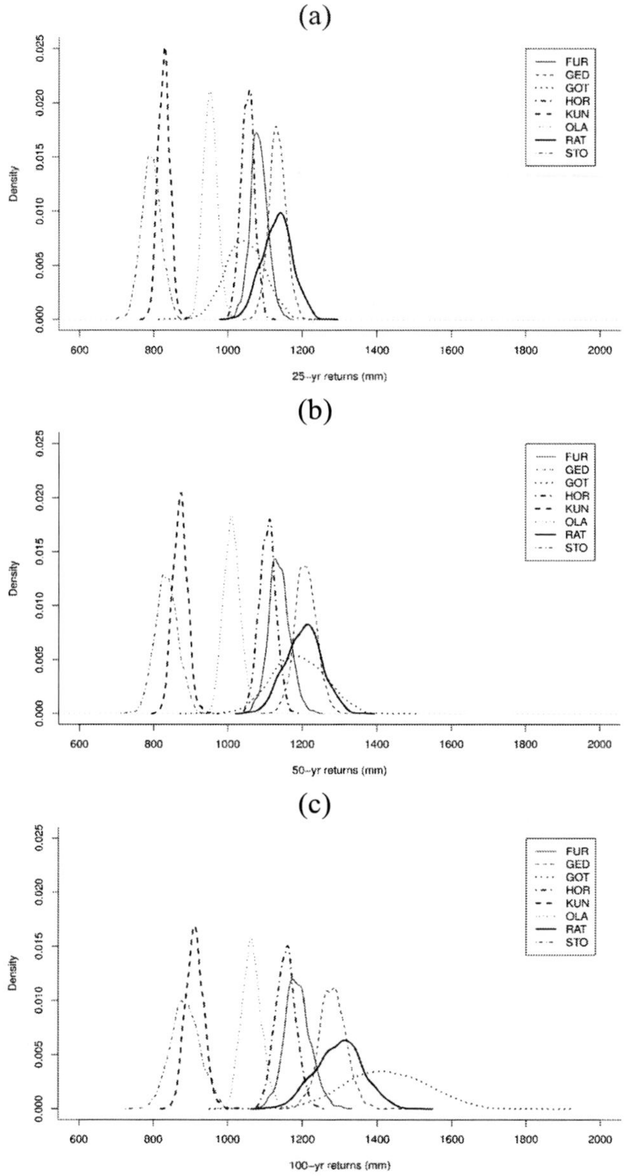

Figure 3. Predictive distributions for 25(a), 50(b)- and 100(c)-years.

of return periods and further combined with a time series clustering procedure in order to obtain a description of the relationship between return values at different locations.

The results show distinct groupings of return levels, with lowest returns for the stations located in the Baltic Proper, and highest return values at both extremities of the Baltic area (in the Bothnian Sea, on the eastern side, and the Kattegat and Danish straits, on the western side). This spatial distribution suggests the influence of atmospheric events on extreme sea-levels, possibly associated with the effect of persistent winds over the North Sea and the Baltic entrance driving water in and out of the Baltic, depending on the wind direction.

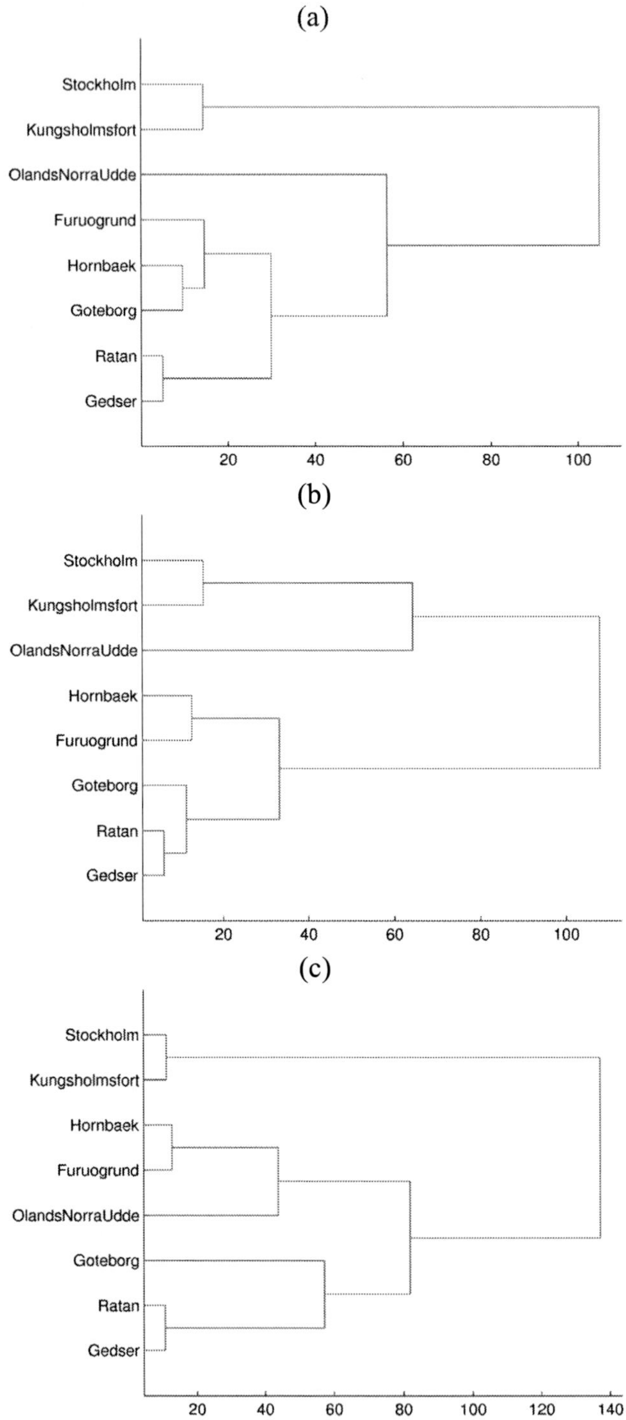

Figure 4. Dendrogram for 25(a), 50(b)- and 100(c)-years return values based on the average linkage method.

Table 3. Bayesian estimates for the return values of 25-, 50- and 100-years. Posterior standard deviations in parenthesis.

	Return values (mm)		
Location	25	50	100
FUR	1083.20(23.35)	1138.38(28.51)	1187.72(34.16)
GED	1133.54(12.70)	1209.16(28.46)	1280.82(35.10)
GOT	1046.64(51.56)	1190.54(71.15)	1423.4(108.77)
HOR	1052.97(18.22)	1107.12(22.31)	1155.89(26.90)
KUN	827.17(16.16)	871.56(20.07)	912.22(24.40)
OLA	952.50(18.43)	1010.66(22.13)	1064.47(26.26)
RAT	1132.72(40.36)	1202.83(48.01)	1297.56(62.18)
STO	792.56(25.64)	834.52(30.38)	886.65(39.22)

A clear advantage of the approach applied in this work is that clustering is performed on the distribution of return values rather than on the return values themselves, enabling a more complete description of sea-level extremes and corresponding uncertainties.

Acknowledgments

The three authors are supported by "Acções Integradas Luso-Espanholas" under the grants E-83/09 and HP2008-008. S.M. Barbosa is also supported by FCT through the Compromisso para a Ciência Programme Ciência 2008. Tide gauge data kindly provided by DMI (K. Madsen), SMHI (T. Hammarklint) and UHSLC.

References

[1] Barbosa SM. (2008). Quantile trends in Baltic sea-level. *Geophysical Research Letters* **35**, L22704.

[2] Bijl W, Flather RA, de Ronde JG, Schmith T. (1999). Changing storminess? an analysis of long-term sea level data sets. *Climate Research* **11**, 161–172.

[3] Castillo E, Hadi AS, Balakrishnan N, Sarabia JM. (2004). *Extreme Value and Related Models with Applications in Engineering and Science*. John Wiley & Sons: Chichester.

[4] Coles SG. (2001). *An Introduction to Statistical Modeling of Extreme Values*. Springer-Verlag: London.

[5] Coles SG, Powell EA. (1996). Bayesian methods in extreme value modelling: a review and new developments. *International Statistical Review* **64**, 119–136.

[6] Egozcue JJ, Pawlowsky-Glahn V, Ortego MI. (2005). Wave-height hazard analysis in Eastern Coast of Spain - Bayesian approach using generalized Pareto distribution. *Advances in Geosciences* **2**, 25–30.

[7] Ekman M. (1988). The world's longuest continued series of sea level observations. *Pure and Applied Geophysics* **127**, 73–77.

[8] Ekman M. (1996). A consistent map of the postglacial uplift of Fennoscandia. *Terra Nova* **8**, 158–165.

[9] Ekman M. (1999). Climate changes detected through the world's longest sea level series. *Global and Planetary Change* **21**, 215–224.

[10] Everitt BS, Landau S, Leese M. (2001). *Cluster Analysis*. Arnold: London.

[11] Hansen L. (2007). Hourly values of sea level observations from two stations in Denmark Hornbæk 1890-2005 and Gedser 1891-2005. *DMI Technical Report No. 07-09*.

[12] Jolliffe IT, Philipp A. (2009). Some recent developments in cluster analysis. *Physics and Chemistry of the Earth* (to appear) DOI:10.1016/j.pce.2009.07.014.

[13] Katz RW, Parlange MB, Naveau P. (2002). Statistics of extremes in hydrology. *Advances in Water Resources* **25**, 1287–1304.

[14] Katz RW, Brush GS, Parlange MB. (2005). Statistics of extremes: modelling ecological disturbances. *Ecology* **86**, 1124–1134.

[15] Geweke J. (1992). Evaluating the accuracy of sampling-based approaches to the calculation of posterior moments. *Bayesian Statistics 4* (J. M. Bernardo, J. O. Berger, A. P. Dawin and A. F. M. Smith, eds.). Oxford: University Press, 169–193, (with discussion).

[16] Meehl GA, Easterling KT, Changnon S, Pielke R, Changnon D, Evans J, Groisman PY, Knutson TR, Kunkel KE, Mearns LO, Parmesan C, Pulwarty R, Root T, Sylves RT, Whetton P, Zwiers F. (2000). An introduction to trends in extreme weather and climate events: observations, socioeconomic impacts, terrestrial ecological impacts, and model projections. *Bulletin of the American Meteorological Society* **81**, 413–416.

[17] Mendez FJ, Menendez M, Luceno A, Losada IJ. (2007). Analyzing monthly extreme sea levels with a time-dependent GEV model. *Journal of Atmospheric and Oceanic Tecnology* **24**, 894–911.

[18] Mergensen KL, Robert CP. Guihenneuc-Jouyaux C. (1999). MCMC convergence diagnostic: a review. *Bayesian Statistics 6* (J. M. Bernardo, J. O. Berger, A. P. Dawin and A. F. M. Smith, eds.). Oxford: University Press, 415–440, (with discussion).

[19] Raftery AE, Lewis S. (1992). How many interactions in the Gibbs sampler? *Bayesian Statistics 4* (J. M. Bernardo, J. O. Berger, A. P. Dawin and A. F. M. Smith, eds.). Oxford: University Press, 763–773.

[20] Reiss RD, Thomas M. (2007). *Statistical Analysis of Extreme Values: with Applications to Insurance, Finance, Hydrology and Other Fields*. Birkhauser: Basel.

[21] Scotto MG, Guedes Soares C. (2007). Bayesian inference for long-term prediction of significant wave height. *Coastal Engineering* **54**, 393-400.

[22] Scotto MG, Alonso AM, Barbosa SM. (2010). Clustering time series of sea levels: extreme value approach. *Journal of Waterway, Port, Coastal, and Ocean Engineering* **136**, 215-225.

[23] Smith AFM, Roberts GO. (1993). Bayesian computation via the Gibbs sampler and related Markov chain Monte Carlo methods (with discussion). *Journal of the Royal Statistical Society. Series B* **55**, 3–23.

[24] Smith RL. (1986). Extreme value theory based on the r largest annual events. *Journal of Hydrology* **86**, 27–43.

[25] Smith RL. (1989). Extreme value analysis of environmental time series: an approach to trend detection in ground-level ozone. *Statistical Science* **4**, 367–393.

[26] Stephenson A. Tawn JA. (2004). Bayesian inference for extremes: accounting for the three extremal types. *Extremes* **7**, 291–307.

[27] Stigebrandt A. (1984). Analysis Of an 89-year-long sea-level record from the kattegat with special reference to the barotropically driven water exchange between the Baltic and the sea. *Tellus* **36A**, 401–408.

[28] Tawn JA. (1988). An extreme value theory model for dependent observations. *Journal of Hydrology* **101**, 227–250.

[29] Tsimplis MN, Blackman DL. (1997). Extreme sea-level distribution and return periods in the Aegean and Ionian seas. *Estuarine, Coastal and Shelf Science* **44**, 79–89.

[30] Unnikrishnan AS, Sundar D, Blackman DL. (2004). Analysis of extreme sea level along the east coast of India. *Journal of Geophysical Research* **109**, C06023.

[31] Weissman J. (1978). Estimation of parameters and lange quantiles based on the k largest observations. *Journal of the American Statistical Association* **73**, 812–815.

[32] Woodworth PL, Blackman DL. (2004). Evidence of systematic changes in extreme high waters since the mid-1970s. *Journal of Climate* **17**, 1190–1197.

In: Sea Level Rise, Coastal Engineering... ISBN 978-1-61728-655-1
Editor: Linda L. Wright, pp. 295-317 © 2011 Nova Science Publishers, Inc.

Chapter 16

Neural-Network Modeling and Data Analysis Techniques in Coastal Hydrodynamic Studies — a Review

Agnieszka Herman*
Institute of Oceanography,
University of Gdańsk, Poland

PACS 05.45-a, 52.35.Mw, 96.50.Fm.

Keywords: coastal hydrodynamics; nonlinear multivariate analysis; artificial neural networks; pattern recognition; nonlinear principal component analysis; nonlinear regression; hybrid models; nearshore currents; SandyDuck97.

1. Introduction

Physical processes in the coastal zone are shaped by numerous factors acting on a broad range of spatial and temporal scales. Shallow-water hydrodynamics is highly nonlinear, which presents a challenge to the available methods of studying and analyzing them. Generally, the difficulties may be broadly grouped into two major categories. The first group relates to issues concerning mathematical description of the processes involved, as well as numerical implementation of the theoretical models. Turbulence, mixing, wave breaking, nonlinear wave–wave interactions are examples of processes that are difficult to measure both in the field and under laboratory conditions. The mathematical models used to describe them are often approximate, semi-empirical and valid in certain (although often not easy to specify) range of conditions. Moreover, even parameterized, highly simplified formulations of nonlinear terms in the governing equations require sophisticated numerical schemes, which make the resulting models computationally expensive. Nonlinear wave–wave interactions in spectral wind-wave models or turbulent mixing in hydrodynamic models are good examples of such processes—algorithms that handle them consume a substantial part of the overall computation time and memory required by the respective models.

*E-mail address: oceagah@univ.gda.pl

This raises the need of searching for alternative methods of accounting for processes that, on the one hand, are crucial for achieving satisfactory quality of the modeling results, and, on the other hand, hard to handle with 'traditional' numerical techniques. It is important especially in models run operationally and used for forecasting purposes; the accuracy of the results and computational effectiveness of a model become then the two most important criteria of its usefulness. In such cases, neural-network and other 'nonphysical' methods provide an attractive alternative to 'physical' ones.

The second category of challenges comprises issues of analyzing, interpreting, and visualizing the (huge amounts of) data resulting from numerical modeling and/or measurements (esp. spatially-resolved remotely sensed data). In hydrodynamic modeling of the coastal zone, computational grids with more than 10^6 nodes are no exception; short time steps of minutes or seconds are often required by stability conditions of numerical schemes. As a result, huge data sets are produced with spatial and temporal resolution higher than would be necessary for a subsequent analysis of a given scientific or engineering problem. The actually useful information remains 'hidden' in the data and must be extracted and transformed into a handleable form. Advanced data-analysis methods are necessary e.g. to extract dominant spatio-temporal patterns in the data, or to formulate and test hypotheses regarding possible cause-and-effect relationships among the variables involved. Both groups of problems present an excellent opportunity for application of modern data analysis and modeling techniques, which enable to go beyond simple linear relationships between the parameters analyzed.

This chapter presents recent developments concerning application of neural-network-based nonlinear modeling and data analysis techniques in coastal engineering, with particular attention to analysis of the hydrodynamics of the coastal zone. The chapter provides a concise description of the theoretical basis of artificial neural network (ANN) methods (Section 3.), as well as examples of their application and a discussion of their strengths and limitations. The methods presented include: gap-filling and interpolation (Section 4.1.); pattern recognition with nonlinear principal component analysis (Section 4.2.); nonlinear regression (Section 5.1.); and hybrid ANN and physically-based models (Section 5.2.). Cross-shore and alongshore current data from a field experiment in North Carolina, USA (Section 2.) are used to illustrate possibilities and limitations of some of these data-analysis and modeling techniques. The chapter concludes with a summary of recent developments and a brief description of some other predictive-learning methods with potential applicability in coastal engineering.

2. Data and Software

In this chapter, apart from previously published results, selected data collected during the SandyDuck'97 (SD97) Coastal Field Experiment will be used to demonstrate some of the analysis methods presented. In particular, cross-shore and alongshore surf-zone currents, measured as a part of the "Surf Zone Waves, Currents, and Morphology" experiment, are used, see [Elgar et al.(1997)]. The experiment, funded by the U.S. Office of Naval Research, was hosted at the Field Research Facility (FRF) of the U.S. Army Engineer Research and Development Center—a sandy beach facing the Atlantic Ocean, located near Duck, North Carolina. The objective of the project was to collect comprehensive field data

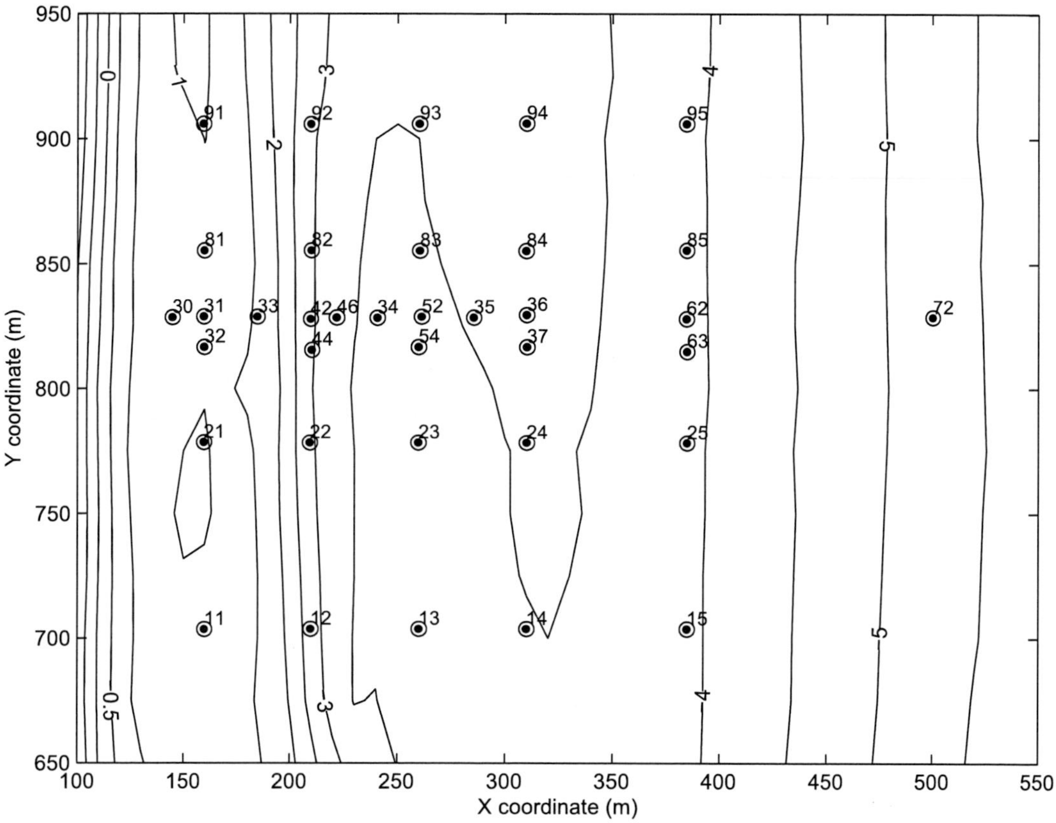

Figure 1. Bathymetry (in meters; measured on 30 September 1997) of the SandyDuck'97 area and location of 36 current meters operated during the experiment.

that could be used to advance numerical models of wave transformation, wave-driven circulation and sediment transport nearshore, including the surf zone. See e.g. [Feddersen et al.(1998), Conley and Beach(2003), Feddersen and Guza(2003), Noyes et al.(2004)] for works using the SD97 data.

Figure 1 shows the bathymetry of the study area and the SD97 measurement array with positions of the 36 individual instruments, consisting of sonar altimeters, pressure gauges and bi-directional current meters deployed between the shoreline and the 5-m depth contour between 1 August and 5 December 1997. In the following, mean cross-shore (U; positive onshore) and alongshore (V; positive southward) velocities from 36 stations are used. At most stations no more than a few percent of the data are missing, and most gaps are relatively short. However, at least one longer gap (more than 100 subsequent observations) is present at most stations, especially those situated in shallow water. Generally, alongshore currents are much stronger than cross-shore currents (compare σ_V and V_{max} with σ_U and U_{max} in Table 1 and panels a,b and c,d in Fig. 2) and their spatial variability is much more uniform, especially in the alongshore direction. Further description of these data is given in the following sections, together with the analysis of the ANN results.

The simulations presented here were performed with NETLAB, a package of pattern-

Table 1. Basic characteristics of SD97 current data

Station	g_{max}	n_{gap}	$\bar{U}$	σ_U	U_{max}	n_{gap}	$\bar{V}$	σ_V	V_{max}
11	220	22.4	−0.10	0.18	0.59	22.4	−0.02	0.39	1.73
12	820	25.3	−0.03	0.05	0.14	27.4	0.02	0.19	0.94
13	240	7.4	−0.03	0.06	0.20	7.4	0.04	0.16	1.27
14	80	5.2	−0.04	0.06	0.14	5.2	0.03	0.17	1.25
15	150	6.2	−0.02	0.04	0.08	6.8	0.05	0.17	1.45
21	930	32.6	−0.08	0.11	0.39	29.7	−0.03	0.39	1.75
22	1120	18.5	−0.03	0.04	0.10	18.5	0.04	0.20	1.28
23	170	5.9	−0.02	0.05	0.14	5.9	0.05	0.21	1.46
24	170	7.0	−0.02	0.05	0.13	7.0	0.04	0.17	1.34
25	180	6.0	−0.03	0.04	0.07	6.0	0.04	0.15	1.30
30	170	19.1	−0.08	0.11	0.28	19.1	0.00	0.39	1.86
31	570	25.4	−0.07	0.11	0.34	25.3	0.01	0.39	1.67
32	310	23.8	−0.09	0.14	0.54	23.8	0.03	0.41	1.90
33	110	5.8	−0.07	0.08	0.16	5.8	0.04	0.30	1.84
34	350	6.4	−0.03	0.04	0.14	3.0	0.05	0.20	1.33
35	80	3.7	−0.03	0.05	0.11	3.7	0.04	0.17	1.30
36	80	4.0	−0.04	0.05	0.09	7.9	0.04	0.13	1.00
37	20	2.7	−0.04	0.07	0.14	6.3	0.08	0.21	1.62
42	770	14.6	−0.04	0.04	0.08	14.6	0.04	0.21	1.39
44	270	6.5	−0.03	0.06	0.09	6.5	0.02	0.21	1.12
46	970	15.4	−0.02	0.03	0.11	15.4	0.02	0.17	1.20
52	100	4.7	−0.03	0.04	0.16	4.6	0.05	0.17	1.25
54	80	3.5	−0.03	0.04	0.19	3.5	0.04	0.16	1.19
62	20	2.3	−0.02	0.04	0.09	2.3	0.05	0.15	1.30
63	1670	19.1	−0.05	0.07	0.18	19.1	0.05	0.21	1.49
72	240	6.1	−0.03	0.04	0.10	5.7	0.06	0.13	1.12
81	740	25.0	−0.07	0.12	0.41	24.3	0.01	0.40	1.72
82	290	8.1	−0.04	0.04	0.11	8.1	0.04	0.23	1.40
83	160	6.1	−0.03	0.05	0.14	9.3	0.05	0.20	1.35
84	160	5.5	−0.04	0.06	0.09	5.5	0.05	0.19	1.52
85	160	6.6	−0.02	0.04	0.09	6.6	0.06	0.18	1.56
91	4010	57.6	−0.08	0.11	0.35	57.6	−0.03	0.37	1.82
92	200	10.7	−0.04	0.06	0.14	10.7	0.04	0.25	1.62
93	290	10.1	−0.03	0.06	0.13	10.1	0.05	0.19	1.33
94	170	6.7	−0.03	0.06	0.10	6.7	0.05	0.18	1.42
95	310	9.2	−0.03	0.05	0.19	9.3	0.05	0.17	1.47

g_{max} – maximum gap length, n_{gap} – fraction of missing data
$\bar{U}$, σ_U, U_{max} – mean, standard deviation and maximum of U (m/s)
$\bar{V}$, σ_V, V_{max} – mean, standard deviation and maximum of V (m/s)

recognition Matlab® scripts developed by [Nabney(2004)] and freely available at `http://www.ncrg.aston.ac.uk/netlab/index.php`. The Nonlinear Principal Component Analysis (NLPCA), presented in Section 4.2., was performed with the NeuMATSA ('Neuralnets for Multivariate And Time Series Analysis', version 5.0.1) package by [Hsieh(2008)], downloadable from `http://www.ocgy.ubc.ca/projects/clim.pred/download.html`.

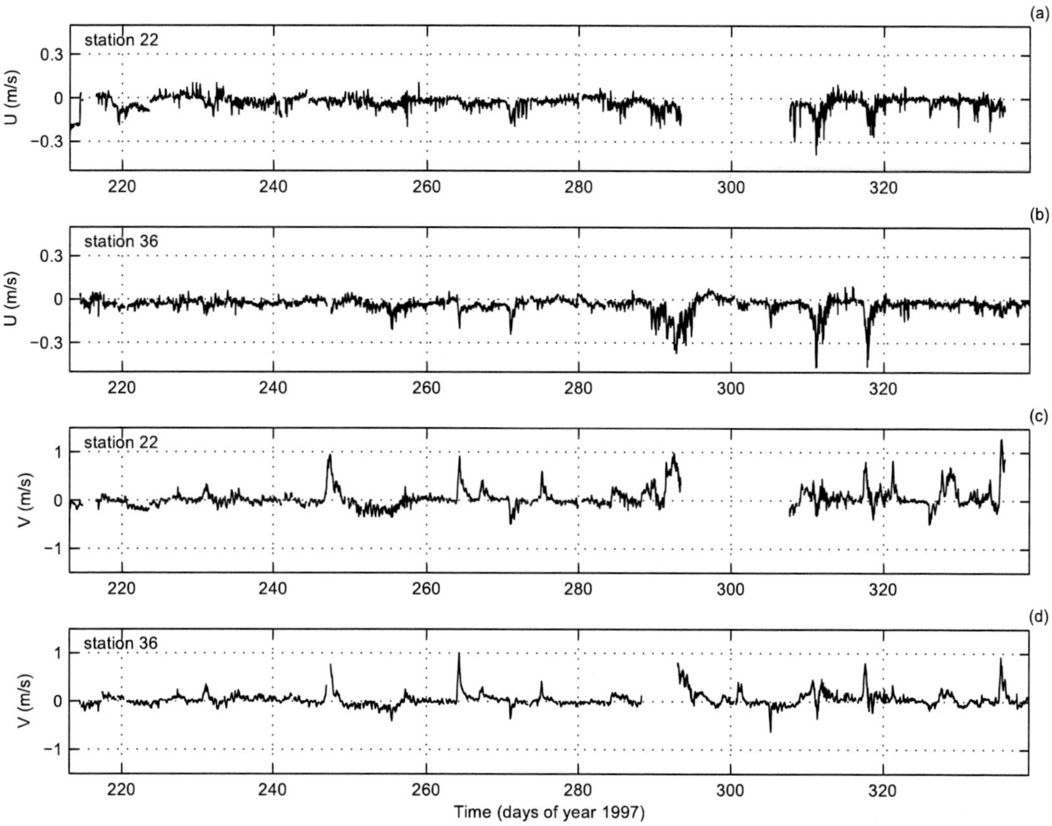

Figure 2. Measured cross-shore (a,b) and alongshore (c,d) current velocity (in m/s) at stations 22 (a,c) and 36 (b,d). Note different vertical scales in (a,b) and (c,d).

3. Theory of Artificial Neural Networks

Neural networks are particularly useful for problems that are difficult to handle with 'traditional' methods. Typical application fields include classification, pattern recognition, identification, nonlinear multivariate regression and many others, see, e.g., [Ripley(1996)], [Nabney(2004)], [Tirozzi et al.(2006)], [Hsieh(2009)]. Excellent introduction to ANN methods can be found in [Aleksander and Morton(1990)], [Wu(1993)] and [Hagan et al.(1995)], among others. This section provides only a brief theoretical background of artificial neural networks, which form a basis for modeling and analysis techniques described further in this chapter.

Artificial neural networks (ANNs) are inspired by biological nervous systems, in which interactions between and cooperation among a large number of simple elements (neurons) results in a powerful, highly efficient system, capable of storing and utilizing experiential knowledge. Similarly as in the brain, the knowledge is acquired by an ANN through a learning process. During learning, interneuron connection strengths (synaptic weights) are adjusted to optimize the ANN performance with respect to the learning data (i.e., sets of 'problems' and their 'solutions' presented to the network). Subsequently, the knowledge

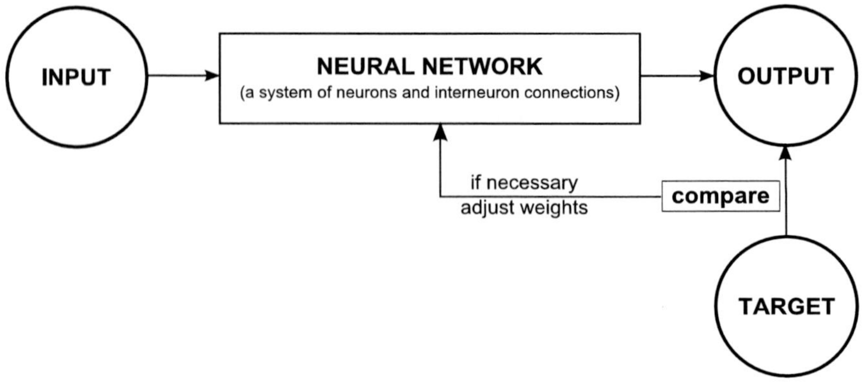

Figure 3. Learning principle of an Artificial Neural Network. A number of input–target pairs is presented to the network. The ANN outputs are compared to the targets and—if the cost function (e.g., root-mean-square differences between the targets and the outputs) is large—the weights are modified in order to minimize it. The procedure is repeated until the cost function decreases below a desired value, or until a prescribed number of iterations is reached.

stored in synaptic weights may be used for prediction. Learning capabilities of any particular ANN are mainly determined by the internal structure of the network, i.e., by the properties and organization of interneuron connections; predictive capabilities depend as well as on the amount and characteristics of data available for training. Figure 3 illustrates the working principle of a typical ANN. Its performance by training is measured with a cost function, which in the simplest case is a root-mean-square error between the target and the ANN output. In some cases, for example when the ANN performance by extreme events is of particular interest, other, more complex forms of cost functions are more appropriate.

A typical artificial neuron is shown in Fig. 4 (some network types, e.g. radial basis networks, are not covered by this scheme; they are not discussed here). Its basic elements are: a vector of adjustable weights $\mathbf{W} = [w_1, w_2, w_3, \ldots, w_{N_i}]$ (where N_i is the number of elements of the input vector), a scalar bias b and a scalar transfer function $f(\mathbf{WX} + b)$. Depending on particular application, many different types of transfer functions are used. Most popular are linear and various types of sigmoid functions, including the hyperbolic-tangent and the logistic function $f(x) = (1 + e^{-x})^{-1}$. From the point of view of learning efficiency, it is desirable that the transfer functions are differentiable. This property is particularly important in backpropagation networks described further. In some ANN applications, neuronal dynamics is additionally controlled by a prescribed threshold value x_t: a given neuron becomes active only if the weighted sum of inputs $\mathbf{WX}$ received by this neuron exceeds x_t; otherwise the output y is set to zero.

An integral component of an ANN is a learning rule, i.e., a method for modifying weights and biases of neurons building that network. Very generally, there are two categories of learning rules. The first one, called unsupervised learning, is applicable to situations where no target outputs are available. Thus, weights and biases are adjusted based on a set of inputs presented to the network. A typical application domain of unsupervised learning is clustering of data. However, most ANNs make use of the second learning

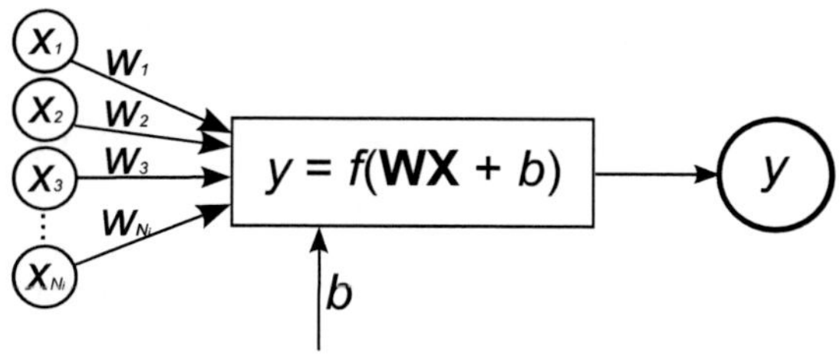

Figure 4. Basic components of an artificial neuron. The input vector $\mathbf{X} = [x_1, x_2, x_3, \ldots, x_{N_i}]$ is multiplied by the vector of weights $\mathbf{W} = [w_1, w_2, w_3, \ldots, w_{N_i}]$. Optionally, if the neuron has a bias, its value b is added to the product $\mathbf{WX}$ to form an argument of the transfer function f. The result of this function y is the scalar returned by the neuron as output.

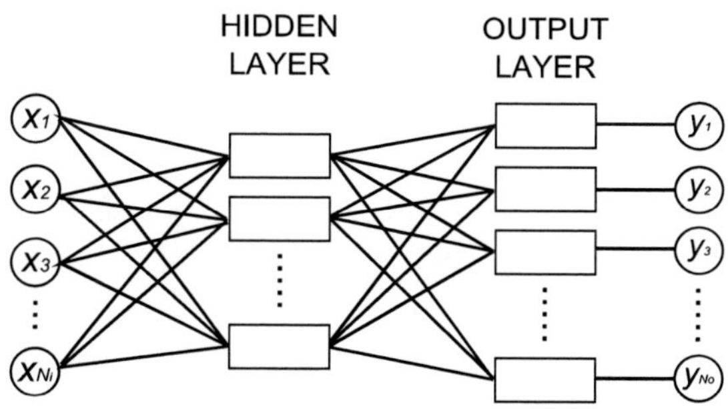

Figure 5. Multi-layer perceptron (MLP) with a single hidden layer—a typical network architecture used in many coastal engineering and environmental applications. Each rectangle symbolizes a neuron shown in Fig. 4. Generally, MLP can contain an arbitrary number of hidden layers.

method—supervised learning, as shown in Fig. 3. In particular, this type of learning is used in multiple layer perceptrons (MLPs), also known as feed-forward neural networks or back-propagation neural networks. MLP is the most popular ANN architecture in coastal engineering and environmental applications, see e.g. [Hsieh and Tang(1998)]. As shown in Fig. 5, MLP has a clearly defined layer structure, with interneuron connections going from one layer to another. It can be shown that MLP with a single hidden layer can approximate any continuous function with any desirable accuracy. There are many learning algorithms applicable to this ANN type, but different modifications of error back-propagation (hence the alternative name of MLPs), as well as conjugate-gradient and quasi-Newton methods, are most often used ([Ripley(1996)], [Nabney(2004)]).

In the next sections, the following concise notation of the MLP networks will be used:

$$\mathcal{N}(\mathbf{X}, \mathbf{N}_h, \mathbf{Y}), \tag{1}$$

where $\mathbf{X}$ and $\mathbf{Y}$ denote the input and output vectors, respectively (with lengths N_i and N_o), and the elements of vector $\mathbf{N}_h = [N_{h,1}, \ldots, N_{h,N_l}]$ contain the number of neuron of each hidden layer (the number of hidden layers being N_l). In all MLPs, there are connections between each pair of neurons of neighboring layers and there are no connections between non-neighboring layers. Conjugate-gradient algorithm is used as a learning method.

4. Application of Neural Network Methods in Data Analysis

Data analysis is a very broad term covering a large group of problems. The aim of data analysis is highlighting important information contained in the data, understanding relationships between their components, checking the quality of the data, removing errors, as well as preparing the data for a subsequent use, e.g. in modeling. Broad range of goals leads to multiple approaches and techniques used to realize them, and here only a small subset will be discussed, in which ANN methods are well established: gap-filling and pattern recognition.

4.1. Interpolation and Gap-filling

Many popular data analysis techniques are applicable only to 'complete' data sets without missing values. Moreover, some methods assume that the data are uniformly distributed over time (and, possibly, space). Similar requirements must be met by measured data used as input (boundary conditions) for numerical models. Yet, in oceanography and coastal engineering, measured data often contain gaps and/or are available at nonuniform time intervals, so that gap-filling and interpolation in space and/or time becomes an inevitable first step preceding further analysis. In that case, ANNs provide an attractive alternative to other methods (polynomial interpolation, splines, spectral methods and so on), mainly because they can be applied without making any *a priori* assumptions concerning properties of the interpolated data.

In coastal engineering, ANNs were used e.g. by [Hsieh and Pratt(2001)] for gap-filling in water level data. [Bhattacharya et al.(2003), Bhattacharya and Solomantine(2006)] and [Makarynskyy et al.(2005a)] applied ANNs to remove missing values from time series of wave parameters. Depending on the available data, a univariate or a multivariate gap-filling may be considered. In the simplest approach, applicable for single time series, only data from this time series are used as input for reconstruction. If time series of other quantities and/or at different locations are available, they may be used as an additional source of information. Often (but not always, as we will see below) this technique allows to reduce the error of the estimation. Here, the use of ANNs as gap-filling tools is demonstrated for the SD97 current data. For each station $k = 1, \ldots, 36$ and for both current components U and V, two two-layer perceptrons, $\mathcal{N}_{C,k,1}$ and $\mathcal{N}_{C,k,2}$ (with C denoting U or V), are constructed. The networks have the same structure (described in Section 3.), but different variables are used as input:

$$\mathbf{x}_{C,k,1} = [C_k(t - N_t\Delta t), \ldots, C_k(t - \Delta t)], \tag{2}$$

$$\mathbf{x}_{C,k,2} = [C_k(t - N_t\Delta t), \ldots, C_k(t - \Delta t), C_m(t - \Delta t), C_m(t), C_n(t - \Delta t), C_n(t)], \tag{3}$$

Thus, networks $\mathcal{N}_{C,k,1}$ perform a univariate gap-filling, based on the values of C at N_t previous time steps. To the contrary, in networks $\mathcal{N}_{C,k,2}$, data from two additional stations m and n are used as input. In this case, the reconstruction is possible only for time periods, when data at those additional stations are available. Hence, stations 37 and 62, with lowest amount of missing data (n_{gap} in Table 1), were chosen. After numerous tests, the number of hidden neurons N_h was set to 100 for all networks; further increase of N_h did not result in any significant improvement in the quality of the results at none of the analyzed stations.

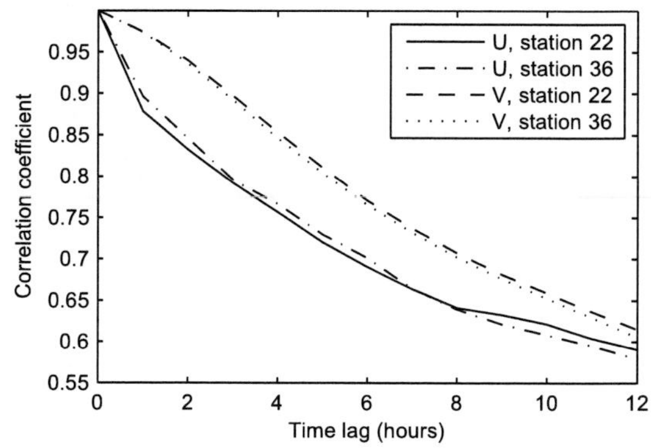

Figure 6. Autocorrelation function for the time series of the cross-shore (U) and alongshore (V) velocity at stations 22 and 36.

Similarly, a common value of $N_t = 4$ was set after a number of tests of the ANNs' performance. Hence, using notation introduced in (1), the ANNs used for the gap-filling task are:

$$\mathcal{N}_{C,k,1} = \mathcal{N}_{C,k,1}(\mathbf{X}_{C,k,1}, [100], C(t)), \quad (4)$$
$$\mathcal{N}_{C,k,2} = \mathcal{N}_{C,k,2}(\mathbf{X}_{C,k,2}, [100], C(t)). \quad (5)$$

In all cases, the input and output vectors were normalized prior to training. 80% of the data were used for training and the remaining 20% for verification of the results.

At all stations, the ANNs' performance was higher for the alongshore current components than for the cross-shore current components. Higher short-time variability of the U time series is one of the main reasons for that behaviour. It is clearly seen in the plots of U and V data at two selected stations, 22 and 36 (Fig. 2). This short-term variability is responsible for different autocorrelation patterns of the U and V time series, as shown in Fig. 6. Moreover, the alongshore-current velocities are strongly spatially correlated, especially in the alongshore direction, i.e., for stations situated in a similar distance from the coast and in a similar water depth. In the case of the cross-shore currents, their spatial variability is significantly higher, which is an important fact from the point of view of multivariate gap-filling with networks $\mathcal{N}_{C,k,2}$. Generally, the ANNs' performance is higher in regions with slowly varying water depth and lower over steep bottom (compare stations 36 and 22 in Fig. 7; see also Fig. 8)—although those differences can be hardly explained with autocorrelation patterns, which are almost the same at both stations (Fig. 6). Spatial distribution of correlation coefficient between the test data and the ANN results reflects the bathymetry of the study area and—in the case of $\mathcal{N}_{C,k,2}$ networks—the correlation between U or V at a given station and at stations used as ANN input (Fig. 8).

In the present approach, each gap is filled with data in an element-by-element manner, i.e., if the gap length is greater than 1, the ANN output is used as input in estimation of subsequent missing values. Hence, the uncertainty of the ANN prediction increases and the quality of the results decreases with increasing gap length. The rate of this decrease depends on the autocorrelation characteristics of the time series analyzed. The SD97 current data provide a good example of that dependence: whereas the alongshore current velocities

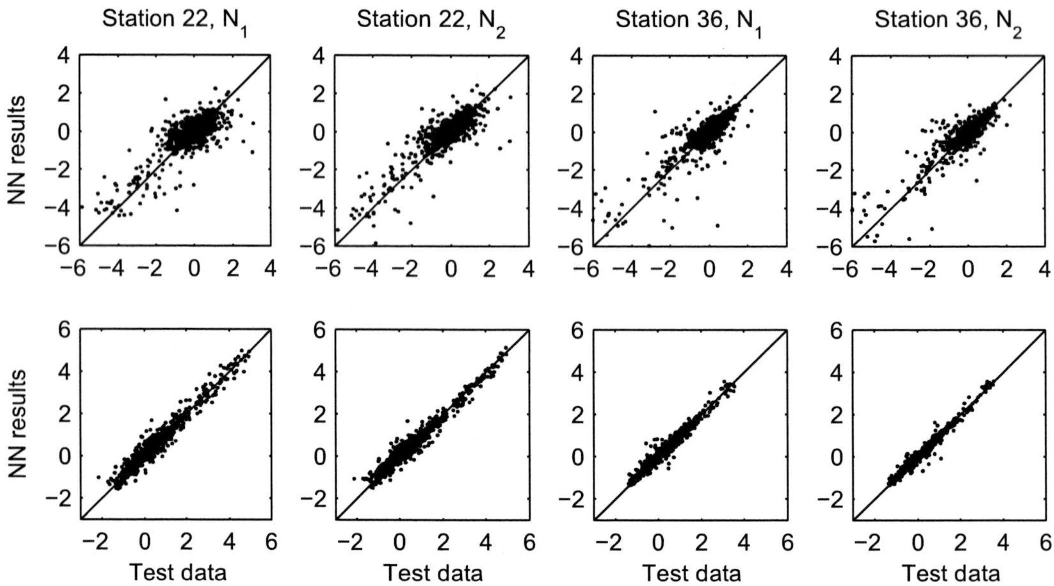

Figure 7. Results of the ANN gap filling for the cross-shore (upper panels) and alongshore (lower panels) current components at stations 22 and 36. The values are normalized. In each case, results obtained with two networks, $\mathcal{N}_{C,k,1}$ (equation 4) and $\mathcal{N}_{C,k,2}$ (equation 5), are shown.

could be reconstructed with high accuracy at all stations (with correlation coefficients r between 0.973 and 0.997 and standard deviation of differences σ_d between 0.17 and 0.26; values obtained for $N_i = 4$), the analogous reconstruction of the cross-shore currents was by far less successful (r and σ_d in the range 0.772–0.970 and 0.20–0.85, respectively). Importantly, in the case of the cross-shore current reconstruction, the quality of the results does not improve (and at some stations even decreases) with increasing N_i for $N_i > 4$. For larger gap lengths, not only the correlation and root-mean-square error, but also the bias of the ANN reconstruction becomes important, because it may lead to positive or negative trends in the reconstructed data. In the analyzed case the normalized bias was at all stations lower than 10^{-3}, insignificant in comparison to previously described sources of error.

4.2. Pattern Recognition

Let $\mathbf{X} = [X_{pt}]_{p=1,\ldots,P,t=1,\ldots,T}$ denote a data set containing P variables, each given at the same time points $t = 1,\ldots,T$. The time series $X_{1t}, X_{2t},\ldots,X_{Pt}$ may denote the same quantity measured/modeled in different spatial locations, different quantities given at the same location or a mixture of both—for the discussion below it is not important. It may be assumed, without loss of generality, that each time series has a zero mean.

In most situations, variables $X_{1t}, X_{2t},\ldots,X_{Pt}$ do not evolve independently, but they are linearly or nonlinearly correlated. This correlation exhibits itself as patterns in the P-dimensional space $\mathcal{S}$ occupied by the data. In other words, the data points in $\mathcal{S}$ are not randomly scattered, but instead are clustered around a subspace of $\mathcal{S}$, which has dimension lower than P. The goal of pattern-recognition (or feature-extraction) task—as it is under-

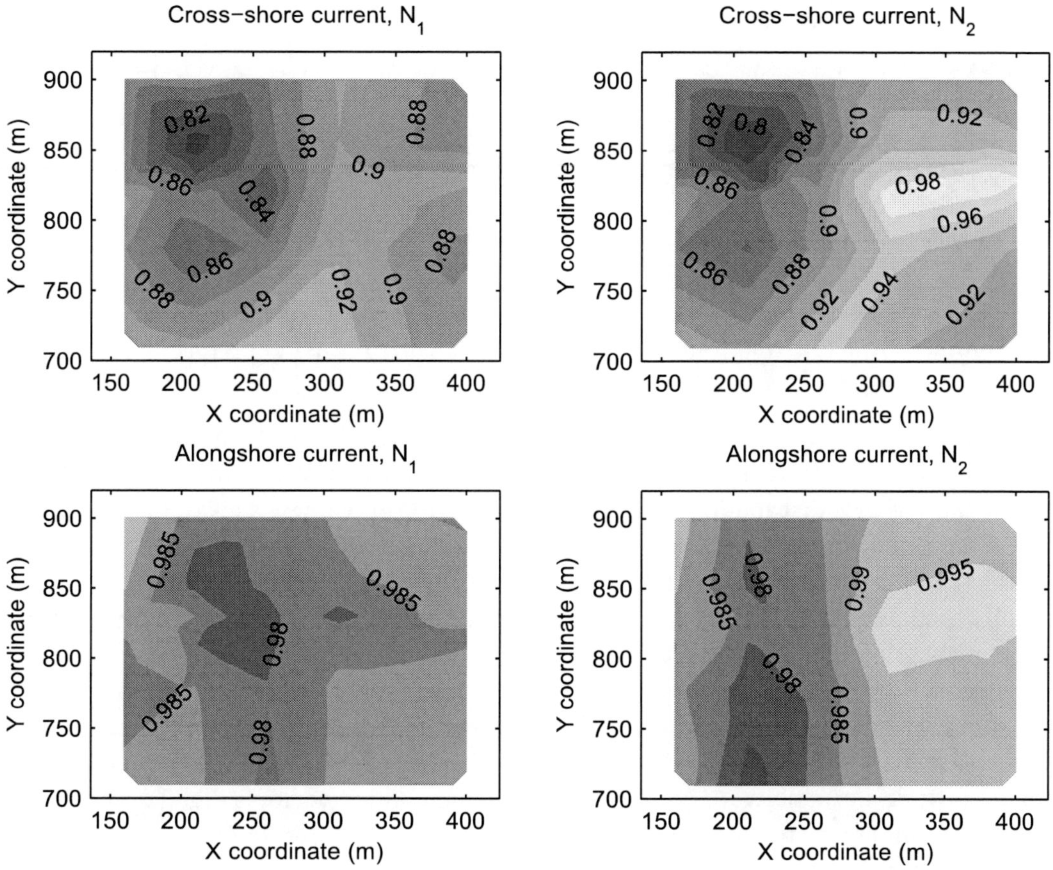

Figure 8. Spatial distribution of the correlation coefficient between the test data and the ANN results obtained with network $\mathcal{N}_{C,k,1}$ (left; equation 4) and $\mathcal{N}_{C,k,2}$ (right; equation 5). Upper/lower panels show the results for the cross-shore/alongshore currents.

stood in earth sciences and related fields—is to extract this lower-dimensional variability 'hidden' in the (possibly noisy) data. This may be done for several purposes. Firstly, if only the low-dimensional variability is retained, the data set becomes smaller and easier to handle, which may be important from a practical point of view. Secondly, pattern recognition may be viewed as a method of signal detection, allowing to remove random noise from the data. Thirdly, patterns resulting from a feature-extraction procedure are sometimes interpreted in terms of physical processes that may have produced those patterns. Thus, the dominant modes of variability revealed during the feature extraction process are given meaning—an undertaking which is often controversial and must be treated with caution.

4.2.1. Nonlinear Principal Component Analysis—Theory

In meteorology and oceanography, linear Principal Component Analysis (PCA) is undoubtedly the most established and well-studied pattern-recognition method. It has been used for more than half a century and its description can be found in a number of books,

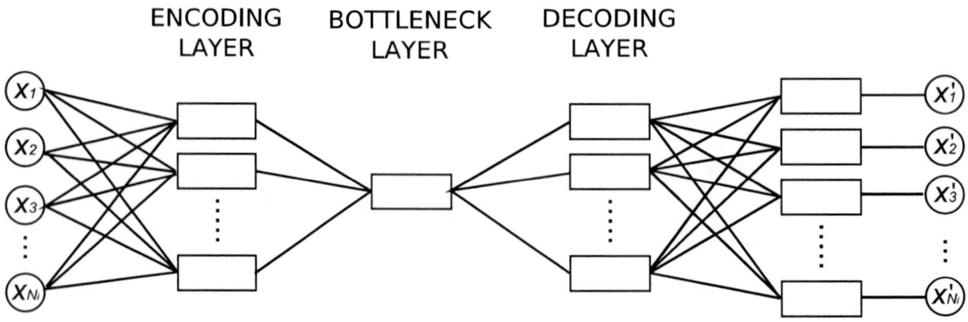

Figure 9. Structure of a four-layer autoassociative MLP used to perform NLPCA.

e.g. in [Preisendorfer(1988)]. In PCA, by solving an eigenvalue problem of the covariance matrix of the original data matrix $\mathbf{X}$, a new linear coordinate system is obtained that better reflects the variance distribution in these data. Typically, only a few axes of this coordinate system are sufficient to reproduce majority of the variance and only these axes are retained, leading to a substantial dimensionality reduction. A result of PCA is a set of time-invariant (usually spatial) modes and corresponding principal components (PCs) describing time evolution of these modes. In coastal hydrodynamic modeling PCA was used e.g. by [van der Merwe et al.(2007)] or [Herman et al.(2007), Herman et al.(2009)].

Nonlinear generalization of PCA was introduced by [Krammer(1991)], who termed it Non-Linear PCA (NLPCA). The methodological basis of NLPCA is an autoassociative neural network, i.e., a network with identical input and target vectors (Fig. 3). The goal of training an autoassociative network is to reproduce its input $\mathbf{X}$, or, in other words, to minimize the mean-square-error E between $\mathbf{X}$ and the output $\mathbf{X}'$. The most important element of the four-layer ANN designed by [Krammer(1991)], through which the compression of the data is achieved, is a so-called bottleneck layer containing only one neuron ($N_{h,2} = 1$, as shown in Fig. 9) with a linear transfer function. The bottleneck layer is preceded by an encoding layer and followed by a decoding layer, each containing the same number of neurons $N_{h,1} = N_{h,3} = N_d$ with hyperbolic-tangent transfer functions (if they are replaced with linear functions, the network performs a linear PCA). Hence, an MLP used for NLPCA may be shortly written as $\mathcal{N}(\mathbf{X}, [N_d, 1, N_d], \mathbf{X})$. After the analyzed data set is presented to a trained network, the time series of states of the bottleneck neuron provide the first nonlinear PC (NLPC) and the vectors $\mathbf{X}'$ build a curve in the space spanned by the data points—the corresponding nonlinear mode. The fraction of variance represented by this NLPC/mode pair equals $1 - \langle ||\mathbf{X} - \mathbf{X}'||^2 \rangle / \langle ||\mathbf{X} - \overline{\mathbf{X}}||^2 \rangle$, where $\langle \cdot \rangle$ denotes averaging over the whole sample, $|| \cdot ||$ is a norm in the space spanned by $\mathbf{X}$ and $\overline{\mathbf{X}}$ is an average of $\mathbf{X}$. Further NLPC/mode pairs can be obtained by training another NLPCA network with $\mathbf{X} - \mathbf{X}'$ as input.

The form of the nonlinear mode depends on the number N_d of encoding/decoding neurons. For $N_d = 1$, the mode is a straight line as in the linear PCA. Curved solutions are obtained for $N_d > 1$, with the maximum possible number of bending points increasing with increasing N_d. For example, for $N_d = 2$ at most 2 bending points (Z-shaped curves) are possible ([Christiansen(2005)]). Thus, the choice of N_d directly influences the complexity of the NLPCA solution. Generally, the higher the number of encoding neurons, the lower the error E (NLPC passes through the clusters of the data points), but this parameter alone is not a sufficient criterion for selecting optimal NLPCA solutions from an ensemble of trained networks. Overfitting is a serious problem of NLPCA. Zigzag curves obtained with high N_d values may be characterized by arbitrarily low E, but produce very misleading results. In particular, NLPCA may suggest the existence of false multimodality in the data, as described by [Christiansen(2005)]. Another problem is that very closely located data points may be projected very far apart on the NLPC mode curve ([Hsieh(2007)]). There are some remedies to these shortcommings, e.g. weight penalty terms or an inconsistency index introduced by [Hsieh(2001)] and [Hsieh(2007)], respectively, but nevertheless the choice of an optimal set of parameters for NLPCA must be always made after an in-depth, critical analysis of the results. Excellent review of the theory of NLPCA may be found in [Hsieh(2004)]. [Rattan and Hsieh(2005)] developed an extension of NLPCA to handle complex-valued data, e.g. vector fields.

Even though some ambiguity exists in the interpretation of NLPCA results, it proved very useful in a number of issues in earth sciences and coastal engineering. Among others, NLPCA was used by [Monahan(2000)], who demonstrated its applicability to study the nonlinear variability of the Lorenz attractor, by [Monahan(2001)] to analyze the 500 hPa geopotential height field variability, and by [Hsieh(2001)] and [Rattan and Hsieh(2004)] for an analysis of tropical Pacific sea surface temperature and wind variability, respectively. From a coastal engineering point of view, more interesting are the works of [Ruessink et al.(2004)] and [Herman(2007)], as they describe NLPCA applications to analysis of processes in the coastal zone. Both authors used an extension of NLPCA, called circular NLPCA (NLPCA.cir), appropriate for analyzing time-periodic signals, see [Kirby and Miranda(1996), Hsieh(2001)] for details. [Ruessink et al.(2004)] applied NLPCA.cir to study nonlinear behavior of nearshore bathymetric data from three locations in the Netherlands, Japan and the USA. The focus of their work lied on the spatio-temporal characteristics of the sandbar migration. Later, [Herman(2007)] used NLPCA.cir for an analysis of water level and current data from the German Wadden Sea. The data sets were obtained with a numerical hydrodynamic model, as described in [Herman et al.(2007)], and subject to a linear PCA prior to the NLPCA.cir analysis in order to preliminary reduce the dimensionality of the NLPCA input. It is a common practice, used also by [Hsieh(2001), Hsieh(2004), Ruessink et al.(2004)] and others. Finally, it is worth noticing that, if correlated patterns between two different data sets are of interest, a ANN-based extension of the canonical correlation analysis (CCA), called NLCCA can be used, see [Hsieh(2004)].

4.2.2. Nonlinear Principal Component Analysis—Example

Here, possibilities of NLPCA are demonstrated for the SD97 alongshore current data V. Because—as already discussed—the alongshore variability of V is very small (data from pairs of stations equally distanced from the coast are linearly correlated at the level ¿0.95), the following analysis concentrates on the cross-shore variability of V. To this end, a subset of stations lying along the straight line between stations 30 and 72 (Fig. 1) is selected. Hence, the number of variables $P = 11$, which is small enough so that the (normalized) current velocity components can be used directly as input to NLPCA, without prior PCA analysis. This shall simplify the interpretation of the results. The 4740 data points were separated into the training (85%) and validation (15%) data set. For each combination of

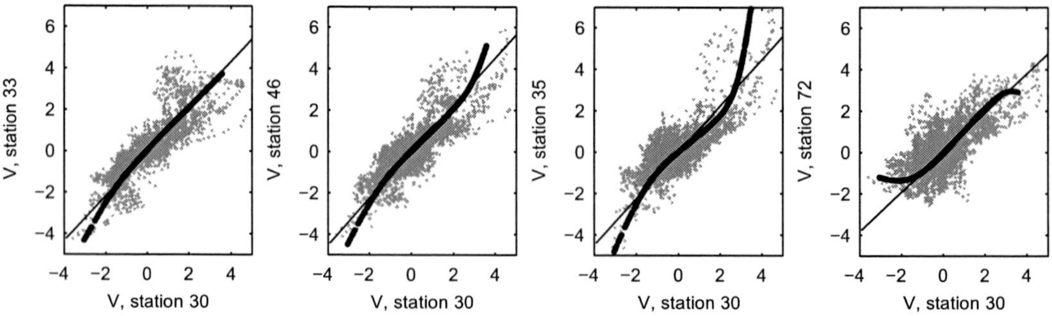

Figure 10. Results of the NLPCA analysis of alongshore currents (obtained with $N_d = 3$ and $w_p = 0.1$), projected onto planes of selected station pairs. Gray dots show the individual data points, black points – the first nonlinear mode, black line – the first linear PC. Values have been normalized for visualization purposes.

$N_d \in \{2,3,4\}$ and wealth penalty $w_p \in \{1, 0.1, 0.01, 0\}$, an ensemble of 10 neural networks was trained and from each ensemble the ANN with the lowest cost function was selected. The results for a few stations along the analyzed profile are shown in Fig. 10. In terms of the reconstructed variance, only a minor gain could be reached in comparison to the linear PCA (straight lines in Fig. 10). The fraction of the reconstructed variance was 0.856 and 0.869 for the first linear and nonlinear PC, respectively. However, these numbers are dominated by a very large number of small values (within ± 2 standard deviations). For strong currents—which are arguably the most interesting part of the data record—the first NLPC provides a much better description of their variability than the first linear PC does. Clearly, in these situations nonlinearities in the cross-shore variability of alongshore currents become important (Fig. 10). A remarkable feature of this variability is the asymmetry of the cross-shore profile of V at positive and negative current speeds, as shown in Fig. 11. At the negative extreme of the first NLPC (corresponding to a southward flow), the cross-shore variability of V is much more uniform than at the opposite (positive, corresponding to a northward flow) extreme, where a strong shallow-water maximum around station 31 develops. The differences between the two cases are smallest at the base of the steep slope separating the shallowest stations from those situated in deeper water. Contrary to NLPCA, linear PCA always produces ideally symmetric patterns, which may represent some kind of 'average' of two quite dissimilar extreme states of the system.

4.2.3. Other ANN Pattern Recognition Methods

Apart from the approaches described above, pattern recognition is often understood as a classification/clustering task, during which the data samples collected in the matrix **X** should be divided into a number of categories (classes) based on some predefined criteria. In studies of coastal hydrodynamics, dealing mainly with continuously changing fields, clustering is not a frequently applied task, but, nevertheless, for some class of problems it is useful. And here, too, ANNs provide an attractive alternative to more classical methods (hierarchical or k-means clustering, Gaussian mixture models, discriminant analysis and so on). A group of problems in which (automatic) classification is an important issue concerns

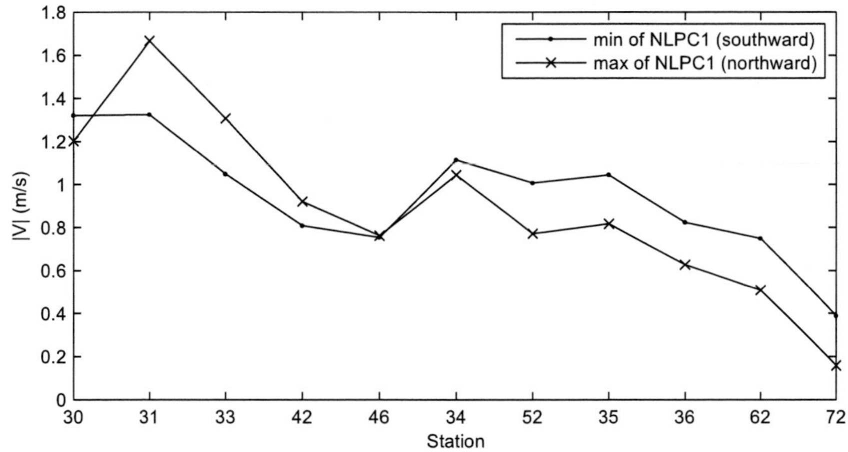

Figure 11. Cross-shore variability of alongshore currents (in m/s) between stations 30 and 72, for minimum and maximum values of the first NLPC.

interpretation and analysis of remotely sensed data (e.g., segmentation of multispectral images). ANN classification was used by [Eom(1999)] for determination of different sea-ice types from radar images, by [Maciolowska et al.(1998), Dung and Stepnowski(2000)] for classification of bottom sediments and by [Topouzelis et al.(2007)] for separating oil spills from other similar phenomena in radar backscatter images.

5. Application of ANN Methods in Modeling

Generally, the purpose of a modeling task as discussed in this section is to find a relationship f between a set of 'known' variables $\mathbf{X}$ and another set of variables $\mathbf{Y}$: $\mathbf{Y} = f(\mathbf{X}, \mathbf{P})$, where $\mathbf{P}$ denotes a vector of model parameters. The transformation f may be purely mathematical, or it may be constructed based on physical laws governing the dependence of $\mathbf{Y}$ on $\mathbf{X}$. The first case is usually broadly termed *regression* and it is one of the most often used tasks accomplished with the help of ANN methods, as discussed in Section 5.1. In the second case, f has a form of a set of physically meaningful equations, which (usually transformed into a discrete form) build up a *physically-based (numerical) model.* Spectral wind-wave models or hydrodynamic circulation models are typical examples. Of course, those types of modeling approaches can be combined into hybrid models—and this is where ANNs, although they do not take any physical processes into account, can be used, as described further in Section 5.2. In both cases, an essential phase in the construction of the model is to adjust the model parameters $\mathbf{P}$ so that, for some test data set $(\mathbf{X}_T, \mathbf{Y}_T)$, the root-mean-square error $\langle ||\mathbf{Y}_T - f(\mathbf{X}_T, \mathbf{P})||^2 \rangle$, or another, more complicated cost function, is minimized.

The simplest ANN-based modeling approaches are similar to the gap-filling techniques described in Section 4.1. For example, [Makarynskyy(2004)] used measured wave data from two locations at the Irish coast to predict future wave parameters at those stations. He trained a separate ANN for each location and for each forecasted parameter, with past values of a given time series used as input. The same technique was used by [Deo and Naidu(1999)] and [Makarynskyy et al.(2005b)]. Similarly, past water level data were applied by [Lee and Jeng(2002)] and [Lee(2004)] to predict tidal elevations at future time steps. [Tsai et al.(2002)] used ANN techniques for measured wave parameters from three nearby stations to remove missing data and to forecast future wave conditions at those stations. Water level data from a number of stations was also used for ANN tidal predictions by [Huang et al.(2003)].

5.1. Nonlinear Regression

Nonlinear regression by means of ANNs is particularly useful in situations where fast and efficient prediction of the variables of interest is more important than gaining insight into the nature of relationships between the modeled quantities. There are many examples of ANN regression modelling in coastal engineering and related subjects. Typical applications include water-level prediction from wind, atmospheric pressure and tidal data (e.g., [Lee(2008), Lee(2009)]) or wind-wave modeling from meteorological data (e.g., [Deo et al.(2001), Kalra et al.(2005), Rao and Mandal(2005)]). [Altunkaynak and Özger(2004)] combined ANN techniques with Kalman filtering in order to simulate wave heights from measured wind-speeds. Similarly, [Deo and Jagdale(2003)] designed an ANN to model the breaking-wave height and depth from deep-water wave height, period and the bottom slope. In most cited cases, the performance of the ANNs was better than obtained with routinely used empirical formulae.

A great majority of those works is based on MLP neural-network type (with radial basis networks being a relatively often used alternative). An interesting group of studies combines ANN techniques with fuzzy inference systems (FIS), see [Jang(1993), Mahjoobi et al.(2008)]. In this approach, called ANFIS, ANNs are used to tune parameters of fuzzy if–then rules, describing relationships between the input and output data sets. Additionally, in GA-ANFIS algorithms, a genetic-algorithm part of the model may be used to vary the number of those rules. An ANFIS model was used for example by [Kazeminezhad et al.(2005)] to predict wave parameters from wind data in Lake Ontario. [Zanaganeh et al.(2009)] did a similar study for Lake Michigan with a GA-ANFIS model and demonstrated its superiority over ANFIS and other widely-used empirical formulae. Another approach to nonlinear modeling, called Nonlinear Principal Prediction Analysis (NLPPA) and developed by [Cannon(2006)], is an extension of NLCCA with more than one neuron in the bottleneck layer. It is a kind of compromise between the goal to minimize the cost function and the goal to reduce the dimensionality of the data of interest. Recently, [Cannon(2007)] combined NLPPA with an analog model (see e.g. [Zorita and von Storch(1999)]) and demonstrated possibilities of the combined model for synthetic data and for synoptic downscaling.

5.2. Numerical Wave and Current Modeling

Spatial and temporal variability of hydro- and morphodynamic processes in the coastal zone is routinely simulated with physics-based numerical models that solve a given set of governing equations on a prescribed (two- or three-dimensional, structured or unstructured) grid of points. The distribution of the points and the time step of the model must be adjusted to the characteristic spatial and temporal scales of variability of the bathymetry of the study area and of the external forcing (wind, atmospheric pressure, water levels at the open boundary, and so on), as well as to the constraints resulting from the type of equations being solved and the numerical schemes being used (stability criteria etc.). In a typical application, a set of over 10^6 equations must be solved at each model time step, which requires huge computational resources and makes some modeling tasks prohibitively time- and memory-consuming or even impossible. This is particularly true in forecasting. At present, most widely used circulation and wave models can be run a few times faster than 'the real life'. This eliminates from practical use probabilistic forecasting methods, which require simulating of hundreds or even thousands of scenarios, see e.g. [van der Merwe et al.(2007)].

One of possible solutions to this problem is to replace the computationally most expensive parts of the models—usually corresponding to physically meaningful terms in the governing equations—with computationally more efficient algorithms. Spectral wind-wave modeling provides a good example of how this can be done. In a SWAN (Simulating WAves Nearshore) model setup of [Herman et al.(2009)], covering parts of the German Wadden Sea, a curvilinear grid of more than $3 \cdot 10^5$ points was used. In nonstationary simulations run in a parallel mode on a cluster of 16 processors, the algorithm for calculating the quadruplet wave–wave interactions (Discrete Interaction Approximation; [Hasselmann et al.(1985)]) consumed over 27% of the total computation time. This number can be regarded as typical for a shallow water wind-wave model application. In view of this fact—and serious systematic shortcomings of DIA and similar quadruplet parameterizations—the NNIA (Neural-Network Interaction Approximation) proposed by [Krasnopolsky and Chevallier(2003)] and [Tolman et al.(2005)] seems a promising direction of research that may provide an alternative to earlier quadruplet wave–wave interaction algorithms. A preliminary version of NNIA, appropriate for single-peaked spectra, has been implemented in the WaveWatchIII model. It is a few times slower than DIA, but still 10^5 times faster than the original 'exact' approximation and—most importantly—an order of magnitude more accurate (in terms of root-mean-square error) than DIA (see Table 3 in [Krasnopolsky and Fox-Rabinovitz(2006b)]).

In a similar way, [Krasnopolsky et al.(2002)] developed an ANN surrogate replacing the UNESCO equation of state in oceanic circulation models. [Krasnopolsky et al.(2005)] replaced longwave radiation components—computationally most expensive parts of atmospheric climate models, occupying up to 50% of the total calculation time—with fast ANN equivalents. The concept of hybrid modeling, together with a formulation of developmental framework and validation criteria for such models, is given in [Krasnopolsky and Fox-Rabinovitz(2006b), Krasnopolsky and Fox-Rabinovitz(2006a)].

Instead of replacing some physics-based model terms with faster neural-network surrogates, it is possible to construct an ANN equivalent for a whole numerical model. This approach has been successfully used in hydrology and in atmospheric and ocean sciences—see [Krasnopolsky and Chevallier(2003)] and [Cherkassky et al.(2006)] for brief reviews. The works of [van der Merwe et al.(2007)] and [Herman et al.(2007), Herman et al.(2009)] provide good examples for shallow-water circulation and wind-wave modeling. [van der Merwe et al.(2007)] constructed an ANN surrogate for a three-dimensional, unstructured-grid hydrodynamic model of the Columbia River estuary, forced with tidal, river flux, wind and atmospheric-pressure data and producing fields of water levels, current velocities, water temperature and salinity. Similarly, the work of [Herman et al.(2007)] was based on the results of two-dimensional curvilinear hydrodynamic model of parts of the German Wadden Sea. Later, [Herman et al.(2009)] used the spectral wave model SWAN to simulate wind-wave variability in the same area. In all three studies, the first step in developing an ANN surrogate was a linear PCA of the data sets of interest in order to reduce dimensionality of these data. PCA results where then used to train two-layer MLP networks (with hyperbolic-tangent transfer functions of the hidden layer and linear transfer functions of the output layer) that could replace the computationally expensive models. In all cases, a few orders of magnitude speed reduction was achieved without a significant loss of the quality of the results. E.g., the surrogate ANN model of [van der Merwe et al.(2007)] could reproduce salt-intrusion and freshwater-plume dynamics in the analyzed estuary with a satisfactory accuracy (limited rather by the PCA than by the capabilities of the ANN model).

6. Conclusion

Apart from ANN-based data-analysis and modeling techniques, there are a number of predictive-learning methods not covered in this review. Among those, evolutionary algorithms deserve attention, as their use in engineering and related problems has increased in recent years. Usage of GA methods in modeling was already mentioned in Section 5.1. A promising branch of genetic methods is Gene Expression Programming (GEP; [Ferreira(2005)]). A frequently pointed-out advantage of these methods is that they provide explicit analytical formulae describing relationships between the analyzed data sets, which makes those relationships easier to analyze and to understand. However, these advantages are often only apparent. In fact, genetic methods share a number of limitations with ANNs and other predictive-learning techniques, including overfitting and complex, incomprehensible formulae. Also, more than one different 'good' approximation to the analyzed problem may exist. One undisputable advantage is that sensitivity analysis of the results is simpler and more straightforward.

ANN methods are sometimes—rather pejoratively—called black-box models, with an intention to stress that they do not provide understanding of processes shaping the relationships between the analyzed variables. Or, at least, with huge numbers of parameters involved in building those relationships, ANNs do not explain them in a way comprehensible for us, humans. However, in some situations this kind of approach seems unavoidable (and, in fact, in this respect it is not very different from a classical statistics, like multivariate regression and so on). Moreover, the knowledge of the underlying processes is often very useful, or even necessary, by selecting the ANN structure, its parameters etc. Another (justified) criticism towards ANNs is related to difficulties in estimating uncertainties of the modeling results. Even though some progress has recently been made in these matters ([Nabney(2004)]), estimation of the accuracy of the results remains a serious problem, especially in engineering applications. All those shortcomings in mind, with increasing need for reliable and efficient operational models, the ANN methods are becoming an attractive—and sometimes indispensable—supplementary or alternative tool in a broad range of scientific and engineering applications.

References

[Aleksander and Morton(1990)] Aleksander, I. and H. Morton, 1990: *An Introduction to Neural Computing*. London: Chapman and Hall.

[Altunkaynak and Özger(2004)] Altunkaynak, A. and M. Özger, 2004: Temporal significant wave height estimation from wind speed by perceptron Kalman filtering. *Ocean Engng*, **31**, 1245–1255.

[Bhattacharya et al.(2003)] Bhattacharya, B., D. Shrestha, and D. Solomantine, 2003: Neural networks in reconstructing missing wave data in sedimentation modelling. *Proc. XXXth IAHR Congress, Thessaloniki, Greece*, Vol. D, 209–216.

[Bhattacharya and Solomantine(2006)] Bhattacharya, B. and D. Solomantine, 2006: Machine learning in sedimentation modelling. *Neural Networks*, **19 (2)**, 208–214.

[Cannon(2006)] Cannon, A., 2006: Nonlinear principal predictor analysis: Application to the Lorenz system. *J. Climate*, **19**, 579–589.

[Cannon(2007)] Cannon, A., 2007: Nonlinear analog prediction analysis: A coupled neural network/analog model for climate downscaling. *Neural Networks*, **20**, 444–453.

[Cherkassky et al.(2006)] Cherkassky, V., V. Krasnopolsky, D. Solomatine, and J. Valdes, 2006: Computational intelligence in earth sciences and environmental applications: Issues and challenges. *Neural Networks*, **19**, 113–121.

[Christiansen(2005)] Christiansen, B., 2005: The shortcommings of nonlinear principal component analysis in identifying circulation regimes. *J. Climate*, **20**, 378–379.

[Conley and Beach(2003)] Conley, D. and R. Beach, 2003: Cross-shore sediment transport partitioning in the nearshore during a storm event. *J. Geophys. Res.*, **108 (C3)**, 3065, doi:10.1029/2001JC001230.

[Deo and Jagdale(2003)] Deo, M. and S. Jagdale, 2003: Prediction of breaking waves with neural networks. *Ocean Engng*, **30**, 1163–1178.

[Deo et al.(2001)] Deo, M., A. Jha, A. Chaphekar, and K. Ravikant, 2001: Neural networks for wave forecasting. *Ocean Engng*, **28**, 889–898.

[Deo and Naidu(1999)] Deo, M. and C. Naidu, 1999: Real time wave forecasting using neural networks. *Ocean Engng*, **26**, 191–203.

[Dung and Stepnowski(2000)] Dung, T. and A. Stepnowski, 2000: Sea bottom recognition using multistage fuzzy neural network operating on multi-frequency data. *Acta Acoustica*, **9**, 830–837.

[Elgar et al.(1997)] Elgar, S., T. Herbers, W. O'Reilly, and R. Guza, 1997: Surf zone waves, currents, and morphology. Online: `http://www.frf.usace.army.mil/sandyduck/SandyDuck.stm`.

[Eom(1999)] Eom, K., 1999: Fuzzy clustering approach in unsupervised sea-ice classification. *Neurocomputing*, **25**, 149–166.

[Feddersen and Guza(2003)] Feddersen, F. and R. Guza, 2003: Observations of nearshore circulation: alongshore uniformity. *J. Geophys. Res.*, **108 (C1)**, 3006, doi:10.1029/2001JC001293.

[Feddersen et al.(1998)] Feddersen, F., R. Guza, S. Elgar, and T. Herbers, 1998: Alongshore momentum balances in the nearshore. *J. Geophys. Res.*, **103 (C8)**, 15 667–15 676.

[Ferreira(2005)] Ferreira, C., 2005: *Gene Expression Programming: Mathematical Modeling by an Artificial Intelligence*, Studies in Computational Intelligence, Vol. 21. 2nd ed. ed., Springer Verlag, 478 pp.

[Hagan et al.(1995)] Hagan, M., H. Demuth, and M. Beale, 1995: *Neural Network Design*. PWS Publishing Company, Boston, 736 pp.

[Hasselmann et al.(1985)] Hasselmann, S., K. Hasselmann, J. Allender, and T. Barnett, 1985: Computations and parameterizations of the nonlinear energy transfer in a gravity-wave spectrum, Part II: parameterizations of the nonlinear energy transfer for application in wave models. *J. Phys. Oceanogr.*, **15**, 1378–1391.

[Herman(2007)] Herman, A., 2007: Nonlinear principal component analysis of the tidal dynamics in a shallow sea. *Geophys. Res. Lett.*, **34**, L02 608, doi:10.1029/2006GL027 769.

[Herman et al.(2007)] Herman, A., R. Kaiser, and H. Niemeyer, 2007: Modelling of a medium-term dynamics in a shallow tidal sea, based on combined physical and neural network methods. *Coastal Modelling*, **17**, 277–299.

[Herman et al.(2009)] Herman, A., R. Kaiser, and H. Niemeyer, 2009: Wind-wave variability in a shallow tidal sea – spectral modelling combined with neural network methods. *Coastal Engng*, **56**, 759–772.

[Hsieh and Pratt(2001)] Hsieh, B. and T. Pratt, 2001: Field data recovery in tidal system using Artificial Neural Networks (ANNs). Coastal and hydraulics engineering technical notes chetn-iv-38, U.S. Army Corps of Engineers. 10 pp.

[Hsieh(2001)] Hsieh, W., 2001: Nonlinear principal component analysis by neural networks. *Tellus*, **53A**, 599–615.

[Hsieh(2004)] Hsieh, W., 2004: Nonlinear multivariate and time series analysis by neural network methods. *Rev. Geophys.*, **42**, RG1003, doi:10.1029/2002RG000 112.

[Hsieh(2007)] Hsieh, W., 2007: Nonlinear principal component analysis of noisy data. *Neural Networks*, **20**, 434–443.

[Hsieh(2008)] Hsieh, W., 2008: *Neuralnets for Multivariate and Time Series Analysis (NeuMATSA): A User Manual.* Dep. Earth and Ocean Sciences, University of British Columbia, online: www.ocgy.ubc.ca/~william/Pubs/NN.manual2008.pdf, 20 pp.

[Hsieh(2009)] Hsieh, W., 2009: *Machine Learning Methods in the Environmental Sciences: Neural Networks and Kernels.* Cambridge Univ. Press, 364 pp.

[Hsieh and Tang(1998)] Hsieh, W. and B. Tang, 1998: Applying neural network models to prediction and data analysis in meteorology and oceanography. *Bull. Am. Meteorol. Soc.*, **79**, 1855–1870.

[Huang et al.(2003)] Huang, W., C. Murray, N. Kraus, and J. Rosati, 2003: Development of a regional neural network for coastal water level predictions. *Ocean Engng*, **30**, 2275–2295.

[Jang(1993)] Jang, J.-S., 1993: ANFIS: Adaptive-Network-Based Fuzzy Inference systems. *IEEE Trans. Syst. Man Cybernet.*, **23**, 665–685.

[Kalra et al.(2005)] Kalra, R., M. Deo, R. Kumar, and V. Agarwal, 2005: Artificial neural network to translate offshore satellite data to coastal locations. *Ocean Engng*, **32**, 1917–1932.

[Kazeminezhad et al.(2005)] Kazeminezhad, M., A. Etemad-Shahidi, and S. Mousavi, 2005: Application of fuzzy inference system in the prediction wave parameters. *Ocean Engng*, **32**, 1709–1725.

[Kirby and Miranda(1996)] Kirby, M. and R. Miranda, 1996: Circular nodes in neural networks. *Neural Computation*, **8**, 390–402.

[Krammer(1991)] Krammer, M., 1991: Nonlinear principal component analysis using autoassociative neural networks. *AIChE Journal*, **37**, 233–243.

[Krasnopolsky et al.(2002)] Krasnopolsky, V., D. Chalikov, and H. Tolman, 2002: A neural network technique to improve computational efficiency of numerical oceanic models. *Ocean Modelling*, **4**, 363–383.

[Krasnopolsky and Chevallier(2003)] Krasnopolsky, V. and F. Chevallier, 2003: Some neural network applications in environmental sciences. Part II: advancing computational efficiency of environmental numerical models. *Neural Networks*, **16**, 335–348.

[Krasnopolsky and Fox-Rabinovitz(2006a)] Krasnopolsky, V. and M. Fox-Rabinovitz, 2006a: Complex hybrid models combining deterministic and machine learning components for numerical climate modeling and weather prediction. *Neural Networks*, **19**, 122–134.

[Krasnopolsky and Fox-Rabinovitz(2006b)] Krasnopolsky, V. and M. Fox-Rabinovitz, 2006b: A new synergetic paradigm in environmental numerical modeling: Hybrid models combining deterministic and machine learning components. *Ecological Modelling*, **191**, 5–18.

[Krasnopolsky et al.(2005)] Krasnopolsky, V., M. Fox-Rabinovitz, and D. Chalikov, 2005: New approach to calculation of atmospheric model physics: Accurate and fast neural network emulation of longwave radiation in a climate model. *Monthly Weather Rev.*, **133**, 1370–1383.

[Lee and Jeng(2002)] Lee, T. and D. Jeng, 2002: Application of artificial neural networks in tide-forecasting. *Ocean Engng*, **29**, 1003–1022.

[Lee(2004)] Lee, T.-L., 2004: Back-propagation neural network for long-term tidal predictions. *Ocean Engng*, **31**, 225–238.

[Lee(2008)] Lee, T.-L., 2008: Neural network prediction of a storm surge. *Ocean Engng*, **33**, 483–494.

[Lee(2009)] Lee, T.-L., 2009: Predictions of typhoon storm surge in Taiwan using artificial neural networks. *Advances Engng Software*, **40**, 1200–1206.

[Maciolowska et al.(1998)] Maciolowska, J., A. Stepnowski, and D. Dung, 1998: Fish schools and seabed identification using neural networks and fuzzy logic classifiers. *Proc. 4th European Conf. Underwater Acoustics*, 275–280.

[Mahjoobi et al.(2008)] Mahjoobi, J., A. Etemad-Shahidi, and M. Kazeminezhad, 2008: Hindcasting of wave parameters using different soft computing methods. *Appl. Ocean Res.*, **30**, 28–36.

[Makarynskyy(2004)] Makarynskyy, O., 2004: Improving wave predictions with artificial neural networks. *Ocean Engng*, **31**, 709–724.

[Makarynskyy et al.(2005a)] Makarynskyy, O., D. Makarynska, E. Rusu, and A. Gavrilov, 2005a: Filling gaps in wave records with artificial neural networks. *Proc. Maritime Transportation and Exploitation of Ocean and Coastal Resources*, iMAM 26-30.09.2005, Lisbon, Portugal.

[Makarynskyy et al.(2005b)] Makarynskyy, O., A. Pires-Silva, D. Makarynska, and C. Ventura-Soares, 2005b: Artificial neural networks in wave predictions at the west coast of Portugal. *Computers and Geosciences*, **31**, 415–424.

[Monahan(2000)] Monahan, A., 2000: Nonlinear principal component analysis by neural networks: theory and application to the Lorenz system. *J. Climate*, **13**, 821–835.

[Monahan(2001)] Monahan, A., 2001: Nonlinear principal component analysis: Tropical Indo-Pacific sea surface temperature and sea level pressure. *J. Climate*, **14**, 219–233.

[Nabney(2004)] Nabney, I., 2004: *NETLAB. Algorithms for pattern recognition*. Advances in Pattern Recognition, 4th Ed., Springer-Verlag, UK, 420 pp.

[Noyes et al.(2004)] Noyes, T., R. Guza, S. Elgar, and T. Herbers, 2004: Field observations of shear waves in the surf zone. *J. Geophys. Res.*, **109 (C01031)**, doi:10.1029/2002JC001761.

[Preisendorfer(1988)] Preisendorfer, R., 1988: *Principal Component Analysis in Meteorology and Oceanography*. Elsevier, 425 pp.

[Rao and Mandal(2005)] Rao, S. and S. Mandal, 2005: Hindcasting of storm waves using neural networks. *Ocean Engng*, **32**, 667–684.

[Rattan and Hsieh(2004)] Rattan, S. and W. Hsieh, 2004: Nonlinear complex principal component analysis of the tropical Pacific interannual wind variability. *Geophys. Res. Lett.*, **31**, L21 201, doi:10.1029/2004GL020446.

[Rattan and Hsieh(2005)] Rattan, S. and W. Hsieh, 2005: Complex-valued neural networks for nonlinear complex principal component analysis. *Neural Networks*, **18**, 61–69.

[Ripley(1996)] Ripley, B., 1996: *Pattern Recognition and Neural Networks*. Cambridge University Press, 403 pp.

[Ruessink et al.(2004)] Ruessink, B., I. van Enckevort, and Y. Kuriyama, 2004: Non-linear principal component analysis of nearshore bathymetry. *Marine Geology*, **203**, 185–197.

[Tirozzi et al.(2006)] Tirozzi, B., S. Puca, S. Pittalis, A. Bruschi, S. Morucci, E. Ferraro, and S. Corsini, 2006: *Neural Networks and Sea Time Series. Reconstruction and Extreme-Event Analysis*. Modeling and Simulation in Science, Engineering and Technology, Birkhäuser, Boston - Basel - Berlin, 179 pp.

[Tolman et al.(2005)] Tolman, H., V. Krasnopolsky, and D. Chalikov, 2005: Neural network approximations for nonlinear interactions in wind wave spectra: direct mapping for wind seas in deep water. *Ocean Modelling*, **8**, 253–278.

[Topouzelis et al.(2007)] Topouzelis, K., V. Karathanassi, P. Pavlakis, and D. Rokos, 2007: Detection and discrimination between oil spills and look-alike phenomena through neural networks. *ISPRS J. Photogrammetry and Remote Sensing*, **62**, 264–270.

[Tsai et al.(2002)] Tsai, C.-P., C. Lin, and J.-N. Shen, 2002: Neural network for wave forecasting among multi-stations. *Ocean Engng*, **29**, 1683–1695.

[van der Merwe et al.(2007)] van der Merwe, R., T. Leen, Z. Lu, S. Frolov, and A. Baptista, 2007: Fast neural network surrogates for very high dimensional physics-based models in computational oceanography. *Neural Networks*, **20**, 462–478.

[Wu(1993)] Wu, J.-K., 1993: *Neural Networks and Simulation Methods*. CRC, New York, 456 pp.

[Zanaganeh et al.(2009)] Zanaganeh, M., S. Mousavi, and A. Etemad-Shahidi, 2009: A hybrid genetic algorithm-adaptive network-based fuzzy inference system in prediction of wave parameters. *Engng Appl. Artificial Intelligence*, **22**, 1194–1202.

[Zorita and von Storch(1999)] Zorita, E. and H. von Storch, 1999: The analog method as a simple statistical downscaling technique: Comparison with more complicated methods. *J. Climate*, **12**, 2474–2489.

INDEX

A

B

C

D

E

F

G

H

I

J

K

L

M

N

O

P

Q

R

S

T

U

V

W

Y

Z